A guide to Feynman diagrams in the many-body problem

Richard D. Mattuck

SECOND EDITION

DOVER PUBLICATIONS, INC., *New York*

Published in Canada by General Publishing Company, Ltd., 30 Lesmill Road, Don Mills, Toronto, Ontario.
Published in the United Kingdom by Constable and Company, Ltd., 3 The Lanchesters, 162–164 Fulham Palace Road, London W6 9ER.

This Dover edition, first published in 1992, is an unabridged and unaltered republication of the 1976 second edition of the work first published in 1967 by The McGraw-Hill Book Company, New York.

Manufactured in the United States of America
Dover Publications, Inc., 31 East 2nd Street, Mineola, N.Y. 11501

Library of Congress Cataloging-in-Publication Data

Mattuck, Richard D.
    A guide to Feynman diagrams in the many-body problem / Richard D. Mattuck.—2nd ed.
        p.   cm.
    Originally published: New York : McGraw-Hill, c1976.
    Includes bibliographical references and index.
    ISBN 0-486-67047-3 (pbk.)
    1. Many-body problem.   2. Feynman diagrams.   I. Title.
[QC174.17.P7M37   1992]
530.1'44—dc20   92-4895
                                                                                    CIP

Preface to the second edition

I was delighted by the extreme reactions to the first edition of this book. One reviewer called it a 'pedagogical jewel ... useful as a crutch for poorly-prepared students', while another felt that it was primarily for people who were 'well-prepared and courageous'. The preface to the Russian edition referred to the pinball game analogy on p. 29 as some sort of world's record in popularization, but an English critic complained that the pinball picture was 'highly offensive' and 'had no place in a serious work of science'. A student told me that at his university, the book was known as 'Feynman Diagrams for Idiots', while other students felt that it was only for people exceptionally well-grounded in quantum mechanics. One critic stated that the 'possibilities for classroom use should be rather wide', but others claimed that the book was useless, since no detailed calculations were carried out in it.

In short, the first edition is too elementary and too advanced. Therefore, the purpose of the second edition is to make the book more advanced and more elementary. Toward this end, on the elementary side, a zeroth and first chapter have been added which are on the pre-kindergarten or nursery school level. This gives a view of the entire field based almost purely on pictures, cartoons, and virtual movies, with essentially no mathematics.

On the more advanced side, I have added to many chapters a new section showing in mathematical detail how typical many-body calculations with Feynman diagrams are carried out. For example, chapter 3 contains the detailed calculation of the energy and lifetime of an electron in an impure metal. In chapter 9, the single pair-bubble approximation is used to compute the quasi particle lifetime diagrammatically. The pair-bubble integrations are done in detail in chapter 10 and the results employed to obtain the form of the effective interaction in an electron gas, and the plasmon dispersion law in chapter 13. Chapter 14 contains the calculation of the finite temperature pair-bubble.

A number of new exercises have been added, some of which give the student the opportunity to carry out simpler many-body calculations himself. For example, Ex. 10.7 requires solving the K-matrix equation in ladder approximation, computing the integrals and showing that the hole lines give a negligible contribution in the low density case.

A new chapter on the quantum field theory of phase transitions has been added. It includes, on the kindergarten level, an analysis of the 'staring crowd' transition (see p. 290) and on the more advanced level, the diagrammatic calculation of the magnetization and transition point for the ferromagnetic phase. There are also new chapters on the Kondo problem and on the renormalization group.

I have also written several new appendices. Appendix L reviews the analytic properties of propagators, which I make considerable use of at various points in the text. Appendix M shows the relation between the equation of motion and Feynman diagram methods for calculating the propagator. Appendix N gives the basic ideas of the 'reduced graph' method, used in connection with the Kondo problem.

In preparing the second edition, special thanks are due to Stud. Scient. Nikolai Nissen for pointing out better methods for carrying out many of the calculations, and for carefully reading and criticizing the new material.

I am also very grateful to my colleague Dr Ulf Larsen for the many fruitful and stimulating discussions of many-body theory we have had during the last five years, for his help in working out the chapter on the Kondo problem, and for weeding out many of the inaccuracies which had crept into the book.

I would like to thank Professor P. W. Atkins of Lincoln College, Oxford, for pointing out how the book could be modified to make it of more value to chemists.

I am much indebted to my cousin's son, David Lustbader, B.A., for his aid in improving chapter Zero, and to my own son, Allan, for help in pasting together the thousands of pieces of paper which were the raw material for the second edition. And I want to express my gratitude to my students, whose unending stream of questions forced me to replace fuzziness by clarity throughout the book.

And finally, a word of thanks to the many people who, by telling me how much they enjoyed the first edition (one wrote: 'Please allow me to express my gratitude for a ray of sunshine that you have cast into the windowless office of a second year graduate student in the form of your book on Feynman diagrams'), gave me the inspiration and fortitude to sweat my way through the production of the second edition.

Copenhagen, 1974

Preface to the first edition

This book is written for laymen, i.e., for experimental physicists and for those theoreticians who don't mind getting caught reading something easy.

Most laymen are aware that many-body theory is very much in vogue these days, and that it is producing a wealth of fundamental results in all fields of physics. Unfortunately, the subject is notoriously difficult, and the only previously available books on it are written on such a high level that they are completely inaccessible to the average experimenter or non-specialist theoretician.

The purpose of this book is to help bridge the pedagogical gap by providing an easy introduction to just one aspect of many-body theory, i.e., the method of Feynman diagrams. Since the word 'easy', along with its cousins 'elementary', 'introductory', or 'for five-year olds', has been applied to some pretty formidable physics literature in the past, I had better make clear how it is used here. It means first that, as far as I know, the present book is simpler than anything else which has been written in the modern many-body field. This establishes an upper bound on 'easy'. The lower bound is fixed by the system illustrated on p. 29. This is the classical example I have invented to introduce the main ideas of the subject. The whole first half of the book is derived essentially by analogy to this example.

Since this book does not fit into any of the usual categories, it may help to prevent misunderstanding if I state clearly what it is not. It is not a many-body 'textbook' in itself; it is simply an elementary introduction to the textbooks which already exist in the field. It does not prepare the student to plunge into the latest literature; it can only give him a glimpse of what this literature is about. It does not train students to do many-body calculations any more than a music-appreciation course trains students to compose music; it can, however, help them to grasp the elegance and significance of these calculations.

In short, it is not a text for the usual 'elementary course in many-body theory', because such a course would have as its purpose the bringing of beginners to the point where they would be able to do calculations and solve real problems in the field. It is rather intended primarily as a 'home study' book for non-specialists trying to get some idea of what Feynman diagrams in many-body physics are all about. In addition, it could serve as a reference in courses on solid state and nuclear physics which make some use of the many-body techniques. And, finally, it can be used (by those who like to start with something simple) as a supplementary reference in a many-body course.

Now a word about the organization of the book. Measured on a scale established by the other literature in the field, it is divided into three parts: kindergarten, elementary, and intermediate.

Chapters 1–6 constitute the kindergarten part. This provides an introduction to the major concepts of the field on a level somewhere between 'Donald Duck' and the 'American Journal of Physics'. The quantum diagram technique is developed by analogy to a transparent classical case. It is first applied in detail to trivial one-particle systems; this gives the reader a feeling for the method by showing him how it works on problems he can easily solve by elementary quantum mechanics. The many-body diagrams are presented using the same simple-minded approach. There is also a short introduction to second quantization, but this is optional, and no essential use of it is made in the first part of the book.

The kindergarten part may be read as a book in itself by people who just want to learn enough so they no longer tremble with awe when a many-body theoretician covers the blackboard with Feynman diagrams.

Chapters 7–16 constitute the elementary part. The topics here, ranging from second quantization to superconductivity, are standard for most of the other many-body books. But they are covered on a much lower, more restricted level. This means essentially that, first of all, the only physical properties of systems which I discuss are the energies of the ground and excited states, and that, secondly, there is no discussion of the analytic properties of propagators. I have instead concentrated exclusively on giving the reader a feeling for the diagrams themselves, their physical significance, and the various summation techniques for manipulating them. A novel feature of the pedagogical technique here is that all calculations are done completely diagrammatically up to the point where the diagram solution is translated into integrals; at this point, I simply state the numerical result and refer the reader to the appropriate book or paper for the details of the integrations.

The appendices A–J are the intermediate part of the book. They begin with a brief summary of Dirac formalism and include a more or less rigorous derivation of the rules for diagrams.

There are a few short exercises at the end of each chapter, and the answers to the exercises appear at the end of the book.

Note: Optional reading is enclosed in double brackets: [[ ]].

This book grew out of a series of lectures I gave to the Solid State Physics Study Group at the University of Copenhagen during 1962–5. Of the many people at the university who have aided me during this period, I wish especially to thank Professor H. Højgaard Jensen, both for giving me the opportunity to get into the many-body field, and for the many helpful and stimulating conversations I have had with him. I am also very grateful to Professor M. Pihl for his criticism of the manuscript in its early stages and for his encouragement.

I would like to acknowledge the many valuable suggestions for improving the manuscript which were made by Professor D. J. Thouless of the University of Birmingham and Dr. A. W. B. Taylor of the University of Liverpool.

Among my colleagues, special thanks are due, first, to Lic. Börje Johansson of NORDITA, for the innumerable long and lively discussions of many-body theory we have had together, and for his reading and criticism of the entire manuscript and, second, to Civ. Ing. F. Greisen of Danmarks Tekniske Højskole, for the extraordinary care with which he read the complete first draft of the book, pointing out countless errors which would otherwise have gone unnoticed and making many worthwhile suggestions. I am also indebted to Mag. P. Laut for his extremely valuable criticism.

In addition, I want to acknowledge the many helpful conversations I have had with my other colleagues, in particular Mag. C. Fogedby, Dr. B. Easlea (University of Sussex), Mag. O. Bundsgaard (Danmarks Tekniske Højskole), Mag. P. Mogensen, Dr. Antonina Kowalska (University of Krakow), Mag. P. Voetmann Christiansen, Mag. E. Brun Hansen, Mag. H. Nielsen, Dr. D. Kobe, Mag. H. Smith, Mag. F. Berg Rasmussen, and Mag. O. P. Hansen. I also would like to thank Professor G. E. Brown, whose lectures on many-body theory at the Niels Bohr Institute (1960–1) constituted my initiation into the subject. I wish to express my gratitude to Academic Press, Inc., for its kind permission to reprint the article 'Phonons From a Many-Body Viewpoint', which appeared in *Annals of Physics* 27, p. 216 (1964), and to Mrs Vera Rothenberg and Mrs Elin Hallden for their fine work in typing the first draft of the book. And, finally, many thanks are due to my brother, Professor Arthur Mattuck (M.I.T.), for extremely helpful suggestions, and to Stud. Mag. Alice Mattuck, for her careful reading and criticism of the jokes in the manuscript.

Richard D. Mattuck

Contents

CONTENTS

Chapter 0

The Many-Body Problem for Everybody

0.0 What the many-body problem is about

The many-body problem has attracted attention ever since the philosophers of old speculated over the question of how many angels could dance on the head of a pin. In the angel problem, as in all many-body problems, there are two essential ingredients. First of all, there have to be many bodies present—many angels, many electrons, many atoms, many molecules, many people, etc. Secondly, for there to be a problem, these bodies have to interact with each other. To see why this is so, suppose the bodies did not interact. Then each body would act independently of all the others, so that we could simply investigate the behaviour of each body separately. In other words, without interaction, instead of having one many-body problem, we would have many one-body problems. Thus, interactions are essential, and in fact the many-body problem may be defined as *the study of the effects of interaction between bodies on the behaviour of a many-body system.*

(It might be noted here, for the benefit of those interested in exact solutions, that there is an alternative formulation of the many-body problem, i.e., how many bodies are required before we have a problem? G. E. Brown points out that this can be answered by a look at history. In eighteenth-century Newtonian mechanics, the three-body problem was insoluble. With the birth of general relativity around 1910 and quantum electrodynamics in 1930, the two- and one-body problems became insoluble. And within modern quantum field theory, the problem of zero bodies (vacuum) is insoluble. So, if we are out after exact solutions, no bodies at all is already too many!)

The importance of the many-body problem derives from the fact that almost any real physical system one can think of is composed of a set of interacting particles. For example, nucleons in a nucleus interact by nuclear forces, electrons in an atom or metal interact by Coulomb forces, etc. Some examples are shown schematically in Fig. 0.1. Furthermore, it turns out that in the calculation of physical properties of such systems—for example, the energy levels of the atom, or magnetic susceptibility of the metal—interactions between particles play a very important role.

It should be clear from the variety of systems in Fig. 0.1 that the many-body problem is *not* a branch of solid state, or nuclear, or atomic physics, etc. It deals rather with *general* methods applicable to *all* many-body systems.

1

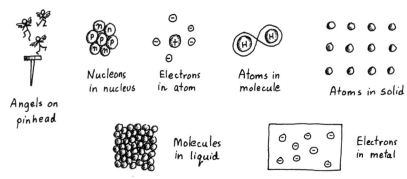

Angels on
pinhead

Nucleons
in nucleus

Electrons
in atom

Atoms in
molecule

Atoms in solid

Molecules
in liquid

Electrons
in metal

Fig. 0.1 *Some Many-body Systems*

The many-body problem is an extraordinarily difficult one because of the incredibly intricate motions of the particles in an interacting system. In Fig. 0.2 we contrast the simple behaviour of non-interacting particles with the complicated behaviour of interacting ones. Because of the complexity of the many-body problem, not much progress was made with it for a long time. In fact one of the preferred methods for solving the problem was simply to ignore it, i.e., pretend there were no interactions present. (Surprisingly enough, in some cases this 'method' produced good results anyway, and one of the great mysteries was how this could be possible!)

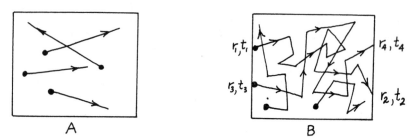

A

B

Fig. 0.2 A. *Non-interacting Particles*
B. *Interacting Particles*

Another of the early approaches to the problem, and one which is still used extensively today is the *canonical transformation* technique, described in appendix $\mathscr{A}$. This involves transforming the basic equations of the many-body system to a new set of coordinates in which the interaction term becomes small. Although considerable success has been achieved with this technique, it is not as systematic as one would like, and this sometimes makes it difficult to apply. It was this lack of a systematic method which kept the many-body field in its cradle well up into the 1950s.

The situation changed radically in 1956–7. In a series of pioneering papers, it was shown that the methods of *quantum field theory*, already famous for its success in elementary particle physics, provided a powerful, unified way of attacking the many-body problem. The new key opened many doors, and in rapid succession the idea was applied to nuclei, electrons in metals, ferromagnets, atoms, superconductors, plasmas, molecules—virtually everything in sight.

From that time on, much of the most exciting and fundamental research into the nature of matter has been based on the quantum field theory method. One of the things emerging from this research is a new simple picture of matter in which systems of interacting real particles are described in terms of approximately non-interacting fictitious bodies called 'quasi particles' and 'collective excitations'. Another thing is new results for calculated physical quantities which are in excellent agreement with experiment—for example, energy levels of light atoms, binding energy of nuclear matter, Fermi energy and effective electron mass in a variety of metals.

In this introductory chapter, we will give a physical picture of quasi particles and collective excitations. Then in the next chapter we show qualitatively how to describe quasi particles and calculate their properties by means of the quantum field theoretical technique known as the method of *Feynman diagrams*.

0.1. Simple example of non-interacting fictitious bodies

As mentioned at the beginning, one of nature's little surprises is that many-body systems often behave as if the bodies of which they are composed hardly interact at all! The reason for this is that the 'bodies' involved are not real but *fictitious*. That is, the system composed of *strongly* interacting *real* bodies acts *as if* it were composed of *weakly* interacting (or non-interacting) *fictitious* bodies. We consider now a very simple example of how this can occur.

Suppose we have two masses, m_1 and m_2 held together by a strong spring as shown in Fig. 0.3. That is, our system here consists of two strongly coupled real bodies. If this contraption is tossed up in a gravitational field, the motion of each body considered separately is very complicated because of the strong interaction (spring force) between the bodies.

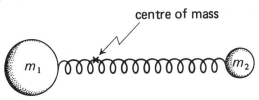

Fig. 0.3 *Two-body System*

However, we can break up the complicated motion into two independent simple motions: motion of the centre of mass and motion about the centre of mass. The centre of mass moves exactly as if it were an independent body of mass $m_1 + m_2$, so it is one of the non-interacting fictitious bodies here. The other fictitious body is a body of mass $m_1 m_2/(m_1 + m_2)$—the so-called 'reduced mass'—which moves independently relative to the centre of mass. Thus the system acts as if it were composed of two non-interacting fictitious bodies: the 'centre of mass body' and the 'reduced mass body'. (See appendix $\mathscr{A}$, eqs. $(\mathscr{A}.11)\rightarrow(\mathscr{A}.14)$ for details.)

0.2 Quasi particles and quasi horses

The above two-body example is easy enough to understand, but finding the weakly interacting fictitious bodies in a set of *many* strongly interacting real bodies is a bit harder. We consider first the fictitious bodies called 'quasi particles'. These arise from the fact that when a real particle moves through the system, it pushes or pulls on its neighbours and thus becomes surrounded by a 'cloud' of agitated particles similar to the dust cloud kicked up by a galloping horse in a western. The real particle plus its cloud is the quasi particle (Fig. 0.4).

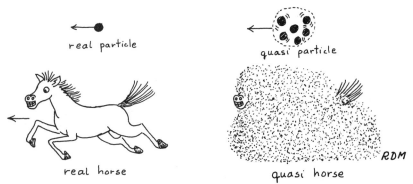

Fig. 0.4 *Quasi Particle Concept*

Just as the dust cloud hides the horse, the particle cloud 'shields' or 'screens' the real particles so that quasi particles interact only weakly with one another. The presence of the cloud also makes the properties of the quasi particle different from that of the real particle—it may have an '*effective mass*' different from the real mass, and a '*lifetime*'. These properties of quasi particles are directly observable experimentally.

It should be remarked that the quasi particle is in an excited energy level of the many-body system. Hence it is referred to as an '*elementary excitation*' of the system. (See appendix $\mathscr{A}$, §$\mathscr{A}.2$.) We now consider some examples of quasi particles.

1 *Quasi ion in a classical liquid*

Imagine that we have an electrolyte solution composed of an equal number of positive and negative ions moving about and colliding with each other as illustrated in Fig. 0.5. Let us focus our attention on a typical (+) ion in the

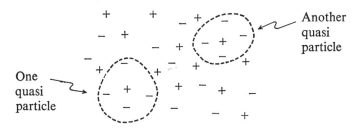

Fig. 0.5 *Quasi Particles in a Liquid of Positive and Negative Ions*

system. As this ion moves, on account of the strong Coulomb interaction, it will attract (−) ions to it. Some of these (−) ions will stick to the (+) for a while, then fall off due to collisions, then be replaced by other (−) ions, etc. Thus, on the average, because of the interaction, this typical (+) ion (and therefore every (+) ion) will be surrounded by a 'coat' or 'cloud' of (−) ions as shown in Fig. 0.5 inside the dotted lines. And of course each (−) ion will similarly have a coat of (+) ions. This coat of opposite charge will shield the ion's own charge so that its interaction with other similarly shielded ions will be much weaker than in the unshielded case. Thus the ions wearing their coats will act approximately independently of each other and constitute the quasi particles of this particular system. Many different types of systems of interacting particles may be described in this manner, and in general we have

$$\text{real particle} \quad + \quad \begin{array}{c}\text{'coat' or 'cloud'}\\ \text{of other particles}\end{array} = \quad \text{quasi particle.} \qquad (0.1)$$

Sometimes this same equation is stated in a more powerful terminology coming from quantum field theory:

$$\text{'bare' particle} \quad + \quad \begin{array}{c}\text{'clothing'}\\ \text{or 'cloud'}\end{array} = \quad \begin{array}{c}\text{'dressed' or 'clothed'}\\ \text{or 'physical' or}\\ \text{'renormalized' particle.}\end{array} \qquad (0.2)$$

For example, in quantum electrodynamics a 'bare' electron interacting with a field of photons acquires a cloud of virtual photons around it, converting it into the 'dressed' electron. In a similar manner, the interaction between real particles is called the 'bare' interaction, while the weak interaction between quasi particles is referred to as the 'effective' or 'dressed' or 'renormalized' interaction.

It should be noted that each bare particle is simultaneously the 'core' of a quasi particle and a transient 'member' of the cloud of several other quasi particles. Therefore, if we try to visualize the whole system here as composed of quasi particles, we have to be careful, since each particle will have been counted more than once. For this reason, the quasi particle concept is valid only if one talks about a few quasi particles at a time, i.e., few in comparison with the total number of particles. In order to avoid this problem and concentrate attention on just a single quasi particle at a time, it is convenient to define quasi particles in terms of an experiment in which one adds an extra particle to the system, and observes the behaviour of this extra particle as it moves through the system. This is shown in Fig. 0.6 for a (+) ion.

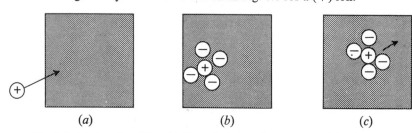

(a) (b) (c)

Fig. 0.6 *Moving Quasi Ion. (a) Extra (+) Ion Shot into Liquid. (b) (+) Ion Acquires Cloud of (−) Ions, Turning it into Quasi Ion. (c) Quasi Ion Moves Through System*

With this intuitive picture in mind, it is possible to guess at some of the properties of quasi particles. First, because there is in general still a small interaction left between quasi particles, a quasi particle of momentum **p** will only keep this momentum for an average time τ_p. This can be understood from Figs. 0.6 and 0.5. If the quasi ion in Fig. 0.6 (*b*) has momentum **p**, it will propagate undisturbed an average time τ_p before undergoing a collision with another quasi ion in the system (that is, a quasi ion which *belongs* to the system, like those shown in Fig. 0.5, *not* one which we shoot into the system) which scatters it out of momentum state **p**. Hence

$$\text{quasi particles have a lifetime, } \tau_p. \tag{0.3}$$

The lifetime must be reasonably long for us to say that the quasi particle approximation is a good one. It can also be seen that because of the average coat of particles on its back, the quasi particle may have an 'effective' or 'renormalized' mass which is different from that of the bare particle. (The effective mass concept is not always applicable however.) This implies that free quasi particles (i.e., not in an externally applied field) have a new energy law

$$\epsilon' = \frac{p^2}{2m^*} \quad \text{instead of} \quad \epsilon = \frac{p^2}{2m} \tag{0.4}$$

where m^* is the effective mass. The difference

$$\epsilon_{\substack{\text{quasi} \\ \text{particle}}} - \epsilon_{\substack{\text{bare} \\ \text{particle}}} = \epsilon_{\text{self}} \qquad (0.5)$$

is called the '*self-energy*' of the quasi particle. This comes from the interpretation that the bare particle interacts with the many-body system, creating the cloud, and the cloud in turn reacts back on the particle, disturbing its motion. Thus the particle is, in a sense, interacting with itself via the many-body system, and changing its own energy.

2 *Quantum system: quasi electron in electron gas*

The 'electron gas' is a simple model often used to describe many-body effects in metals. It consists of a box containing a large number of electrons interacting by means of the Coulomb force. In addition, there is a uniform, fixed, positive charge 'background' put into the box in order to keep the whole system electrically neutral. In the ground state, the electrons are spread out uniformly in the box, as shown schematically in Fig. 0.7.

Fig. 0.7 '*Electron Gas*': *Interacting Electrons Spread Out Uniformly in Box, plus Uniform, Fixed, Positive Charge Background*

Suppose now that we have a single, well-localized electron which we shoot into the electron gas (Fig. 0.8). Because of the repulsive Coulomb interaction between electrons, this extra electron repels other electrons away from it, so

Fig. 0.8 *Extra Electron Shot into Electron Gas*

we get an 'empty space' near the extra electron, and repelled electrons further away (Fig. 0.9). The empty space has positive charge, since the positive charge background is exposed in this region. This empty region may be viewed in a more detailed or 'microscopic' way as composed of 'holes' in the electron gas. That is, the extra electron has 'lifted out' electrons from the uniform charge distribution in its vicinity, thus creating 'holes' in this charge distribution, and has 'put down' these lifted-out electrons further away. This is shown in Fig. 0.10. Because of the exposed positive background, these holes have positive charge.

Fig. 0.9 *Extra Electron Pushes Other Electrons Away, Creating 'Empty' Region in its Immediate Vicinity*

Fig. 0.10 *'Microscopic' View of Fig. 0.9 Showing Electrons Lifted out from Vicinity of the Extra Electron, thus Creating 'Holes'*

The above definition of hole in the sense of 'empty place' is the one commonly used in solid state physics. However, later on we shall re-define things so that the hole becomes an 'anti-particle' analogous to those of elementary particle physics (see §4.2).

The holes and lifted out electrons are constantly being destroyed by interaction with the extra electron and with the other electrons in the system, and new holes and lifted out electrons take their place. The sum of these microscopic processes, which go on all the time, is Fig. 0.9. Thus Fig. 0.9 may be visualized as an extra electron surrounded by a 'cloud' of constantly changing holes and lifted out electrons. This combination is called the *quasi electron*.

The quasi electron moves or 'propagates' through the system as shown in Fig. 0.11.

We now notice that the positive hole cloud immediately around the extra electron partially shields the electron's own negative charge. Hence, if we have two quasi electrons as shown in Fig. 0.12, and these are far enough

Fig. 0.11 *Quasi Electron Propagates Through System*

Fig. 0.12 *Two Quasi Electrons Interact only Weakly Because of Shielding*

apart so that their clouds do not overlap very much, then we see that because of the shielding the two quasi electrons will interact only weakly. That is, quasi electrons act nearly independently of one another. This is why metals generally behave as if their electrons were independent: it is not real electrons but rather quasi electrons we are looking at.

3 *Single electron in a metal*

Actually, the simplest quantum example of the quasi particle idea occurs not in a true many-body system, but rather in a system containing one particle moving in an external potential, i.e., a conduction electron in a metal. In a perfect metal the positive ions form a regular periodic lattice (we ignore lattice vibrations for the moment) so that the electron moves in a periodic force field due to the attractive Coulomb interaction between the ions and the electron (see Fig. 0.13a). In an imperfect metal, the periodicity is spoiled by the presence of a more or less random distribution of some impurity ions in the lattice, or the presence of some displaced ions (Fig. 0.13b).

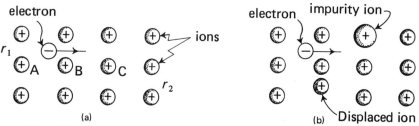

Fig. 0.13 (*a*) *Conduction Electron in Perfect Metal.* (*b*) *Imperfect Metal*

Since the lattice here is assumed fixed, there is no 'moving cloud' of lattice ions following the electron. Nevertheless, it turns out that even these stationary lattice ions are capable of 'clothing' the electron, and we find that for a perfect lattice, there is an effective mass, m^*, and an infinite lifetime. Addition of imperfections causes the lifetime to become finite.

4 Quasi nucleon

Despite powerful short-range forces between nucleons in a nucleus, they behave in many respects as if they were independent of each other, as is indicated by the success of the nuclear shell model. The nearly independent particles here are not the nucleons themselves, but the nucleons each surrounded by a cloud of other nucleons, i.e., the quasi nucleons.

5 Bogoliubov quasi particles ('bogolons')

These are the elementary excitations in a superconductor. We include them here since they are called quasi particles, but actually their structure is quite different from the 'particle plus cloud' picture described above. They consist of a linear combination of an electron in state $(+k, \uparrow)$ and a 'hole' in $(-k, \downarrow)$.

0.3 Collective excitations

As we have seen, the quasi particle consists of the original real, individual particle, plus a cloud of disturbed neighbours. It behaves very much like an individual particle, except that it has an effective mass and a lifetime. But there also exist other kinds of fictitious particles in many-body systems, i.e., 'collective excitations'. These do not centre around individual particles, but instead involve collective, wavelike motion of *all* the particles in the system simultaneously. Here are some examples:

1 Plasmons

If a thin metal foil is bombarded with high energy electrons, it is possible to set up sinusoidal oscillations in the density of the electron gas in the foil. This is known as a 'plasma wave', and it has a frequency ω_p and a wavelength λ_p (see Fig. 0.14a). The plasma wave may be visualized as built up of 'holes' in the low-density regions and extra electrons in the high-density regions as shown in Fig. 0.14(b). Just as light waves are quantized into units having energy $E = \hbar\omega$ called photons, plasma waves are quantized into units with energy $E_p = \hbar\omega_p$ called *plasmons*.

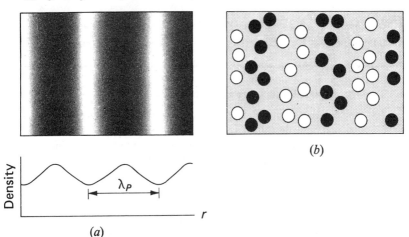

(a)

Fig. 0.14 (a) *Plasma Wave in Electron Gas.* (b) *Particle-hole Picture of Plasma Wave*

2 *Phonons*

Sound waves are sinusoidal oscillations in the crystal lattice of a solid. They are quantized into collective excitations called 'phonons'. (See appendix $\mathscr{A}$.)

3 *Magnons*

In ferromagnets there are regular fluctuations in the density of spin angular momentum known as 'spin waves'. The collective excitation here is the spin wave quantum known as the 'magnon'.

4 *Nuclear quanta*

In nuclei, one finds various vibrational and rotational motions; the associated quanta are the collective excitations in this case.

In the next chapter, we will describe in a very qualitative way how to find the properties of quasi particles and collective excitations by means of 'propagators' and 'Feynman diagrams'.

Further reading

Appendix $\mathscr{A}$
Patterson (1964).
Pines (1963), chap. 1.

Chapter 1

Feynman Diagrams, or how to Solve the Many-Body Problem by means of Pictures

1.1 Propagators—the heroes of the many-body problem

We have seen that many-body systems consisting of strongly interacting real particles can often be described as if they were composed of weakly interacting fictitious particles: quasi particles and collective excitations. The question now is, how can we calculate the properties of these fictitious particles—for example, the effective mass and lifetime of quasi particles? There are various ways of doing this (see appendix $\mathscr{A}$) but the hero roles in the treatment of the many-body problem are played by quantum field theoretical quantities known as *Green's functions* or *propagators*. These are essentially a generalization of the ordinary, familiar undergraduate Green's function. They come in all sizes and shapes—one particle, two particle, no particle, advanced, retarded, causal, zero temperature, finite temperature—an assortment to suit every situation and taste.

There are three reasons for the immense popularity propagators are enjoying these days. First of all, they yield in a direct way the most important physical properties of the system. Secondly, they have a simple physical interpretation. Thirdly, they can be calculated in a way which is highly systematic and 'automatic' and which appeals to one's physical intuition.

The idea behind the propagator method is this: the detailed description of a many-body system requires in the classical case the position of each particle as a function of time, $\mathbf{r}_1(t)$, $\mathbf{r}_2(t)$, ..., $\mathbf{r}_N(t)$, or in the quantum case, the time-dependent wave function of the whole system, $\Psi(\mathbf{r}_1.\mathbf{r}_2,...,\mathbf{r}_N,t)$. A glance at Fig. 0.2B shows that this is an extremely complicated business. Fortunately, it turns out that in order to find the important physical properties of a system it is not necessary to know the detailed behaviour of each particle in the system, but rather just the *average* behaviour of one or two typical particles. The quantities which describe this average behaviour are the *one-particle propagator* and *two-particle propagator* respectively, and physical properties may be calculated directly from them.

Consider the one-particle propagator first. It is defined as follows: We put a particle into the interacting system at point r_1 at time t_1 and let it move through the system colliding with the other particles for a while (i.e., let it 'propagate' through the system). Then the one-particle propagator is the probability (or in quantum systems, the probability *amplitude*—see §3.1) that the particle will be observed at the point r_2 at time t_2. (Note that instead of putting the particle in at a definite point, it is sometimes more convenient to put it in with definite momentum, say p_1, and observe it later with momentum p_2.) The single-particle propagator yields directly the energies and lifetimes of quasi particles. It also gives the momentum distribution, spin and particle density and can be used to calculate the ground state energy.

Similarly, the two-particle propagator is the probability amplitude for observing one particle at r_2, t_2 and another at r_4, t_4 if one was put into the system at r_1, t_1 and another at r_3, t_3 (see Fig. 0.2B). This also has a wide variety of talents, giving directly the energies and lifetimes of collective excitations, as well as the magnetic susceptibility, electrical conductivity, and a host of other non-equilibrium properties.

There is also another useful quantity, the 'no-particle propagator' or so-called 'vacuum amplitude' defined thus: We put no particle into the system at time t_1, let the particles in the system interact with each other from t_1 to t_2, then ask for the probability amplitude that no particles emerge from the system at time t_2. This may be used to calculate the ground state energy and the grand partition function, from which all equilibrium properties of the system may be determined.

1.2 Calculating propagators by Feynman diagrams: the drunken man propagator

There are two different methods available for calculating propagators. One is to solve the chain of differential equations they satisfy—this method is discussed briefly in appendix M. The other is to expand the propagator in an infinite series and evaluate the series approximately. This can be carried out in a general, systematic, and picturesque way with the aid of *Feynman diagrams*.

Just to get an idea of what these diagrams are, consider the following simple example (see Fig. 1.1). A man who has had too much to drink, leaves a party at point 1 and on the way to his home at point 2, he can stop off at one or more bars—Alice's Bar (A), Bardot Bar (B), Club Six Bar (C), ..., etc. He can wind up either at his own home 2, or at any one of his friends' apartments, 3, 4, etc. We ask for the probability, $P(2, 1)$, that he gets home. This probability, which is just the propagator here (with time omitted for simplicity), is the sum of the probabilities for all the different ways he can propagate from 1 to 2 interacting with the various bars.

The first way he can propagate is 'freely' from 1 to 2, i.e., without stopping at a bar. Call the probability for this free propagation $P_0(2, 1)$.

The second way he can propagate is to go freely from 1 to bar A (the probability for this is $P_0(A, 1)$), then stop off at bar A for a drink (call the probability for this $P(A)$), then go freely from A to 2 (probability $= P_0(2, A)$). Assume for simplicity that the three processes here are independent. Then the total probability for this second way is the product of the probabilities for each process taken separately, i.e., $P_0(A, 1) \times P(A) \times P_0(2, A)$. (This is like the case in coin-tossing: since each toss is independent, the probability of first tossing a head, then a tail, equals the probability of tossing a head times the probability of tossing a tail.)

Fig. 1.1 *Propagation of Drunken Man*

(Reproduced with the kind permission of *The Encyclopedia of Physics*)

The third way he can propagate is from 1 to B to 2, with probability $P_0(B, 1)P(B)P_0(2, B)$. Or he could go from 1 to C to 2, etc., or from 1 to A to B to 2, or from 1 to A, come out of A, go back into A, then go to 2, and so on. The total probability, $P(2, 1)$ is then given by the sum of the probabilities for each way, i.e., the infinite series:

$$P(2, 1) = P_0(2, 1) + P_0(A, 1)P(A)P_0(2, A) + P_0(B, 1)P(B)P_0(2, B) + \cdots$$
$$+ P_0(A, 1)P(A)P_0(B, A)P(B)P_0(2, B) + \cdots. \tag{1.1}$$

This is an example of a 'perturbation series', since each interaction with a bar 'perturbs' the free propagation of the drunken man.

Now, such a series is a complicated thing to look at. To make it easier to read, we follow the journal 'Classic Comics' where difficult literary classics are translated into picture form. Let us make a 'picture dictionary' to associate diagrams with the various probabilities as in Table 1.1. Using this dictionary, series (1.1) can be drawn thus:

$$\begin{array}{c}\Vert^2_1\end{array} = \begin{array}{c}|^2_1\end{array} + \fbox{A} + \fbox{B} + \cdots + \begin{array}{c}A\\A\end{array} + \begin{array}{c}B\\A\end{array} + \cdots + \begin{array}{c}A\\A\\A\end{array} + \cdots. \quad (1.2)$$

Since, by dictionary Table 1.1, each diagram element stands for a factor, series (1.2) is completely equivalent to (1.1). However it has the great advantage that it also reveals the physical meaning of the series, giving us a 'map' which helps us to keep track of all the sequences of interactions with bars which the drunken man can have in going from 1 to 2.

Table 1.1 *Diagram dictionary for drunken man propagator*

Word	Picture	Meaning
$P(2, 1)$		probability of propagation from 1 to 2
$P_0(s, r)$		probability of free propagation from r to s
$P(X)$		probability of stopping off at bar X for a drink

The series may be evaluated approximately by selecting the most important types of terms in it and summing them to infinity. This is called *partial summation*. For example, suppose the man is in love with Alice, so that $P(A)$ is large, and all the other $P(X)$'s are small. Then Alice's bar diagrams will dominate, and the series (1.2) may be approximated by a sum over just repeated interactions with Alice's Bar:

(1.3)

Using the above dictionary, this can be translated into functions:

$$P(2,1) \approx P_0(2,1) + P_0(A,1)P(A)P_0(2,A) +$$
$$+ P_0(A,1)P(A)P_0(A,A)P(A)P_0(2,A) + \cdots. \quad (1.4)$$

Assume for simplicity that all $P_0(s,r)$ are equal to the same number, c, i.e., $P_0(2,1) = P_0(2,A) = P_0(A,1) = P_0(A,A) = c$. Then series (1.4) becomes

$$P(2,1) = c + c^2 P(A) + c^3 P^2(A) + \cdots$$
$$= c\{1 + cP(A) + [cP(A)]^2 + [cP(A)]^3 + \cdots\}. \quad (1.5)$$

The series in brackets is geometric and can be summed exactly to yield $1/(1-cP(A))$, so that

$$P(2,1) = c \times \left(\frac{1}{1-cP(A)}\right) = \frac{1}{c^{-1} - P(A)} \quad (1.6)$$

which is the solution for the propagator in this case.

Note that since each diagram element stands for a factor, we could have done calculation (1.5), (1.6) completely diagrammatically:

(1.7)

The partial summation method is extremely useful in dealing with the strong interactions between particles in the many-body problem, and it is the basic method which will be used throughout this book.

1.3 Propagator for single electron moving through a metal

The example here is just like the previous one, except that instead of a propagating drunken man interacting with various bars, we have a propagating electron interacting with various ions in a metal. A metal consists of a set of positively charged ions arranged so they form a regular lattice, as in Fig. 0.13A or a lattice with some irregularities, as in Fig. 0.13B. An electron interacts with these ions by means of the Coulomb force. The single particle propagator here is the sum of the *quantum mechanical probability amplitudes* (see §3.1) for all the possible ways the electron can propagate from point r_1 in the crystal at time t_1, to point r_2 at time t_2, interacting with the various ions on the way. These are: (1) freely, without interaction; (2) freely from r_1, t_1 (= ' 1 ' for short) to the ion at r_A at time t_A, interaction with this ion, then free propagation from the ion to point 2; (3) from 1 to ion B, interaction at B, then from B to 2, etc. Or we could have the routes $1-A-A-2$, $1-A-B-2$, etc. We can now use the dictionary in Table 1.1 to translate this into diagrams, provided the following changes are made: change 'probability' to 'probability amplitude', and change the meaning of the circle with an X to 'probability amplitude for an interaction with the ion at X'. When this is done, the series for the propagator can be translated immediately into exactly the same diagrams as in the drunken man case! That is, (1.2) is also the propagator for an electron in a metal, provided that we just use a quantum dictionary to translate the lines and circles into functions. The series can be partially summed, and from the resulting propagator we obtain immediately the energy of the electron moving in the field of the ions.

1.4 Single-particle propagator for system of many interacting particles

We will now indicate in a qualitative way how the single-particle propagator may be calculated in a system of many interacting particles. The argument is general, but we may think in terms of the electron gas as illustration. The propagator will be the sum of the probability amplitudes for all the different ways the particle can travel through the system from r_1, t_1 to r_2, t_2. First we have free propagation without interaction. Another thing which can happen is shown in the 'movie', Fig. 1.2, which depicts a 'second-order' propagation process (i.e., a process with *two* interactions). (It should be mentioned here that unlike the drunken man case, the processes involved in Fig. 1.2 are not real physical processes, but rather 'virtual' or 'quasi physical', since they do not conserve energy, and they may violate the Pauli exclusion principle. The reason for this is that, as we shall see later on, the sequence in Fig. 1.2 (or the corresponding diagram (1.9)) is simply a convenient and picturesque way of describing a certain second-order term which appears in the perturbation expansion of the propagator. Hence Fig. 1.2 and diagram (1.9) are in reality *mathematical expressions* so we have to be careful not to push their physical interpretation too far (see §4.6).)

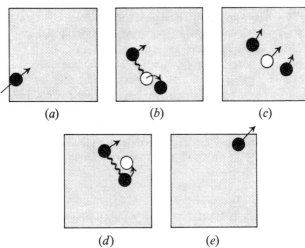

(a) (b) (c)

(d) (e)

Fig. 1.2 '*Movie*' *of Second-order Propagation Process in Many-body System*

(a) At time t_1, extra particle enters system.

(b) At time t, extra particle interacts (wavy line) with a particle in the system, lifting it out of its place, thus creating a 'hole' in the system.

(c) The extra particle, plus the 'hole' and the 'lifted-out' particle ('particle–hole pair') travel through the system.

(d) At time t', the extra particle interacts with the 'lifted-out' particle, knocking it back into the hole, thus destroying the particle–hole pair.

(e) At time t_2, the extra particle moves out of the system.

To represent this sequence of events diagrammatically, let us imagine that time increases in the upward-going direction and we use the following diagram elements:

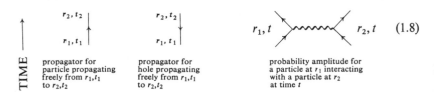

TIME →

| propagator for particle propagating freely from r_1, t_1 to r_2, t_2 | propagator for hole propagating freely from r_1, t_1 to r_2, t_2 | probability amplitude for a particle at r_1 interacting with a particle at r_2 at time t |

(1.8)

(Note that the hole is drawn as a particle moving backward in time. The reason for this is in §4.2.) Then the probability amplitude for the above sequence of events can be represented by the diagram

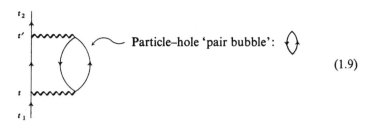

$$(1.9)$$

The piece of diagram:

$$(1.10)$$

is called a 'self-energy part' because it shows the particle interacting with itself via the particle–hole pair it created in the many-body medium. Diagram (1.9) may be evaluated by writing a free propagator factor for each directed line, and an amplitude factor for each wiggly line (see Chapter 4, Table 4.3), analogous to the drunken man case.

Another sequence of events which can occur involves only one interaction (i.e., a 'first-order' process). It is a quick-change act in which the incoming electron at point r interacts with another electron at point r' and changes place with it. This is analogous to billiard ball 1 striking billiard ball 2 and transferring all its momentum to 2. The first-order process and its analogy are shown in Fig. 1.3. The sequence may be drawn diagrammatically

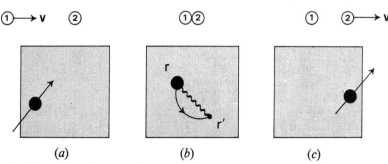

Fig. 1.3 *Movie of First-order Process (Lower Drawing) and its Analogy (Upper Drawing)*

(a) Extra particle enters at time t_1.

(b) At time t, the particle is at point r. It interacts with a particle at r' and changes place with it.

(c) Extra particle leaves at time t_2.

as in (1.11):

$$ t_2 $$

$$ \text{(1.11)} $$

t

t_1

'Open oyster' diagram
(closed oyster is in (1.19))

The diagrams in (1.8)–(1.11) are called *Feynman diagrams* after their inventor, Richard P. Feynman who employed them in his Nobel prize-winning work on quantum electrodynamics. They are used extensively in elementary particle physics.

The total single particle propagator is the sum of the amplitudes for all possible ways the particle can propagate through the system. This will include the above processes, repetitions of them, plus an infinite number of others. Thus we find

$$ \text{(1.12)} $$

(Note: the interpretation of the 'bubble' diagram, just after the open oyster, will be discussed in chapter 4.)

We can see the direct connection between the one-particle propagator and the quasi particle by looking at all the diagrams at a particular time t_0 (dashed line):

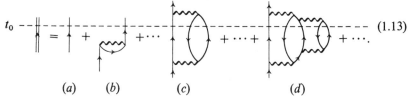

t_0 (1.13)

(a) (b) (c) (d)

At t_0, we see that various situations may exist: there may be just the bare particle (a), or there may exist two particles plus one hole created by the second-order sequence (c), or three particles plus two holes in (d), etc. That is, the diagrams show all the configurations of particles and holes which may be kicked up by the bare particle as it churns through the many-body system. If we now compare with the picture of the quasi particle in Fig. 0.10, we see that *the diagrams reveal the content of the ever-changing cloud of particles and holes surrounding the bare particle and converting it into a quasi particle.*

Just as in the drunken man case, the propagator here may be calculated approximately by doing a partial sum. For example, we can sum over all diagrams containing repeated open oyster parts since they constitute a geometric series (cf. (1.7)):

$$(1.14)$$

For the electron gas, this is the 'Hartree-Fock' approximation. We can also include 'ring' diagrams in the sum, i.e., diagrams in which the self-energy parts are composed of rings of particle–hole pair bubbles (these are the most important in a high-density electron gas):

$$(1.15)$$

This sum can be carried out and yields the so-called 'random phase approximation' or 'RPA', which is extremely useful in analysing the properties of metals.

Note that the essential thing involved in the above partial sums is the structure or *topology* of the diagrams, i.e., how the various lines are connected to each other. Thus we could sum (1.14) because each diagram consisted of single lines connecting the same repeated part. This diagram topology is the key to the quantum field theoretical method in the many-body problem.

1.5 The two-particle propagator and the particle-hole propagator

The two-particle propagator is the sum over the probability amplitudes for all the ways two particles can enter the system, interact with each other and with the particles in the system, then emerge again. The diagram series for it is (note that the dots on the diagram for the two-particle propagator show the points at which directed lines emerge):

$$(1.16)$$

A partial sum over all 'ladder' diagrams here:

$$(1.17)$$

is called 'ladder' approximation, and is very useful in describing nuclear matter, and low-density systems.

The 'particle–hole' propagator, given by

$$(1.18)$$

may be used to find the energy and lifetime of collective excitations, e.g. plasmons.

1.6 The no-particle propagator ('vacuum amplitude')

The ground state energy of a many-body system may be obtained directly from the no-particle propagator, or 'vacuum amplitude'. This is the sum of amplitudes for all the ways the system can begin at time t_1 with no extra or lifted-out particles, or holes in it (this is the undisturbed or 'Fermi vacuum' state), have its particles interact with each other, and wind up at t_2 with no extra or lifted-out particles, or holes. The simplest process is where nothing at all happens—the system just sits there. A first-order process occurs in which two particles change places with each other as shown in the following diagram

(1.19)

'Oyster' diagram

A more complicated process is shown in Fig. 1.4. The vacuum amplitude may thus be represented by the following diagram series:

(1.20)

where '1' is for the nothing-at-all process and (d) is the picture for Fig. 1.4. (The 'double bubble' diagram, (c), is discussed in chapter 5.)

The vacuum amplitude series gives us a vivid picture of the ground state of the many-body system as a sort of 'virtual witches' brew', constantly seething, with particles and holes boiling up, bubbling, and colliding, as in Fig. 1.5.

In conclusion, we see that Feynman diagrams have many appealing features, besides their utility as a calculational tool. One thing which was already pointed out in §1.2 is the fact that they show directly the physical meaning of the perturbation term they represent. Another thing is that they reveal at a glance the structure of very complicated approximations by showing which sets of diagrams have been summed over. In this way, they have introduced a new language into physics, and one often sees phrases like 'ladder approximation' or 'ring approximation' even in articles in which no diagrams appear. And finally, one cannot be immune to the Klee-like charm of the diagrams. Includ-

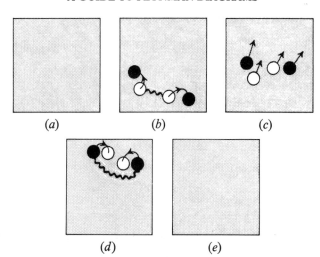

Fig. 1.4 *Virtual Movie of Second-order Vacuum Amplitude Process*
(*a*) Vacuum.
(*b*) At time t_1 interaction between two particles in system causes two particles to be lifted out, forming two holes.
(*c*) The two particle–hole pairs propagate freely through the system.
(*d*) Both pairs annihilated at time t'.
(*e*) Vacuum.

Fig. 1.5 *Modern View of a Many-body System in its Ground State*

ing in their ranks, in addition to the above, such characters as the 'necklace', the 'potato' and the 'tadpole', plus infinite numbers yet unnamed, they constitute what might indeed be called 'perturbation theory in comic-book form.'

Chapter 2

Classical Quasi Particles and the Pinball Propagator

2.1 Physical picture of quasi particle

We saw in §0.2 that the quasi particle is one type of elementary excitation in a many-body system, and that physically it consists of a particle surrounded by a cloud of other particles. The concept was illustrated by examples ranging from the quasi electron to the quasi horse. We also saw how quasi particles may be described by means of propagators, which are calculated with the aid of Feyman diagrams. Here we start with a brief review of the quasi particle idea, then go on to describe the form of the propagator for a classical quasi particle. The partial sum method of calculating the classical propagator is discussed in detail with the aid of a pinball machine example.

For concreteness, let us think in terms of the classical quasi ion in Fig. 0.6 which consists of a bare ion plus a coat of oppositely charged ions surrounding it. This picture led us to the general definition

$$\text{real particle} + \frac{\text{'coat' or 'cloud'}}{\text{of other particles}} = \text{quasi particle} \qquad (2.1)$$

or

$$\text{'bare' particle} + \frac{\text{'clothing'}}{\text{or 'cloud'}} = \frac{\text{'dressed' or 'clothed'}}{\text{or 'renormalized'}} \qquad (2.2)$$

It may be remarked that if we perform a 'Gedanken' calculation and imagine that the transformation in appendix ($\mathscr{A}.9$) were carried out, we see that the quasi particle co-ordinate $\mathbf{R}_i$ will involve the real particle co-ordinate $\mathbf{r}_i$, plus the co-ordinates $\mathbf{r}_j(j \neq i)$ of all the other particles in the system. The $\mathbf{r}_j(j \neq i)$ then evidently describe the shifting cloud, so it is therefore proper to call the cloud a part of the quasi particle.

We saw also that because of the small interactions between quasi particles,

$$\text{quasi particles have a lifetime, } \tau_p, \qquad (2.3)$$

and because of their coat of other particles, quasi particles have a new energy

$$\epsilon = \frac{p^2}{2m^*} \qquad (2.4)$$

law where m^* is the effective mass. Finally, we defined the self-energy, ϵ_{self} by

$$\epsilon_{\substack{quasi \\ particle}} - \epsilon_{\substack{bare \\ particle}} = \epsilon_{self}. \tag{2.5}$$

2.2 The classical quasi particle propagator

Quasi particles in a system may be tracked down by means of the single particle Green's function or 'propagator'. Let us see what this is in the classical case. Imagine we have a many-body system, and we consider the motion of one particle in it under the influence of a constant external force $\mathbf{F}$ applied to it as shown in Fig. 2.1. Suppose the particle begins at $\mathbf{r}_1$ at time t_1.

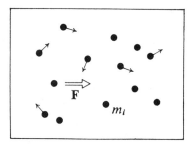

Fig. 2.1 *Many-body System*

If there are no collisions with other particles, the movement or 'propagation' of the particle to the point $\mathbf{r}_2$ at time t_2 is described by

$$\mathbf{r}_2 - \mathbf{r}_1 = \frac{1}{2}\left(\frac{\mathbf{F}}{m}\right)(t_2 - t_1)^2. \tag{2.6}$$

But in the interacting case, collisions take place, and the particle will follow a highly irregular path not described by (2.6). The best one can do in this situation is to talk about the *probability* of the particle going from one point to another. This leads us to define the *classical propagator*:

$P(\mathbf{r}_2, t_2, \mathbf{r}_1, t_1) =$ probability density ($=$probability per unit volume)
that if a particle at rest is put into the system at point
$\mathbf{r}_1$ at time t_1, then it will be found at $\mathbf{r}_2$ at later
time t_2.   (2.7)

It will be convenient, when we later take the Fourier transform, to have P defined also for $t_2 < t_1$:

$$P(\mathbf{r}_2, t_2, \mathbf{r}_1, t_1) = 0, \quad \text{for } t_2 < t_1. \tag{2.8}$$

In Fig. 2.2 is a graph showing a qualitative picture of this propagator in the interacting and non-interacting cases. Probability density is plotted on the vertical axis, and t_2 and an arbitrary component of $\mathbf{r}_2$ on the horizontal axes. In the absence of interactions, P will be a surface which is zero everywhere except on the line $\mathbf{r}_2 - \mathbf{r}_1 = \frac{1}{2}(\mathbf{F}/m)(t_2 - t_1)^2$, where it equals ∞, i.e., the Dirac δ-function:

$$P_0(\mathbf{r}_2, t_2, \mathbf{r}_1, t_1) = \delta\left[(\mathbf{r}_2 - \mathbf{r}_1) - \frac{1}{2}\left(\frac{\mathbf{F}}{m}\right)(t_2 - t_1)^2\right]. \qquad (2.9)$$

This propagator in the absence of interactions is called the *free propagator*.

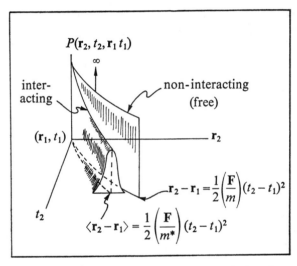

Fig. 2.2 *The Classical Propagator (Schematic—Only One Component of $\mathbf{r}_2$ Shown)*

If interactions between particles are now allowed to occur, this surface will spread out, as shown qualitatively. If we examine $\langle \mathbf{r}_2 - \mathbf{r}_1 \rangle$, the position of the maximum value of P in the interacting case, we see that for some types of interaction we might find that

$$\langle \mathbf{r}_2 - \mathbf{r}_1 \rangle = \frac{1}{2}\left(\frac{\mathbf{F}}{m^*}\right)(t_2 - t_1)^2 \quad \text{for } P = \text{maximum.} \qquad (2.10)$$

If this is true, then $\langle \mathbf{r}_2 - \mathbf{r}_1 \rangle$ behaves as the co-ordinate of a quasi particle of effective mass m^*. Look now at the maximum height of P as a function of t_2. Because of the 'spreading out' of the particle position, P_{max} will first fall infinitely rapidly from its value of ∞ at $t_2 = t_1$, then more slowly. If this slower decay is exponential:

$$P_{max}(\mathbf{r}_2, t_2, \mathbf{r}_1, t_1) \propto e^{-(t_2 - t_1)/\tau}, \qquad (2.11)$$

then τ may be identified as the quasi particle lifetime; it clearly must be fairly large if the quasi particle picture is to be useful. Thus, if we calculate P and find that it shows the above behaviour, then the system is describable in terms of quasi particles and their lifetime and effective mass may be determined.

2.3 Calculation of the propagator by means of diagrams

The actual calculation of the propagator P is quite complicated, but it is easy to illustrate all the principles involved with the aid of a simple analogue example in which the many-body system is replaced by a set of fixed scattering centres. (The system considered here is essentially the same as the drunken man case in chapter 1, but it will be treated in much more detail.)

The example involves the particle accelerator in Fig. 2.3. A pinball is injected at the point r_1, at time t_1 and propagates through the system, being scattered at the various centres. We ask for the probability $P(r_2, t_2, r_1, t_1)$ that the particle reaches the point r_2 at time t_2.

The scattering mechanism is assumed to be such that (1) if the pinball strikes the shaded circle at animal A, then there is probability $P(A)$ that it is scattered and $1 - P(A)$ that it will go straight through without scattering, (2) the probability distribution of pinball paths and velocities after scattering at A must be independent of the pinball path and velocity before scattering— that is, the pinball loses its 'memory' of how it got to A.

(There are many ways in which the above properties can be approximately realized. For example, the shaded circle could be a round peg which is pushed up so that it protrudes above the playing board surface a fraction $P(A)$ of the time, and is pulled in so that it is flush with the surface (hence cannot scatter the pinball) the rest of the time. Or we could have an immovable peg (i.e., always protruding) within the shaded circle, having a diameter such that the ratio of the peg diameter to that of the circle $= P(A)$. The loss of memory could be achieved by attaching a 'shuffling' device to each peg—like for example rapidly rotating spokes. The choice of method and the 'Rube Goldberg' details are, however, left as an exercise to the reader. They are of no importance for our discussion!)

For the sake of simplicity, let us leave time out of the argument to begin with, and consider just $P(r_2, r_1)$; this is the probability that if the particle begins at r_2 it will finish at r_2 regardless of the time. From the definition of probability, $P(r_2, r_1)$ is the sum of the probabilities for all the different ways the particle can go through the machine which begin at r_1 and wind up at r_2. For example, it could go 'directly' from r_1 to r_2 (i.e., without being scattered on the way) or it could go from r_1 to the giraffe, be scattered off the giraffe and fall to r_2. Or it could scatter from the giraffe to the monkey to r_2. Or it could scatter twice on the giraffe before falling to r_2. And so on.

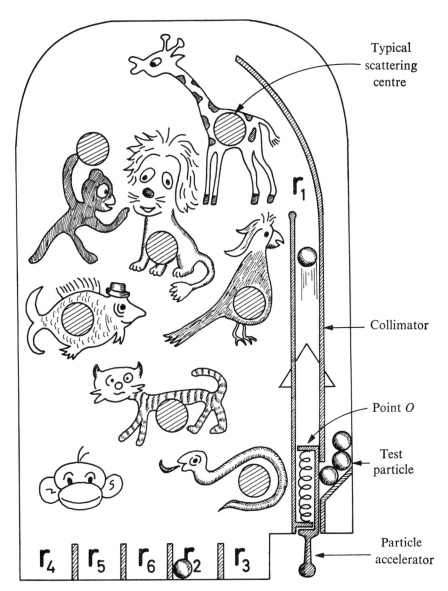

Fig. 2.3 *Classical Analogue Machine to Illustrate the Single-particle Propagator*

We first calculate the probability that the pinball will follow any particular path through the system. Let $P_0(\mathbf{r}_j, \mathbf{r}_i)$ = probability that if the pinball leaves $\mathbf{r}_i$ then it travels to $\mathbf{r}_j$ without being scattered by an animal en route ('free propagator'). The simplest path the pinball can follow is from $\mathbf{r}_1$ to $\mathbf{r}_2$ without scattering; this has probability $P_0(\mathbf{r}_2, \mathbf{r}_1)$. Another path is from $\mathbf{r}_1$ to the giraffe at $\mathbf{r}_G$ (probability = $P_0(\mathbf{r}_G, \mathbf{r}_1)$)), scattering at the giraffe (probability = $P(G)$), then from $\mathbf{r}_G$ to $\mathbf{r}_2$ (probability = $P_0(\mathbf{r}_2, \mathbf{r}_G)$)). Because the pinball loses its memory after the scattering at $\mathbf{r}_G$, these probabilities are independent of each other, and the joint probability for the whole path is just the product of the probabilities for each part of the path:

$$P\{(\mathbf{r}_1 \rightarrow \mathbf{r}_G), (\text{scattered at } \mathbf{r}_G), (\mathbf{r}_G \rightarrow \mathbf{r}_2)\} = P_0(\mathbf{r}_G, \mathbf{r}_1) P(G) P_0(\mathbf{r}_2, \mathbf{r}_G). \quad (2.12)$$

(Note that a process in which the pinball goes from $\mathbf{r}_1$ to $\mathbf{r}_G$, is not scattered at $\mathbf{r}_G$, and continues to $\mathbf{r}_2$, is not included in (2.12), but in the free propagator, $P_0(\mathbf{r}_2, \mathbf{r}_1)$.) The probabilities for the other paths are calculated in a similar fashion.

The total probability, $P(\mathbf{r}_2, \mathbf{r}_1)$, is just the sum of the probabilities for the various paths. Thus we find

$$P(\mathbf{r}_2, \mathbf{r}_1) = P_0(\mathbf{r}_2, \mathbf{r}_1) + P_0(\mathbf{r}_G, \mathbf{r}_1) P(G) P_0(\mathbf{r}_2, \mathbf{r}_G) + P_0(\mathbf{r}_M, \mathbf{r}_1) P(M) P_0(\mathbf{r}_2, \mathbf{r}_M) +$$

$$+ P_0(\mathbf{r}_G, \mathbf{r}_1) P(G) P_0(\mathbf{r}_G, \mathbf{r}_G) P(G) P_0(\mathbf{r}_2, \mathbf{r}_G) + \cdots \quad (2.13)$$

where G = giraffe, M = monkey, etc. What we have here is evidently just a perturbation expansion of the propagator, in which the $P(A)$'s play the same sort of role that the matrix elements of the perturbation, V_{kl}, play in quantum mechanical perturbation expansions.

In order to make series (2.13) easier to interpret, we draw a 'picture dictionary' to associate diagrams with the various probabilities as shown in Table 2.1.

Table 2.1 *Diagram dictionary for the pinball propagator*

Word	Picture
$P(\mathbf{r}_j, \mathbf{r}_i)$	
$P_0(\mathbf{r}_j, \mathbf{r}_i)$	
$P(A)$	

Then the series (2.13) may be drawn thus:

$$\text{(2.14)}$$

Equations (2.14) and (2.13) are of course completely equivalent to each other, being in one-to-one correspondence by the dictionary Table 2.1. But the picture has the advantage of revealing the physical meaning of (2.13), showing directly the particle shooting out from r_1, undergoing various sequences of collisions and coming finally to r_2. It presents in a vivid and systematic way the total probability as the sum of the probabilities associated with all the possible paths or 'histories' the particle can have as it goes through the system. Note that it is possible to interpret the $r_1, r_2, r_G, \ldots$ on the diagrams as being points in real space if we just re-draw the diagrams so the points lie as in Fig. 2.3 thus:

$$\text{(2.15)}$$

It is important to observe that in terms of diagrams, 'the sum of the probabilities for all the different ways the particle can go from r_1 to r_2, interacting with the various scatterers' may be translated into 'the sum of all possible different diagrams which can be built up out of labelled circles connected by directed lines, beginning at r_1 and terminating at r_2'. This is because there is just one diagram corresponding to each physical path through the system.

How can this series be evaluated? If we assume that the P_0's are large, say $\sim \frac{1}{2}$ or so, and the various interaction $P(A)$'s are small, say $\sim \frac{1}{10}$, then the higher order diagrams (i.e., terms; note that by order here we mean the total number of interactions) will give successively smaller contributions, and just as in ordinary perturbation theory, we can get an approximate solution by simply summing the series up through the first- or second-order terms. Thus, the zeroth-order approximation would be just the unperturbed case where the particle propagates freely from r_1 to r_2. When we add the possibility of a perturbing (scattering) interaction with the various animals just once each, we get the first-order approximation

$$\approx \quad + \quad (G) \quad + \quad (M) \quad + \cdots + \quad (L) . \qquad (2.16)$$

Allowing two interactions gives the second-order approximation and so on. If, on the other hand, one or more of the interaction terms $P(A)$ is large (i.e., strong scattering at A) this method is not practical, since the series converges too slowly, and the summation must be carried out to extremely high orders to give a good result.

However, there is another kind of approximation we can make in this strong interaction case, an approximation that does not stop at second order, but instead sums over diagrams to infinite order. Suppose, for example, that only P (monkey) is large and all the other $P(A)$'s are small. Then the monkey diagrams will dominate, and the series may be approximated by the sum over just repeated monkeys, thus:

$$\approx \quad + \quad (M) \quad + \quad (M) \atop (M) \quad + \quad (M) \atop (M) \atop (M) \quad + \cdots . \qquad (2.17)$$

Translating each element of the diagrams into the appropriate probability, it is easy to write down the corresponding series:

$$P(\mathbf{r}_2,\mathbf{r}_1) \approx P_0(\mathbf{r}_2,\mathbf{r}_1) + P_0(\mathbf{r}_M,\mathbf{r}_1) P(M) P_0(\mathbf{r}_2,\mathbf{r}_M) +$$
$$+ P_0(\mathbf{r}_M,\mathbf{r}_1) P(M) P_0(\mathbf{r}_M,\mathbf{r}_M) P(M) P_0(\mathbf{r}_2,\mathbf{r}_M) + \cdots . \quad (2.18)$$

And now we notice that this infinite series is easily summed, since it is just a geometric progression:

$$P(\mathbf{r}_2,\mathbf{r}_1) \approx P_0(\mathbf{r}_2,\mathbf{r}_1) + P_0(\mathbf{r}_M,\mathbf{r}_1) P(M) P_0(\mathbf{r}_2,\mathbf{r}_M) \times$$
$$\times [1 + P(M) P_0(\mathbf{r}_M,\mathbf{r}_M) + P(M)^2 P_0(\mathbf{r}_M,\mathbf{r}_M)^2 + \cdots]$$
$$= P_0(\mathbf{r}_2,\mathbf{r}_1) + \frac{P_0(\mathbf{r}_M,\mathbf{r}_1) P(M) P_0(\mathbf{r}_2,\mathbf{r}_M)}{1 - P(M) P_0(\mathbf{r}_M,\mathbf{r}_M)} . \qquad (2.19)$$

Thus, we have obtained an approximate solution for the propagator $P(\mathbf{r}_2,\mathbf{r}_1)$ which is valid in the strong interaction case.

This new approximation, involving the summation of a perturbation series to infinite order over a selected class of repeated diagrams (i.e., terms) is called '*partial summation*' or '*selective summation*'. It is drastically different from the ordinary perturbation approximation. It goes beyond conventional perturbation theory and can be used in cases where the interaction term is so large that the ordinary low-order perturbation approximation won't work. It is this property which makes the new technique of great value in tackling the strong interactions encountered in the many-body problem. As will be seen shortly, this method of partial summation is the basic procedure underlying the calculation of the quantum mechanical propagator.

The above diagram technique may easily be extended to the time-dependent propagator, $P(\mathbf{r}_2, \mathbf{r}_1, t_2 - t_1)$. (We have written $t_2 - t_1$ since the force is time independent so the propagators can depend only on time differences.) Let $P_0(\mathbf{r}_j, \mathbf{r}_i, t_j - t_i) =$ probability that if the particle leaves the point $\mathbf{r}_i$ at time t_i then it arrives at $\mathbf{r}_j$ at time t_j without undergoing any interaction on the way (this is the 'free propagator'). Let $P(A)$ be the interaction term, assumed instantaneous for simplicity. Then, using the convention that time increases in the positive y direction, the new diagram dictionary is given by Table 2.2 and the diagrammatic expansion becomes

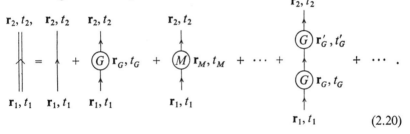

$$(2.20)$$

(Analogous to (2.15), these diagrams may be re-drawn (at least in the one-dimensional case) in a co-ordinate system with t as ordinate, and r as abscissa.) Then, in writing down the corresponding series, it is necessary to remember that t_A, the time at which the scattering from A occurs, may be anywhere between t_1 and t_2, and there is some probability that it occurs at any of these intermediate moments. Thus, the total probability is the sum of all these contributions, and this implies that we must integrate over all intermediate times, t_A. This leads to the series:

$$P(\mathbf{r}_2, \mathbf{r}_1, t_2 - t_1) = P_0(\mathbf{r}_2, \mathbf{r}_1, t_2 - t_1) +$$

$$+ \int_{t_1}^{t_2} dt_G \, P_0(\mathbf{r}_G, \mathbf{r}_1, t_G - t_1) \, P(G) \, P_0(\mathbf{r}_2, \mathbf{r}_G, t_2 - t_G) +$$

$$+ \int dt_M \cdots + \int \int + \cdots + \int \int \int + \cdots + \cdots. \quad (2.21)$$

Table 2.2 *Dictionary for pinball propagator including time*

Word	Diagram
$P(\mathbf{r}_j,\mathbf{r}_i,t_j-t_i)$	$\mathbf{r}_j,t_j$ / $\mathbf{r}_i,t_i$
$P_0(\mathbf{r}_j,\mathbf{r}_i,t_j-t_i)$	$\mathbf{r}_j,t_j$ / $\mathbf{r}_i,t_i$
$P(A)$	(A)

The unpleasant integrals parading through this expression may be removed by noticing that they all have the form of 'folded' products. This means they can be converted into simple products by a Fourier transformation. Suppose we define the transformed propagator, $P_0(\mathbf{r}_j,\mathbf{r}_i,\omega)$ (ω=frequency) by

$$P_0(\mathbf{r}_j,\mathbf{r}_i,t_j-t_i) = \frac{1}{2\pi}\int_{-\infty}^{+\infty} d\omega\, e^{-i\omega(t_j-t_i)} P_0(\mathbf{r}_j,\mathbf{r}_i,\omega) \qquad (2.22)$$

with a similar expression for $P(\mathbf{r}_j,\mathbf{r}_i,\omega)$. Then the first two terms of (2.21) become (note that we can integrate over t_G from $-\infty$ to $+\infty$ because condition (2.8) automatically limits the integral to the region $t_1 \to t_2$):

$$P_0(\mathbf{r}_2,\mathbf{r}_1,t_2-t_1) = \frac{1}{2\pi}\int_{-\infty}^{+\infty} d\omega\, e^{-i\omega(t_2-t_1)} P_0(\mathbf{r}_2,\mathbf{r}_1,\omega)$$

$$\int_{-\infty}^{+\infty} dt_G\, P_0(\mathbf{r}_G,\mathbf{r}_1,t_G-t_1) P(G) P_0(\mathbf{r}_2,\mathbf{r}_G,t_2-t_G) =$$

$$= \int_{-\infty}^{+\infty} dt_G \left[\frac{1}{2\pi}\int_{-\infty}^{+\infty} d\omega'\, e^{-i\omega'(t_G-t_1)} P_0(\mathbf{r}_G,\mathbf{r}_1,\omega')\right] \times$$

$$\times P(G) \times \left[\frac{1}{2\pi}\int_{-\infty}^{+\infty} d\omega\, e^{-i\omega(t_2-t_G)} P_0(\mathbf{r}_2,\mathbf{r}_G,\omega)\right] =$$

$$= \frac{1}{(2\pi)^2} \int\limits_{-\infty}^{+\infty} d\omega \int\limits_{-\infty}^{+\infty} d\omega' P_0(\mathbf{r}_G, \mathbf{r}_1, \omega') \times$$

$$\times P(G) P_0(\mathbf{r}_2, \mathbf{r}_G, \omega) e^{+i(\omega' t_1 - \omega t_2)} \underbrace{\int\limits_{-\infty}^{+\infty} dt_G e^{-it_G(\omega' - \omega)}}_{2\pi\delta(\omega' - \omega)}$$

$$= \frac{1}{2\pi} \int\limits_{-\infty}^{+\infty} d\omega\, e^{-i\omega(t_2 - t_1)} P_0(\mathbf{r}_G, \mathbf{r}_1, \omega) P(G) P_0(\mathbf{r}_2, \mathbf{r}_G, \omega). \quad (2.23)$$

Continuing thus, and finally taking the inverse transform, yields

$$P(\mathbf{r}_2, \mathbf{r}_1, \omega) = P_0(\mathbf{r}_2, \mathbf{r}_1, \omega) + P_0(\mathbf{r}_G, \mathbf{r}_1, \omega) P(G) P_0(\mathbf{r}_2, \mathbf{r}_G, \omega) + \cdots. \quad (2.24)$$

This is just as simple as the series (2.13) for the time-independent case. We can use the partial summation trick on it just as before. Thus inclusion of time in the propagator creates no special difficulties. Note that the Fourier transformed series may be gotten directly by using a revised edition of the 'dictionary', Table 2.2, in which the diagrams are for transforms of the propagators, as in Table 2.3. Hence the diagrams for (2.24) are just the same

Table 2.3 *Fourier transformed pinball dictionary*

Word	Diagram
$P(\mathbf{r}_j, \mathbf{r}_i, \omega)$	
$P_0(\mathbf{r}_j, \mathbf{r}_i, \omega)$	
$P(A)$	

as those in (2.20) provided we erase all the t's and put in all the ω's. Thus, we have

$$(2.25)$$

We shall not actually apply this formalism to the calculation of classical quasi particles—this would take us too far afield. Instead we go on directly to the quantum case.

Exercises

2.1 Write the diagram series for the propagator $P(\mathbf{r}_2, \mathbf{r}_1)$ assuming that the scattering at both the monkey and the lion are large, while all other interactions are small. Include all terms through second order, plus a couple of third-order terms. How many diagrams are there in nth order?

2.2 Translate the first few terms of Ex. 2.1 into functions.

2.3 Evaluate the above propagator by partial summation assuming that all $P_0(\mathbf{r}_i, \mathbf{r}_j) = c$.

2.4 Assuming all free propagators $= c$, generalize the above results to include scattering from all animals.

Chapter 3

Quantum Quasi Particles and the Quantum Pinball Propagator

3.1 The quantum mechanical propagator

In this chapter we are going to solve the simplest existing example of a quantum field theoretical problem. We call it the 'quantum pinball game' since it is the precise quantum analogue of the classical pinball machine just discussed, and in fact gives rise to a diagrammatic series having exactly the same form as (2.25). It is a sub-trivial problem, one which can be solved in a microsecond by elementary quantum mechanics. It takes a little longer to do by diagrams, but like its classical cousin in Fig. 2.3 has the great merit of illustrating all the basic principles without immersing the reader in a morass of mathematics. At the end of the chapter, the diagram method is applied to a non-trivial problem, i.e., finding the energy and lifetime of an electron propagating through a set of randomly distributed scattering centres (e.g., impurity atoms in a metal).

The fundamental difference between the classical propagator, P, and the quantum propagator, G, is that P is a probability, whereas G is a probability *amplitude*, with corresponding probability given by $|G|^2 (= G^* G)$. Thus in the classical case, the total probability for propagation from point 1 to point 2 is just the sum of the probabilities for each propagation process taken separately:

$$P(2,1)_{\text{classical}} = P(\text{process I}) + P(\text{process II}) + \cdots.$$

But in the quantum case, the total probability *amplitude* is the sum of the probability *amplitudes* for each process taken separately

$$G(2,1) = G(\text{process I}) + G(\text{process II}) + \cdots$$

so that the corresponding probability is given by

$$P(2,1)_{\text{quantum}} = G^* G = \underbrace{|G(\text{I})|^2}_{P(\text{I})} + \underbrace{|G(\text{II})|^2}_{P(\text{II})} + \underbrace{G(\text{I})^* \, G(\text{II}) + G(\text{II})^* \, G(\text{I})}_{\text{interference terms}} + \cdots.$$

Because of the characteristic 'interference terms', the quantum probability is not just the sum of the probabilities for the individual processes, in contrast to the classical case.

37

A familiar example of this is the decay of an atom, molecule, or nucleus from a state i to a state f by means of photon emission. Suppose the atom can either decay directly: $i \to f$, or via the intermediate state m: $i \to m \to f$. Then we have (call A the probability amplitude):

$$
\begin{aligned}
P(i \to f) = A^* A &= |A(i \to f) + A(i \to m \to f)|^2 \\
&= |A(i \to f)|^2 + |A(i \to m \to f)|^2 + A^*(i \to f)\, A(i \to m \to f) \\
&\quad + A^*(i \to m \to f)\, A\,(i \to f),
\end{aligned}
$$

which shows the interference between processes $i \to f$ and $i \to m \to f$. (See also Feynman (1965), pp. 19, 20.)

Let us begin by defining the quantum propagator in general, then show what it looks like in the case of free particles and quasi particles. The quantum analogue of the classical propagator is (assuming that the Hamiltonian is time-independent, so that the propagator depends only on time differences):

$$
iG(\mathbf{r}_2, \mathbf{r}_1, t_2 - t_1)_{t_2 > t_1} = iG^+(\mathbf{r}_2, \mathbf{r}_1, t_2 - t_1)
$$

$\qquad\qquad$ = probability amplitude that if at time t_1 we add a particle at point $\mathbf{r}_1$ to the interacting system in its ground state, then at time t_2 the system will be in its ground state with an added particle at $\mathbf{r}_2$. $\qquad$ (3.1)

The i factor is purely for decoration (a matter of convention) and the $+$ superscript denotes $t_2 > t_1$. (The precise meaning of the word 'add' here is discussed in detail in §9.2.) The probability corresponding to the amplitude (3.1) is

$$
P(\mathbf{r}_2, \mathbf{r}_1, t_2 - t_1) = G^+(\mathbf{r}_2, \mathbf{r}_1, t_2 - t_1)^* \, G^+(\mathbf{r}_2, \mathbf{r}_1, t_2 - t_1).
$$

Note that it is not necessarily the 'same' particle which is observed at t_2, since this has no meaning in the systems of identical particles with which we shall generally deal. Note also that a more precise way of saying that the particle is 'at point $\mathbf{r}_1$' is to say that it is 'in the position eigenstate $\delta(\mathbf{r} - \mathbf{r}_1)$'.

The quantity G^+ defined in (3.1) is called a 'retarded' propagator (or Green's function). By definition, it is equal to zero for $t_2 \leqslant t_1$. There is also an 'advanced' propagator, G^-, which is finite for $t_2 \leqslant t_1$; this will be discussed in chapter 4. (See appendix L for other types of retarded and advanced propagators.)

It is actually more convenient to work with an equivalent definition of G in terms of arbitrary single-particle eigenstates, $\phi_k(\mathbf{r})$, instead of position eigenstates. Then we have

$iG^+(k_2, k_1, t_2 - t_1)_{t_2 > t_1}$ = probability amplitude that if at time t_1 we add a particle in $\phi_{k_1}(\mathbf{r})$ to the interacting system in its ground state, then at time t_2 the system will be in its ground state with an added particle in $\phi_{k_2}(\mathbf{r})$.   (3.2)

For $t_2 \leqslant t_1$, G^+ is defined so that:

$$iG^+(k_2, k_1, t_2 - t_1)_{t_2 \leqslant t_1} = 0. \tag{3.3}$$

A convenient choice for $\phi_k(\mathbf{r})$ is the eigenstates of the unperturbed single particle Hamiltonian H_i in Appendix ($\mathscr{A}$.2), which we will call H_0:

$$H_0 = \frac{p^2}{2m} + U(\mathbf{r}) = -\frac{1}{2m}\nabla_r^2 + U(\mathbf{r}) \quad (\hbar \text{ set} = 1)$$

with

$$H_0\phi_k(\mathbf{r}) = \epsilon_k \phi_k(\mathbf{r}). \tag{3.4}$$

If $U(\mathbf{r}) = 0$, then this is just the free particle case:

$$H_0 = -\frac{\nabla_r^2}{2m}, \quad \phi_k(\mathbf{r}) = \frac{1}{\sqrt{\Omega}} e^{i\mathbf{k}\cdot\mathbf{r}}, \quad \epsilon_k = \frac{k^2}{2m} \tag{3.5}$$

where Ω = normalization volume. We shall usually set $\hbar = 1$. Spin has been neglected for simplicity.

(Note regarding notation: In (3.4), the subscript k (or k_1, or k_2, etc.) stands for all the quantum numbers necessary to designate an *arbitrary* energy eigenstate. The *particular* eigenstates will be labelled with p-subscripts thus: $\phi_{p_1}(\mathbf{r}), \phi_{p_2}(\mathbf{r}), \phi_{p_3}(\mathbf{r}), \ldots$, or, for short $\phi_1(\mathbf{r}), \phi_2(\mathbf{r}), \ldots$ (the arrangement is roughly in order of increasing energy). In the special case where $U(\mathbf{r}) = 0$, k is a wave-vector and will be written $\mathbf{k}$ (or $\mathbf{k}, \sigma$ if spin is included).)

Definition (3.2) describes 'propagation' of a particle from state $\phi_{k_1}(\mathbf{r})$ to $\phi_{k_2}(\mathbf{r})$. Note that if $k_1 = k_2$, the particle propagates in time only.

Let us first get the free propagator G_0^+ (no perturbing interaction). Suppose at time t_1 the wave function of the free particle is $\phi_{k_1}(\mathbf{r})$. Then we have:

$$\psi(\mathbf{r}, t_1) = \phi_{k_1}(\mathbf{r}). \tag{3.6}$$

At later time t_2, by the time-dependent Schrödinger equation, we find that the wave function has become

$$\psi(\mathbf{r}, t_2) = \phi_{k_1}(\mathbf{r}) e^{-i\epsilon_{k_1}(t_2 - t_1)} \tag{3.7}$$

where ϵ_{k_1} is the single particle energy of (3.4). The probability amplitude for the particle being in state ϕ_{k_2} at time t_2 is then just the component of $\psi(\mathbf{r}, t_2)$ along ϕ_{k_2} or:

$$\int d^3\mathbf{r}\,\psi(\mathbf{r}, t_2)\,\phi_{k_2}^*(\mathbf{r}) = e^{-i\epsilon_{k_1}(t_2-t_1)}\underbrace{\int d^3\mathbf{r}\,\phi_{k_1}(\mathbf{r})\,\phi_{k_2}^*(\mathbf{r})}_{\delta_{k_2 k_1}}, \tag{3.8}$$

whence, by definition of G^+:

$$\begin{aligned} G_0^+(k_2, k_1, t_2-t_1) &= -i\theta_{t_2-t_1}e^{-i\epsilon_{k_1}(t_2-t_1)}\,\delta_{k_2 k_1} \\ &= \delta_{k_2 k_1}\,G_0^+(k_1, t_2-t_1) \end{aligned} \tag{3.9}$$

where

$$G_0^+(k, t_2-t_1) = \begin{cases} -i\,\theta_{t_2-t_1}e^{-i\epsilon_k(t_2-t_1)}, & \text{for } t_2 \neq t_1 \\ 0, & \text{for } t_2 = t_1 \end{cases} \tag{3.10}$$

and

$$\theta_{t_2-t_1}\begin{cases} = 1, & \text{if } t_2 > t_1 \\ = 0, & \text{if } t_2 < t_1. \end{cases} \tag{3.11}$$

The $\theta_{t_2-t_1}$ factor is put in to take care of the fact that by definition (3.3), $G^+=0$ for $t_2 < t_1$. Note also that $G_0^+=0$ for $t_2=t_1$, by (3.3). (See (9.2), (9.4), end of appendix F.) Note that for fermions, all levels up to ϵ_F ($=$ Fermi energy—see §4.2) are filled, so we can only propagate a particle with $\epsilon_{k_1} > \epsilon_F$.

Just as with the classical pinball propagator, it is convenient to work with the Fourier transform of (3.10) ($\omega=$ frequency or 'energy parameter'):

$$\begin{aligned} G_0^+(k, \omega) &= -i\int_{-\infty}^{+\infty} d(t_2-t_1)\,\theta_{t_2-t_1}e^{i\omega(t_2-t_1)}e^{-i\epsilon_k(t_2-t_1)} \\ &= (-1)\frac{e^{i(\omega-\epsilon_k)(t_2-t_1)}}{\omega-\epsilon_k}\Bigg|_0^\infty = \frac{1}{\omega-\epsilon_k} - \frac{e^{i(\omega-\epsilon_k)\infty}}{\omega-\epsilon_k}. \end{aligned} \tag{3.12}$$

Because of the exponential oscillating at ∞, this function is not well defined. In order to get around this difficulty, we have to slightly modify the expression for the free propagator. This is done by multiplying the propagator by the factor $\exp(-\delta(t_2-t_1))$, where δ is a positive infinitesimal such that $\delta \times \infty = \infty$. Then (3.10) becomes:

$$G_0^+(k, t_2-t_1) = -i\theta_{t_2-t_1}e^{-i(\epsilon_k-i\delta)(t_2-t_1)}. \tag{3.12'}$$

For any finite (t_2-t_1), we have $\delta \times (t_2-t_1)=0$, so this is just (3.10). But for infinite (t_2-t_1), $\delta \times (t_2-t_1)=\infty$ so $G_0^+=0$. When (3.12') is placed in (3.12), we find

$$G_0^+(k, \omega) = \frac{1}{\omega-\epsilon_k+i\delta} - \frac{e^{i(\omega-\epsilon_k+i\delta)\infty}}{\omega-\epsilon_k+i\delta} = \frac{1}{\omega-\epsilon_k+i\delta}. \tag{3.13}$$

In Appendix I, it is shown that the inverse transform, i.e., the Fourier transform of (3.13), yields exactly (3.12').

The above modification of G_0^+ has no physical significance since $t_2 - t_1$ is always finite in any experiment. However, it is mathematically very convenient, because it allows us to work with well-defined integrals.

Note: The usual way of introducing the modified free propagator employs the integral representation of the step function:

$$\theta_{t_2-t_1} = - \int_{-\infty}^{+\infty} \frac{d\omega'}{2\pi i} \frac{e^{-i\omega'(t_2-t_1)}}{\omega' + i\delta}. \tag{3.13'}$$

This is not precisely a true step function but rather a modified step function, which can be seen by evaluating it using exactly the same technique as in appendix I. This yields

$$\theta_{t_2-t_1} = \begin{cases} e^{-\delta(t_2-t_1)}, & \text{for } (t_2 - t_1) > 0 \\ 0, & \text{for } (t_2 - t_1) < 0 \end{cases} \tag{3.13''}$$

which is just (3.11), except when $(t_2 - t_1) \to \infty$, where it goes to zero. Inserting (3.13'') in (3.10) yields just (3.12'). Alternatively, we can place (3.13') in (3.12), integrate over $t_2 - t_1$ first (which gives $2\pi\delta(\omega' - \omega + \epsilon_k)$), then over ω' and immediately obtain (3.13).

In this transformed version, (3.13) it is seen that the free propagator possesses poles at (i.e., infinitesimally close to) $\omega = \epsilon_k$, i.e., at the energy of the added particle in state ϕ_k. This turns out to be quite general, and in fact it may be shown that (see appendix H):

> The poles of $G^+(k,l;\omega)$, the Fourier transform of the single-particle propagator, occur at values of ω equal to the excited state energies of the interacting $(N+1)$-particle system minus the ground energy of the interacting N-particle system. (3.14)

This property accounts for the extraordinary utility of the propagator in many-body theory.

Now consider the propagator in the presence of interaction. Analogous to the classical case in chapter 2, quantum quasi particles act like free particles except that they have a new energy ϵ_k' instead of ϵ_k, and a lifetime τ_k. Therefore we expect that if the added particle behaves as a quasi particle, the single-particle propagator will have the same form as the free propagator except for the replacement of ϵ_k by ϵ_k' and the inclusion of an exponential decay factor with time constant τ_k. One more thing: In a Fermi system, because of the Pauli principle, each state can hold at most one particle. Therefore, if state k is already partially (or fully) occupied, the probability amplitude that we can add an extra particle in state k will be less than 1. Hence we have to multiply by a factor $Z_k \leqslant 1$. This gives us:

$$G^+_{\substack{quasi \\ particle}}(k, t_2 - t_1) = -iZ_k e^{-i\epsilon_{k'}(t_2 - t_1)} e^{-(t_2 - t_1)/\tau_k}. \qquad (3.15)$$

This has the Fourier transform

$$G^+_{\substack{quasi \\ particle}}(k, \omega) = \frac{Z_k}{\omega - \epsilon_k' + i\tau_k^{-1}}. \qquad (3.16)$$

For these expressions to be sensible, it is evident that the lifetime of the quasi particles must be long, so that the width of the energy levels, τ_k^{-1} (see appendix $\mathscr{A}$ after ($\mathscr{A}$.43)) is much less than the values of the energies themselves, i.e.:

$$\tau_k^{-1} \ll \epsilon_k'. \qquad (3.17)$$

(A more exact condition on τ_k is given in (8.21).)

Thus, if G^+ is calculated, and it is found that it has the above form, then the system is describable in terms of the simple quasi particle picture. Such systems are rewarded with the name 'normal'. On the other hand, even if the system turns out to be of the less co-operative 'abnormal' variety where (3.16) does not hold (like for example the one-particle system of (4.39), or the superconducting system of chapter 15), we can still get the excited state energies by means of (3.14).

It is still possible (in the case of normal systems) to interpret the poles of the quasi particle propagator (3.16) as yielding the excited state energies of the system (as in (3.14)), if the energy is regarded as being complex, with ϵ_k' being its real, and $i\tau_k^{-1}$ its (small) imaginary part:

$$\omega_{\text{pole}} = \epsilon_k' - i\tau_k^{-1}. \qquad (3.18)$$

Such complex energies are the same sort of thing we meet in the case of an atom in an excited state, ϕ_n, with energy ϵ_n. In the absence of interaction with other atoms or with radiation, the wave function is

$$\psi_n(t) = \phi_n e^{-i\epsilon_n t}. \qquad (3.18')$$

If weak interactions are turned on, the energy shifts to ϵ_n' and the atom starts to decay out of state ϕ_n. Thus, the approximate wave function may be written

$$\psi_n'(t) \approx \phi_n e^{-i\epsilon_{n'} t} e^{-t/\tau_n} = \phi_n e^{-i(\epsilon_{n'} - i\tau_n^{-1}) t} \qquad (3.18'')$$

which has just the form of (3.18'), but with a complex energy $\epsilon_n' - i\tau_n^{-1}$ replacing the real energy ϵ_n. (See note after (3.70) and also after appendix (H.10).)

3.2 The quantum pinball game

In order to illustrate the principles involved, we will now find the propagator for a simple system consisting of one particle in an external momentum-conserving potential, which turns out to be the exact quantum analogue of the classical pinball game. Of course, in a one-particle system, we cannot have quasi particles in the 'real particle plus cloud of other particles' sense. However, as mentioned in connection with conduction electrons in §0.2, we may rather loosely regard the particle as being clothed by the external potential itself.

The quantum pinball game consists of a single free particle subjected simultaneously to two perturbing potentials V_M, V_L, which are the analogues of two different animal scatterers in the classical pinball game. The unperturbed Hamiltonian, wave functions, and energies are given by (3.5). We take as the perturbation the 'velocity dependent' potential

$$V(\mathbf{p}) = V_M + V_L = Mp^2 + Lp^4 = -M\nabla_r^2 + L\nabla_r^4 \qquad (3.19)$$

where M, L, are real constants, and it is assumed that $M \gg L$.

This odd-looking potential, which has been chosen because of its great mathematical simplicity, may have a traumatic effect on some readers. It is certainly not the sort of potential one meets on the street—those are mostly of the familiar $V(\mathbf{r})$ form. Nevertheless it is quite easy to construct perturbations of the form (3.19) artificially. For example, the Hamiltonian for the centre of mass motion of a free hydrogen atom is $H = p^2/(m + m_e)$ where m = proton mass and m_e = electron mass. This may be broken up into

$$H = \frac{p^2}{2m} - \frac{m_e}{(m_e + m)\, m} p^2,$$

and the second term treated as if it were a 'perturbing potential'. In a similar fashion, a p^4 term can come as a relativistic correction when we expand the relativistic Hamiltonian:

$$H = (m_0^2 c^4 + p^2 c^2)^{\frac{1}{2}} \approx m_0 c^2 + \frac{p^2}{2m_0} - \frac{p^4}{8m_0^3 c^2}.$$

In fact, if we regard this as the relativistic Hamiltonian for the centre of mass motion of a free hydrogen atom, with $m_0 = m + m_e$, then we can write

$$H \approx (m + m_e)c^2 + \frac{p^2}{2m} - \frac{m_e}{(m_e + m)\, m} p^2 - \frac{p^4}{8(m + m_e)^3 c^2},$$

which has just the form (3.19), except for the unimportant constant term. Examples of real velocity-dependent potentials arise in the case of an electron in a magnetic field ($V \propto \mathbf{A} \cdot \mathbf{p}$), and in nuclear physics.

The problem, then, is to find the energy of the free particle when it is perturbed by $V(\mathbf{p})$.

Let us first look at the conventional solution of the problem. Since $M \gg L$, we may at first neglect the L term and have $V(\mathbf{p}) \approx Mp^2$ so

$$H \approx \frac{p^2}{2m} + Mp^2. \tag{3.20}$$

Because the perturbation has the same form as the unperturbed $H_0 = p^2/2m$, the perturbed wave functions are just the old ϕ_k's of (3.5) and the new energy is

$$\epsilon_k' = \left(\frac{1}{2m} + M\right) k^2. \tag{3.21}$$

For purposes of comparison with appendix ($\mathscr{A}$.21), this result may be obtained by means of the trivial 'canonical transformation'

$$H = \underbrace{\frac{p^2}{2m}}_{H_0} + \underbrace{Mp^2}_{H_1} \rightarrow H' = \underbrace{\left(\frac{1}{2m} + M\right) p^2}_{H_0'} + \underbrace{0}_{H_1'} \; . \tag{3.22}$$

Thus, H_0' may be regarded as describing a sort of rudimentary 'quasi particle' having a modified energy dispersion law given by (3.21). (In this simple example, the 'fictitious bodies' of ($\mathscr{A}$.21) and the quasi particles of ($\mathscr{A}$.43) are the same thing.)

Consider next the effect of adding the L term. This also has the same eigenfunctions as H_0 and we find:

$$L\nabla^4 \phi_k = Lk^4 \phi_k \tag{3.23}$$

from which it follows that the total energy of the particle is

$$\epsilon_k'' = \left(\frac{1}{2m} + M\right) k^2 + Lk^4. \tag{3.24}$$

Let us now solve the same trivial problem with the aid of the single-particle propagator and see how we can get the above energies, ϵ_k', ϵ_k'', as 'quasi particle' energies from the poles of the propagator. This requires that we first obtain the perturbation series for the propagator analogous to the series (2.21) for the classical animal game case. We will get this series by the same sort of physically intuitive argument used in the classical case. (The rigorous mathematical way of getting the perturbation series is outlined in §3.4.)

According to the instructions in the definition of the propagator, at time t_1 a particle is introduced into the (in this case, initially empty) system in state

$\phi_{k_1}(\mathbf{r}) = \Omega^{-\frac{1}{2}}\exp(i\mathbf{k}_1 \cdot \mathbf{r})$, and propagates through the system, being scattered zero, one or more times by the external potentials:

$$V_M = Mp^2 \quad \text{or} \quad V_L = Lp^4. \tag{3.25}$$

By definition (3.2) the propagator $iG^+(\mathbf{k}_2, \mathbf{k}_1, t_2 - t_1)$ is just the probability amplitude that the particle will be in the state $\phi_{k_2}(\mathbf{r}) = \Omega^{-\frac{1}{2}}\exp(i\mathbf{k}_2 \cdot \mathbf{r})$ at time t_2. Analogous to the animal case, this amplitude iG^+ is just the sum of the probability amplitudes for all the different ways the particle can go through the system, beginning in state ϕ_{k_1} and winding up in state ϕ_{k_2}.

For example, the simplest way the particle can propagate through the system is freely, without interaction. The probability amplitude for this is just the free propagator $i\delta_{k_1 k_2}G_0^+(\mathbf{k}_1, t_2 - t_1)$ as in (3.9), (3.10). Another way is to enter in ϕ_{k_1} at time t_1, be scattered into state ϕ_{k_2} at time t_M by the potential V_M, then continue freely in ϕ_{k_2} until time t_2. (It may seem peculiar to say that the particle is scattered by the potential V_M at time t_M, or to say that the particle is scattered several times by the potential, when the potential is actually there the whole time. However, this is just a result of the fact that what we are doing in such a perturbation expansion is to decompose the total propagator into primitive components, each component being an instantaneous scattering by the potential. At the end we integrate over all times as shown in (3.28), and sum over all sequences of scattering processes as in (3.30), thus 'putting the propagator back together again'.) The amplitude for this second way will be, by analogy with the classical pinball case, the product of the amplitudes for the independent processes it is composed of. (That these processes are independent can be seen from the fact that a particle which has been scattered into state ϕ_k from state ϕ_l cannot be distinguished from one scattered into ϕ_k from another state $\phi_{l'}$. That is, the particle now in ϕ_k has no 'memory' of how it got there, just as in the classical pinball case.)

The first of these independent processes, free propagation from t_1 to t_M in state ϕ_{k_1}, has amplitude $iG_0^+(\mathbf{k}_1, t_M - t_1)$, according to (3.10). The amplitude for the second process, i.e., scattering from ϕ_{k_1} to ϕ_{k_2} by V_M at time t_M, can be obtained from ordinary time-dependent perturbation theory as follows: Let c_l be the probability amplitude that at time t_0 a system is in state ϕ_l. Then at later time, t, the time rate of change of any particular c_l, say c_p, under the influence of perturbation V, is given by:

$$\dot{c}_p(t) = -i \sum_l V_{pl} c_l e^{i(\epsilon_p - \epsilon_l)(t - t_0)} \tag{3.26}$$

where V_{pl} is the matrix element of V between states ϕ_p, ϕ_l (see, for example, Dicke (1960), Eq. 14–57, with $\hbar = 1$). In the process under consideration, at time $t_0 = t_M$, the system is definitely in state ϕ_{k_1}, so $c_l = \delta_{lk_1}$. The perturbation $V = V_M$. Hence the probability amplitude per unit time that the system undergoes a transition from ϕ_{k_1} to $\phi_p = \phi_{k_2}$, at time t_M (i.e., t here is also equal to t_M) is

$$\dot{c}_{k_2}(t = t_M) = -iV_{M_{k_2 k_1}} = -i \int d^3 \mathbf{r} \, \phi_{k_2}^*(\mathbf{r}) V_M \phi_{k_1}(\mathbf{r}) =$$

$$= +iM \int d^3 \mathbf{r} \, \phi_{k_2}^* \nabla^2 \phi_{k_1} = -iMk_1^2 \delta_{k_2 k_1}. \qquad (3.27)$$

The $\delta_{k_1 k_2}$ shows that the process here conserves momentum so that the particle still has the same momentum after scattering. The amplitude for the last process is $iG_0^+(\mathbf{k}_2, t_2 - t_M)$. Hence the total amplitude is the product

$$\begin{bmatrix} Probability \\ Amplitude \end{bmatrix}_{t_1 \to t_M \to t_2} = i \int_{-\infty}^{+\infty} dt_M \, G_0^+(\mathbf{k}_1, t_M - t_1) \, V_{M_{k_2 k_1}} \, G_0^+(\mathbf{k}_2, t_2 - t_M). \quad (3.28)$$

We have integrated over t_M since the collision with V_M could have occurred at any intermediate time $t_1 < t_M < t_2$. Note that the θ-function in G_0^+ (see (3.10)) automatically restricts the region of integration to $t_1 < t_M < t_2$.

Similarly, there can be an interaction with V_L described by the matrix element

$$-iV_{L_{k_2 k_1}} = -iLk_1^4 \delta_{k_2 k_1} \qquad (3.29)$$

which also conserves momentum. There are also second- and higher-order processes in which the particle collides with V_M and V_L any number of times. This gives us the series expansion for the propagator (set $\mathbf{k}_1 = \mathbf{k}_2 = \mathbf{k}$ because of conservation of momentum here), after cancelling the i's:

$$G^+(\mathbf{k}, t_2 - t_1) = G_0^+(\mathbf{k}, t_2 - t_1) + \int_{-\infty}^{+\infty} dt_M \, G_0^+(\mathbf{k}, t_M - t_1) \, V_{M_{kk}} \, G_0^+(\mathbf{k}, t_2 - t_M)$$

$$+ \int_{-\infty}^{+\infty} dt_L \, G_0^+(\mathbf{k}, t_L - t_1) \, V_{L_{kk}} \, G_0^+(\mathbf{k}, t_2 - t_L) +$$

$$+ \int dt_M dt_M' \cdots + \int dt_M dt_L \cdots + \cdots +$$

$$+ \int dt_M dt_M' dt_M'' \cdots + \cdots. \qquad (3.30)$$

Just as in the classical pinball case, the integrals in the above series may be eliminated by taking the Fourier transform. This yields, analogous to (2.24):

$$G^+(\mathbf{k}, \omega) = G_0^+(\mathbf{k}, \omega) + [G_0^+(\mathbf{k}, \omega)]^2 V_{M\mathbf{kk}} + [G_0^+(\mathbf{k}, \omega)]^2 V_{L\mathbf{kk}} +$$
$$+ [G_0^+]^3 V_{M\mathbf{kk}}^2 + 2[G_0^+]^3 V_{M\mathbf{kk}} V_{L\mathbf{kk}} + [G_0^+]^3 V_{L\mathbf{kk}}^2 +$$
$$+ [G_0^+]^4 V_M^3 + \cdots. \tag{3.31}$$

We now pull the same trick used in the classical case and make a dictionary to translate the above series into diagrams. The primitive diagrams are in Table 3.1. Compare this with Table 2.2, which is in $(\mathbf{r}, t)$-space, and Table 2.3 in $(\mathbf{r}, \omega)$-space. (Equations (3.30), (3.31) could also be written out in $(\mathbf{r}, t)$- and $(\mathbf{r}, \omega)$-space but in the present case this would not be very useful.)

Table 3.1 *Diagram dictionary for quantum pinball propagator*

(k, t)-space		(k, ω)-space	
Word	Diagram	Word	Diagram
$iG^+(\mathbf{k}_2, \mathbf{k}_1, t_2 - t_1)$	$\mathbf{k}_2 t_2$ / $\mathbf{k}_1 t_1$	$iG^+(\mathbf{k}_2, \mathbf{k}_1, \omega)$	ω $\mathbf{k}_2$ / $\mathbf{k}_1$
$iG_0^+(\mathbf{k}, t_2 - t_1)$ $= \theta_{t_2-t_1} e^{-i\epsilon_k(t_2-t_1)}$	$\mathbf{k}$ t_2 / t_1	$iG_0^+(\mathbf{k}, \omega) = \dfrac{i}{\omega - \epsilon_k + i\delta}$	$\mathbf{k}, \omega$
$-iV_{Aml}$	m / (A) / l	$-iV_{Aml}$	m / (A) / l

With this dictionary it is easy to write out the series of diagrams corresponding to (3.30) or (3.31):

$$\tag{3.32}$$

where the lines may be labelled with t's to give (3.30) or ω's for (3.31). This is evidently the sum of all possible different diagrams for this case.

Now, since we assumed that $M \gg L$, all interactions with V_L may be neglected, and the above series may be approximated by

$$
\text{(3.33)}
$$

This is the precise analogue of the partial sum over all monkey diagrams in (2.17). And, as in the monkey case, the summation is easy, since once again it is just a geometric series. Translating (3.33) into words with the aid of Table 3.1 (use $(\mathbf{k}, \omega)$-space), cancelling i's and dropping $(\mathbf{k}, \omega)$'s for brevity yields

$$
\begin{aligned}
G^+(\mathbf{k}, \omega) &\approx G_0^+ + (G_0^+)^2 \, V_{M_{kk}} + (G_0^+)^3 \, V_{M_{kk}}^2 + \cdots \\
&= G_0^+ [1 + G_0^+ \, V_{M_{kk}} + (G_0^+)^2 \, V_{M_{kk}}^2 + \cdots] \\
&= \frac{G_0^+}{1 - G_0^+ \, V_{M_{kk}}} = \frac{1}{(G_0^+)^{-1} - V_{M_{kk}}}, \quad \text{for } |G_0^+ \, V_{M_{kk}}| < 1. \quad \text{(3.34)}
\end{aligned}
$$

This same result may be obtained conveniently in a way which saves a lot of writing by manipulating the diagrams themselves; this is legitimate because in $(\mathbf{k}, \omega)$-space each diagram part stands for a factor. Thus (3.34) may be re-written:

$$
\text{(3.35)}
$$

3.2] QUANTUM QUASI PARTICLES 49

which may be then translated into

$$G^+(\mathbf{k}, \omega) \approx \frac{1}{(G_0^+)^{-1} - V_{M_{kk}}} \tag{3.36}$$

i.e., just the result (3.34). (Note: the little 'stumps' of line connected to $\widehat{M}$ have no value in themselves. They just show where the propagator lines are to be attached!)

(Observe that (3.35) may be written in a very useful alternative form, i.e.

$$\text{(diagram equation)} \tag{3.36'}$$

This may be proved by iteration:

$$\text{(diagram equation)}$$

In $(\mathbf{k}, \omega)$-space, (3.36') may be factored into

$$\text{(diagram equation)}$$

which may be solved algebraically to yield (3.35). However, (3.36') has the advantage of being more general than (3.35) since it may also be used when the diagrams do not factor. For example in $(\mathbf{k}, t)$-space, it yields an integral equation instead of an algebraic one. (See exercise 3.8).)

Finally, we substitute for G_0^+ and for V_M and obtain:

$$G^+(\mathbf{k}, \omega) = \frac{1}{\omega - \epsilon_k + i\delta - V_{M_{kk}}} = \frac{1}{\omega - (\epsilon_k + Mk^2) + i\delta}. \tag{3.37}$$

Comparing this with the quasi particle propagator (3.16), we find

$$\epsilon_k' = \epsilon_k + Mk^2 = \left(\frac{1}{2m} + M\right)k^2$$

$$\tau_k = \frac{1}{\delta} = \infty. \tag{3.38}$$

That is, the interaction with V_M has 'clothed' the particle, turning it from a 'bare' into a 'quasi' particle, having modified energy dispersion law given by ϵ_k' in (3.38) and infinite lifetime. And comparison with (3.21) shows this to be precisely the same result obtained by direct solution of the Schrödinger equation!

On second thought, when we realize that it has taken us three pages to do by diagrams what we did directly in three lines, there appears to be little cause for celebration. We seem to have built an elephant cannon to shoot a horse-fly. Of course this is not true. The quantum pinball game is intended only as a transparent example to introduce the general principles. The big many-body game will come later. Furthermore, at the end of this chapter, in §3.5, we apply the method to a non-trivial one-particle problem: finding the energy and lifetime of an electron in an impure metal.

In this simple example, it is actually possible to do much better than just the partial sum (3.33). We can in fact sum over all the diagrams of (3.32) as follows:

$$\tag{3.39}$$

or, translating:

$$G^+(\mathbf{k}, \omega) = \frac{1}{(G_0^+)^{-1} - (V_{M\mathbf{k}\mathbf{k}} + V_{L\mathbf{k}\mathbf{k}})} = \frac{1}{\omega - \left(\dfrac{k^2}{2m} + Mk^2 + Lk^4\right) + i\delta} \quad (3.40)$$

which gives

$$\epsilon_k'' = \frac{k^2}{2m} + Mk^2 + Lk^4 \quad (3.41)$$

in agreement with (3.24). This shows that we could just as well have taken $V = V_M + V_L$ together from the start and represented them by a single diagram, $\begin{smallmatrix}|\\(V)\\|\end{smallmatrix}$. The potential was broken up into two parts just to make the parallel with the classical pinball game more obvious.

3.3 Disappearance of disagreeable divergences

It is important to note a weakness in the above method. The geometric series in (3.34) converges only for $|G_0^+ V_{M\mathbf{k}\mathbf{k}}| < 1$, which means that

$$\left|\frac{Mk^2}{\omega - \epsilon_k + i\delta}\right| < 1 \quad \text{or} \quad \begin{array}{l} \omega > \epsilon_k + Mk^2 \\ \omega < \epsilon_k - Mk^2. \end{array} \quad (3.42)$$

But to get (3.38) we set $\omega = \epsilon_k + Mk^2$, which is just where the series begins to diverge! This is a typical example of the sort of divergence which plagues the diagram method. The usual household remedy is to assume that the propagator is still valid for ω in the region of divergence. Or, in more fancy language, one assumes that the partial sum result for the propagator may be 'analytically continued' into the divergent region. This might be called 'the Hypothesis of the Disappearance of Disagreeable Divergences'.

In many cases one can justify this by using a different method to get the propagator. We can do just this in the present case. All that is necessary is to take for the unperturbed Hamiltonian of (3.4)

$$H_0' = \frac{p^2}{2m} + Mp^2 \quad (3.43)$$

instead of just the $p^2/2m$ in (3.5). The free propagator for this new H_0' is, by exactly the same argument leading to (3.13) just

$$G_0^{+'}(\mathbf{k}, \omega) = \frac{1}{\omega - \left(\dfrac{k^2}{2m} + Mk^2\right) + i\delta} \quad (3.44)$$

which is precisely the result in (3.37). This shows that the propagator (3.37) is good for all ω.

Another indication that such divergences are largely spurious (Kurki-Suonio (1965)) is that if we do the partial sum in t-space instead of ω-space, the divergence does not occur, at least not in this simple case. Thus, using (3.30) and Table 3.1 and summing just over terms containing the V_M interaction, we find:

$$G^+(\mathbf{k}, t_2 - t_1) = -ie^{-i\epsilon_k(t_2-t_1)}\left[1 + (-iV_{Mkk})\int_{t_1}^{t_2} dt_M + \right.$$

$$+ (-iV_{Mkk})^2 \int_{t_1}^{t_2} dt'_M \int_{t_1}^{t_M'} dt_M + \cdots \right]$$

$$= -ie^{-i\epsilon_k(t_2-t_1)}\left[1 + (-iV_{Mkk})(t_2-t_1) + \frac{1}{2!}(-iV_{Mkk})^2(t_2-t_1)^2 + \right.$$

$$+ \frac{1}{3!}(-iV_{Mkk})^3(t_2-t_1)^3 + \cdots \right]$$

$$= -ie^{-i\epsilon_k(t_2-t_1)}[e^{-iV_{Mkk}(t_2-t_1)}] \qquad (3.45)$$

which is just the Fourier transform of (3.37) and converges for all values of $-iV_{Mkk}(t_2-t_1)$.

3.4 Where the diagram expansion of the propagator really comes from

The results in this chapter were obtained by analogy with the classical pinball case. Since such intuitive arguments may seem like voodoo to some readers, we will now show in a rough way how the diagram expansion of G^+ in this single-particle case can be gotten from the Schrödinger equation. (The derivation for the many-body case is in the Appendices.)

The first thing to realize is that G_0^+ and G^+ are actually Green's functions. Recall that if we have a differential equation of the form

$$L\psi(\mathbf{x}, t) = f(\mathbf{x}, t), \qquad (3.46)$$

where L is a linear differential operator which does not depend explicitly on x or t, then the Green's function, G, associated with this equation is the solution of

$$LG(\mathbf{x} - \mathbf{x}', t - t') = \delta(\mathbf{x} - \mathbf{x}')\,\delta(t - t'). \qquad (3.47)$$

Now the unperturbed Schrödinger equation may be written

$$\left(+\frac{\nabla^2}{2m} + i\frac{\partial}{\partial t}\right)\psi(\mathbf{x}, t) = 0. \qquad (3.48)$$

This has the form (3.46) (with $f(x,t)=0$), so that the associated Green's function obeys

$$\left(+\frac{\nabla^2}{2m}+i\frac{\partial}{\partial t}\right)G(\mathbf{x}-\mathbf{x}',t-t') = \delta(\mathbf{x}-\mathbf{x}')(t-t').$$ (3.49)

Fourier transforming G, we have

$$G(\mathbf{x}-\mathbf{x}',t-t') = \int \frac{d^3\mathbf{k}}{(2\pi)^3} e^{i\mathbf{k}\cdot(\mathbf{x}-\mathbf{x}')} G(\mathbf{k},t-t').$$ (3.50)

Setting this into (3.49) yields

$$\left(-\frac{k^2}{2m}+i\frac{\partial}{\partial t}\right)G(\mathbf{k},t-t') = \delta(t-t').$$ (3.51)

If we now use for G the free propagator in (3.10):

$$G = G_0^+(\mathbf{k},t-t') = -i\theta_{t-t'} e^{-i\epsilon_k(t-t')}$$ (3.52)

and use the fact that

$$\frac{d\theta_x}{dx} = \delta(x), \quad f(x)\delta(x) = f(0)\delta(x),$$ (3.53)

we find that (3.51) is satisfied, showing that G_0^+ is indeed a Green's function.

In a similar way, the Schrödinger equation with a perturbing potential of form $V(\mathbf{\nabla})$ (as in (3.19)),

$$\left[+\frac{\nabla^2}{2m}+i\frac{\partial}{\partial t}-V(\mathbf{\nabla})\right]\psi(\mathbf{x},t) = 0,$$ (3.54)

has the associated Green's function equation (in k-space):

$$\left[-\frac{k^2}{2m}+i\frac{\partial}{\partial t}-V(\mathbf{k})\right]G^+(\mathbf{k},t-t') = \delta(t-t'),$$ (3.55)

where $V(\mathbf{k})$ is the Fourier transform of $V(\mathbf{\nabla})$. The solution to this may be written as an integral equation

$$G^+(\mathbf{k},t-t') = G_0^+(\mathbf{k},t-t')+ \int_{-\infty}^{+\infty} dt'' G_0^+(\mathbf{k},t-t'') V(\mathbf{k}) G^+(\mathbf{k},t''-t'),$$ (3.56)

as can be seen by substituting (3.56) in (3.55) and using (3.51) with $G=G_0^+$. Finally, we obtain the perturbation expansion for G^+ in terms of G_0^+ by iterating (3.56):

$$G^+(\mathbf{k},t-t') = G_0^+(\mathbf{k},t-t')+ \int_{-\infty}^{+\infty} dt'' G_0^+(\mathbf{k},t-t'') V(\mathbf{k}) G_0^+(\mathbf{k},t''-t')+$$

$$+ \int_{-\infty}^{+\infty}\int_{-\infty}^{+\infty} dt'' dt''' G_0^+ V G_0^+ V G_0^+ +\cdots$$ (3.57)

which is just the series (3.30). The translation into diagrams is accomplished immediately by using dictionary Table 3.1.

It may be remarked that the Green's function for the many-body case obeys an equation of type (3.47) but with L a non-linear operator.

3.5 [Energy and lifetime of an electron in an impure metal]

(This section can be skipped on first reading!)

We will now apply the propagator method to a more realistic problem, i.e., an electron in an impure metal. For simplicity, let us pretend that the regularly arranged lattice ions in the metal have been removed, so that all we have left is an electron interacting with a set of N randomly distributed impurity ions (see Fig. 0.13B), which we assume are identical, in a volume Ω. Then, as discussed in §1.3, the propagator will be given by (1.2) or (2.25) with the circles interpreted as scattering from the various ions:

$$(3.58)$$

where the $\mathbf{k}$'s denote momentum eigenstates of a free electron as in (3.5), and i denotes the impurity ion at position $\mathbf{R}_i$.

If the potential well for an impurity at the origin has the form $W(\mathbf{r})$, then an identical ion at point $\mathbf{R}_i$ will have the potential $W(\mathbf{r} - \mathbf{R}_i)$. Hence the matrix element for the transition $\mathbf{k} \rightarrow \mathbf{l}$ at ion i is given by

$$-iV_{lk}(\mathbf{R}_i) = \frac{-i}{\Omega} \int d^3 r\, e^{-i(\mathbf{l}-\mathbf{k})\cdot\mathbf{r}} W(\mathbf{r} - \mathbf{R}_i) = \frac{(-i)}{\Omega} e^{-i(\mathbf{l}-\mathbf{k})\cdot\mathbf{R}_i} W_{lk}$$

where

$$W_{lk} = \int d^3 \mathbf{r}' \, e^{-i(\mathbf{l}-\mathbf{k})\cdot\mathbf{r}'} W(\mathbf{r}') \tag{3.59}$$

The series (3.58) may now be written out in terms of functions as follows (after eliminating the i's and suppressing ω's for brevity, and noting that it is necessary to sum over all values of the intermediate momentum, l):

$$G^+(\mathbf{k}_2, \mathbf{k}_1) = G_0^+(\mathbf{k}_1)\,\delta_{k_1 k_2} + G_0^+(\mathbf{k}_2) \sum_{i=1}^{N} V_{k_2 k_1}(\mathbf{R}_i)\, G_0^+(\mathbf{k}_1) +$$

$$+ G_0^+(\mathbf{k}_2) \left[\sum_{l} \sum_{i=1}^{N} V_{k_2 l}(\mathbf{R}_i)\, G_0^+(l)\, V_{lk_1}(\mathbf{R}_i) \right] G_0^+(\mathbf{k}_1) +$$

$$+ G_0^+(\mathbf{k}_2) \left[\sum_{l} \sum_{j \neq i} V_{k_2 l}(\mathbf{R}_j)\, G_0^+(l) \sum_{i=1}^{N} V_{lk_1}(\mathbf{R}_i) \right] G_0^+(\mathbf{k}_1) + \cdots. \tag{3.60}$$

The above G^+ is for a particular set of $\mathbf{R}_i$'s, i.e., a particular arrangement of impurities in the system, and for each different set of $\mathbf{R}_i$'s, we will get a different value of G^+. Consider now an ensemble consisting of all possible arrangements of impurities. Suppose this ensemble is *random*, i.e., the coordinate for the ith impurity, $\mathbf{R}_i$, is equally likely to be found anywhere in the volume Ω. Let us imagine that we compute $\langle G^+ \rangle$, the average value of G^+ for the ensemble. Clearly, for any specific arrangement, $G^+ \neq \langle G^+ \rangle$. But, as is common in large systems (see Landau and Lifshitz (1959), pp. 5–8), in the limit $N \to \infty$ (with $N/\Omega = $ constant), the ratio of the mean square fluctuation $(\langle G^{+2} \rangle - \langle G^+ \rangle^2)$ to $\langle G^+ \rangle^2$ will go to zero, so that we can take $G^+ = \langle G^+ \rangle$ for all but a negligible number of arrangements (see Kohn and Luttinger (1957), especially Appendix B). Hence our object here will be to calculate $\langle G^+ \rangle$.

The average $\langle G^+ \rangle$ is the sum of the average of each term in the perturbation expansion (3.60). For the second term on the right side of (3.60) we have, noting that free propagators may be factored out when averaging since they are independent of $\mathbf{R}_i$,

$$\left\langle G_0^+(\mathbf{k}_2)\, G_0^+(\mathbf{k}_1) \sum_{i=1}^{N} V_{k_2 k_1}(\mathbf{R}_i) \right\rangle = G_0^+(\mathbf{k}_2)\, G_0^+(\mathbf{k}_1)\, \frac{W_{k_2 k_1}}{\Omega} \left\langle \sum_{i=1}^{N} e^{-i(\mathbf{k}_2 - \mathbf{k}_1)\cdot\mathbf{R}_i} \right\rangle \tag{3.61}$$

The last factor here may be written

$$\left\langle \sum_{i=1}^{N} e^{-i(\mathbf{k}_2 - \mathbf{k}_1)\cdot\mathbf{R}_i} \right\rangle = \sum_{i=1}^{N} \left\langle e^{-i(\mathbf{k}_2 - \mathbf{k}_1)\cdot\mathbf{R}_i} \right\rangle = N \left\langle e^{-i(\mathbf{k}_2 - \mathbf{k}_1)\cdot\mathbf{R}_i} \right\rangle \tag{3.61'}$$

since each of the N terms is identical in form, so that we can just average over one of them and multiply by N. For a random ensemble, the probability of finding the ith impurity atom within volume $d^3 \mathbf{R}_i$ surrounding the point $\mathbf{R}_i$ will be independent of $\mathbf{R}_i$ and equal to $d^3 \mathbf{R}_i / \Omega$. Hence we have

$$\langle e^{-i(\mathbf{k}_2 - \mathbf{k}_1) \cdot \mathbf{R}_i} \rangle = \frac{1}{\Omega} \int d^3 \mathbf{R}_i \, e^{-i(\mathbf{k}_2 - \mathbf{k}_1) \cdot \mathbf{R}_i} = \frac{1}{\Omega} \times \Omega \delta_{k_2 k_1}. \qquad (3.62)$$

(Note: In a one-dimensional box of length L,

$$I \equiv \int\limits_{-L/2}^{+L/2} dx \exp(-ikx) = 2k^{-1} \sin(kL/2).$$

Because of periodic boundary conditions, the wave function at $x = 0$ equals that at $x = L$, i.e., $\exp(ikx) = \exp(ik(x+L))$. Hence $\exp(ikL) = 1$, or $k = 2\pi n/L$ (n = integer). Thus $I = L\delta_{k,0}$. Equation (3.62) is just the three-dimensional version of this, with $\Omega = L^3$. If k is continuous, the integral (3.62) yields $(2\pi)^3 \delta(\mathbf{k}_2 - \mathbf{k}_1)$, which is a Dirac δ-function.)

In the third term of (3.60), which represents two successive scatterings from the *same* impurity, we have, using the same method as above:

$$\sum_l G_0^+(1) \left\langle \sum_i V_{k_2 l}(\mathbf{R}_i) V_{l k_1}(\mathbf{R}_i) \right\rangle = \sum_l G_0^+(1) \frac{W_{k_2 l} W_{l k_1}}{\Omega^2} \left\langle \sum_i e^{-i(\mathbf{k}_2 - 1 + 1 - \mathbf{k}_1) \cdot \mathbf{R}_i} \right\rangle$$

$$= \frac{N}{\Omega^2} \sum_l G_0^+(1) \, W_{k_2 l} W_{l k_1} \delta_{k_2 k_1}. \qquad (3.63)$$

It is convenient at this point to change from a sum over $\mathbf{l}$ to an integral by

$$\sum_l \rightarrow \frac{\Omega}{(2\pi)^3} \int d^3 \mathbf{l}. \qquad (3.64)$$

This is legitimate in the case of a large (i.e., macroscopic) system, since the points $\mathbf{l}$ in $\mathbf{k}$-space are very close to each other. The factor $\Omega/(2\pi)^3$ is the density of points in $\mathbf{k}$-space. To see this, we note that in one dimension, $k = 2\pi n/L$ (n = integer). (See just after (3.62).) Thus there are $L/2\pi$ points per unit length in **k**-space in one dimension. In three dimensions we have $L^3/(2\pi)^3 = \Omega/(2\pi)^3$ points per unit volume in **k**-space. Using (3.64), (3.63) becomes

$$= \frac{N}{\Omega} \int \frac{d^3 1}{(2\pi)^3} G_0^+(1) W_{k_2 l} W_{l k_1} \delta_{k_2 k_1}. \tag{3.64'}$$

The fourth term of (3.60) (two successive scatterings from *different* impurities) contains the average

$$\sum_l G_0^+(1) \left\langle \sum_{i, j \neq i} V_{k_2 l}(\mathbf{R}_i) V_{l k_1}(\mathbf{R}_i) \right\rangle$$

$$= \sum_l G_0^+(1) \frac{W_{k_2 l} W_{l k_1}}{\Omega^2} \left\langle \sum_{i, j \neq i} e^{-i(k_2 - 1) \cdot \mathbf{R}_j} e^{-i(1 - k_1) \cdot \mathbf{R}_i} \right\rangle$$

$$= \sum_l G_0^+(1) \frac{W_{k_2 l} W_{l k_1}}{\Omega^2} N(N-1) \int \frac{d^3 \mathbf{R}_j}{\Omega} \int \frac{d^3 \mathbf{R}_i}{\Omega} e^{-i(k_2 - 1) \cdot \mathbf{R}_j} e^{-i(1 - k_1) \cdot \mathbf{R}_i}$$

$$\approx \left(\frac{N}{\Omega}\right)^2 \sum_l G_0^+(1) W_{k_2 l} W_{l k_1} \delta_{k_2 l} \delta_{l k_1}$$

$$= \left(\frac{N}{\Omega}\right)^2 G_0^+(\mathbf{k}_1) W_{k_1 k_1}^2 \delta_{k_2 k_1} \tag{3.65}$$

Here we have used that for the random distribution, the probability that impurity i is in $d^3 \mathbf{R}_i$, and j is simultaneously in $d^3 \mathbf{R}_j$ is $(d^3 \mathbf{R}_i/\Omega)(d^3 \mathbf{R}_j/\Omega)$. Also, we have assumed $N \gg 1$. Averages of higher order terms are done in a similar way.

With the aid of these results, we can write out the series for the averaged propagator. It helps here to introduce a couple of new diagram conventions. First of all, since $V_{kl}(\mathbf{R}_i)$ does not occur any more we use just an empty circle, for the transition probability amplitude W_{kl}. Secondly, because each group of two or more successive scatterings at the same ion has an associated density factor N/Ω, we connect successive circles representing the same ion by dotted lines. (Note that a single scattering also has this factor associated with it.) Thus, taking the δ-functions into account, and letting $\mathbf{k}_1 = \mathbf{k} = \mathbf{k}_2$, we have for the averaged propagator:

$$\tag{3.66}$$

This may be translated with the dictionary in Table 3.2. Note that in this table, there is no factor Ω^{-1} in front of $\int d^3l/(2\pi)^3$ because all Ω^{-1} factors are already included in the (N/Ω) factor in line 4 of the table. (See e.g. exercise 3.9).

Let us now evaluate (3.66) assuming the most important processes are single scattering, and double scattering by the same impurity. This means that diagrams containing more than two successive scatterings off the same ion (such as for example, the fifth diagram on the right of (3.66)) are neglected. The partial sum may easily be carried out and yields

$$\tag{3.67}$$

(Note that the complete series for $\textcircled{\Sigma}$ is:

$$\text{(3.67')}$$

For small (N/Ω), we only need to keep terms $\propto (N/\Omega)$, i.e., terms representing multiple scattering from a single impurity.) Translating (3.67) into functions

$$\langle G(\mathbf{k}, \omega) \rangle = 1/[\omega - \epsilon_k + i\delta - \sum (\mathbf{k}, \omega)] \qquad (3.68)$$

where

$$\sum (\mathbf{k}, \omega) = \frac{N}{\Omega} W_{kk} + \frac{N}{\Omega} \int \frac{d^3 \mathbf{l}}{(2\pi)^3} \frac{|W_{kl}|^2}{\omega - \epsilon_l + i\delta} \qquad (3.69)$$

and we have used Table 3.2.

Table 3.2 *Diagram dictionary for electron propagating through a system of randomly distributed impurity ions*

Diagram element	Factor
$\left\langle \Big\Vert_{\mathbf{k},\,\omega} \right\rangle$	$i\langle G^+(\mathbf{k}, \omega)\rangle$
$\Big\uparrow_{\mathbf{k},\,\omega}$	$iG_0^+(\mathbf{k}, \omega) = \dfrac{i}{\omega - \varepsilon_k + i\delta}$
$\underset{\mathbf{k}}{\overset{\mathbf{l}}{\bigcirc}}$	$-iW_{lk}$
, etc.	factor $\dfrac{N}{\Omega}$.
intermediate momentum, **l**	$\displaystyle\int \frac{d^3\mathbf{l}}{(2\pi)^3}$

In order to find the new energy and lifetime of the electron, we need the complex pole of (3.68), i.e., the ω which is the solution of

$$\omega - \epsilon_k - \sum (\mathbf{k}, \omega) + i\delta = 0. \qquad (3.70)$$

(Note: If in (3.69) we use the original sum over **l**, i.e., $\sum_l$ in (3.60) instead of $\int d^3\mathbf{l}$ (see (3.64)), we find that, as expected from (3.14), the pole equation (3.70) will have *real* solutions. This can be seen at once by plotting $\sum (\mathbf{k}, \omega) = (N/\Omega) W_{kk} + (N/\Omega) \sum_l |W_{kl}|^2/(\omega - \epsilon_l + i\delta)$ and $y(\mathbf{k}, \omega) = \omega - \epsilon_k$ vs. ω and noting that the poles occur at the intersection of $\sum (\mathbf{k}, \omega)$ and $y(\mathbf{k}, \omega)$. The *complex* solutions of (3.70) arise because we have gone from a sum to an integral. The physical meaning of this is discussed at the end of appendix H.) If W is small, so that $\sum$ is small, then the zeroth-order approximation to ω is $\omega = \epsilon_k$. The first-order approximation may be obtained by setting $\omega = \epsilon_k$ into $\sum (\mathbf{k}, \omega)$ and re-solving for ω, which gives

$$\omega = \epsilon_k + \sum (\mathbf{k}, \epsilon_k) = \underbrace{\epsilon_k + \mathrm{Re} \sum (\mathbf{k}, \epsilon_k)}_{\epsilon_k'} + i\underbrace{\mathrm{Im} \sum (\mathbf{k}, \epsilon_k)}_{-\tau_k^{-1}}. \qquad (3.71)$$

Hence we need to find the real and imaginary parts of $\sum (\mathbf{k}, \epsilon_k)$.

To do this, we imagine δ is finite to start with, then take the limit $\delta \to 0$. Multiplying numerator and denominator of the integrand of $\sum$ in (3.69) by $\omega - \epsilon_l - i\delta$ we find for the real and imaginary parts of $\sum (\mathbf{k}, \epsilon_k)$:

$$\mathrm{Re} \sum (\mathbf{k}, \epsilon_k) = \frac{N}{\Omega} W_{kk} + \lim_{\delta \to 0} \left(\frac{N}{\Omega}\right) \int \frac{d^3 \mathbf{l}}{(2\pi)^3} \frac{|W_{lk}|^2 (\epsilon_k - \epsilon_l)}{(\epsilon_k - \epsilon_l)^2 + \delta^2}$$

$$= \frac{N}{\Omega} W_{kk} + \left(\frac{N}{\Omega}\right) P \int \frac{d^3 \mathbf{l}}{(2\pi)^3} \frac{|W_{lk}|^2}{(\epsilon_k - \epsilon_l)} \qquad (3.72)$$

$$\mathrm{Im} \sum (\mathbf{k}, \epsilon_k) = -\lim_{\delta \to 0} \left(\frac{N}{\Omega}\right) \int \frac{d^3 \mathbf{l}}{(2\pi)^3} |W_{lk}|^2 \frac{\delta}{(\epsilon_k - \epsilon_l)^2 + \delta^2}$$

$$= -\pi \left(\frac{N}{\Omega}\right) \int \frac{d^3 \mathbf{l}}{(2\pi)^3} |W_{lk}|^2 \delta(\epsilon_k - \epsilon_l). \qquad (3.73)$$

In (3.72), P stands for 'principal part'. (We will show why the limit in (3.72) is a principal part by illustrating with a simple case, the function $1/x$. Using the usual definition:

$$P \int_{-a}^{+b} \frac{dx}{x} = \lim_{\delta \to 0} \left\{ \int_{-a}^{-\delta} \frac{dx}{x} + \int_{+\delta}^{+b} \frac{dx}{x} \right\} = \lim_{\delta \to 0} \{\ln(-\delta) - \ln(-a) + \ln b - \ln \delta\} = \ln b/a \qquad (3.74)$$

Using the alternative definition in (3.72):

$$P \int_{-a}^{+b} \frac{dx}{x} = \lim_{\delta \to 0} \int_{-a}^{+b} dx \frac{x}{x^2 + \delta^2} = \lim_{\delta \to 0} \frac{1}{2} \int_{-a}^{+b} \frac{d(x^2)}{x^2 + \delta^2} = \lim_{\delta \to 0} \frac{1}{2} \ln(x^2 + \delta^2)\big|_{-a}^{b} = \ln b/a.) \qquad (3.75)$$

In (3.73) we have used the 'squeezed Lorentzian' definition of the δ-function. The results (3.72, 3.73) are usually obtained with the aid of the so-called 'well-known theorem from complex function theory', (see, e.g., Dennery and Krzywicki (1967), p. 64),

$$\frac{1}{x + i\delta} = P \frac{1}{x} - i\pi\delta(x) \qquad (3.76)$$

which is short for

$$\int \frac{dx\, f(x)}{x + i\delta} = P \int \frac{dx\, f(x)}{x} - i\pi \int dx\, f(x)\, \delta(x). \qquad (3.76')$$

This can be applied in the present case by noting that the integral in (3.69) may be written in the general form

$$\int d^3 l\, \frac{A(l, \ldots)}{B(l, \ldots) + i\delta} = \int d\phi \int d\theta \sin\theta \int dl\, l^2\, \frac{A(l, \theta, \phi, \ldots)}{B(l, \theta, \phi, \ldots) + i\delta}$$

where l, ϕ, θ are the polar coordinates of l and the dots ... refer to all the other variables. Only the l-variable is relevant here. If we let $x = B(l)$ so $l = B^{-1}(x)$ then $\int dl$ may be written in terms of x, allowing (3.76') to be used. Transforming back to l again after this is done yields

$$\int d^3 l\, \frac{A(l, \ldots)}{B(l, \ldots) + i\delta} = P \int d^3 l\, \frac{A(l, \ldots)}{B(l, \ldots)} - i\pi \int d^3 l\, A(l, \ldots)\, \delta[B(l, \ldots)]$$

$$(3.76'')$$

Applying this to $\sum$ with $\omega = \epsilon_k$ gives just (3.72), (3.73).

Hence, placing (3.72, 3.73) in (3.71), we find for the electron energy and lifetime:

$$\epsilon_k' = \epsilon_k + \frac{N}{\Omega} W_{kk} + \left(\frac{N}{\Omega}\right) P \int \frac{d^3 l}{(2\pi)^3} \frac{|W_{lk}|^2}{\epsilon_k - \epsilon_l} \qquad (3.77)$$

$$\tau_k^{-1} = \pi \left(\frac{N}{\Omega}\right) \int \frac{d^3 l}{(2\pi)^3} |W_{lk}|^2\, \delta(\epsilon_k - \epsilon_l) \qquad (3.78)$$

Equation (3.77) is just the result obtained from second-order perturbation theory. Equation (3.78) is what comes out of applying the 'golden rule' for transition probabilities, i.e., τ_k^{-1} is just the transition probability/sec for the electron to jump from state $\mathbf{k}$ to $\mathbf{l}$, $|W_{lk}|^2$, integrated over all final states $\mathbf{l}$, subject to conservation of energy as expressed by the δ-function. (For a review of electrons in disordered systems see Leath (1970). The method above is applied to the case where the impurity distribution is not completely random but has 'short-range order' by Woolley and Mattuck (1972).)

Further Reading

Feynman (1965), chap. 1.
Bjorken (1964), §6.2.
Feynman (1962), p. 168, §2.

Exercises

3.1 Consider a one-dimensional system with an unperturbed Hamiltonian such that $U(x)$ in (3.4) is a square well of width a with infinitely high walls, i.e., $U(x) = 0$ for $0 < x < a$ and $U(x) = \infty$ for $x < 0$, $x > a$. What are the eigenstates $\phi_n(x)$ and energies ϵ_n for this system? Write out the free propagator and its Fourier transform for this system.

3.2 The system in Ex. 3.1 is acted on by a hypothetical external perturbing potential $V(x) = B \times (p^2/2m + U(x))^3$, where $U(x)$ is defined in Ex. 3.1. Calculate the transition amplitude from single particle eigenstate $\phi_n(x)$ (calculated in Ex. 3.1) to $\phi_m(x)$ under the influence of the perturbation.

3.3 Write out the diagram series for the propagator of the system in Ex. 3.1, with the perturbation in Ex. 3.2 and evaluate it by summing to infinite order (assume the propagating particle is the only particle present).

3.4 What is the quasi particle energy dispersion law and lifetime in the above system?

3.5 Carry out the Fourier transform of the first-order terms in (3.30) and show that you get the corresponding terms in (3.31).

3.6 Use (3.53) to verify that G_0^+ satisfies the equation of the Green's function (3.51).

3.7 We have a random distribution of ions with a potential such that $W_{kl} = Wf_k f_l$, where $f_p = 1$ for $|\mathbf{p}| < a$ and $f_p = 0$ for $|\mathbf{p}| > a$. Show that the energy and reciprocal lifetime of an electron propagating in this sytem are

$$\epsilon_k' = \frac{k^2}{2m} + \left(\frac{N}{\Omega}\right) W + \left(\frac{N}{\Omega}\right)\frac{mW^2}{\pi^2}\left[-a + \frac{k}{2}\ln\left(\frac{a+k}{a-k}\right)\right] \quad \text{for } k < a$$

$$= \frac{k^2}{2m} \quad \text{for } k > a, \text{ where } k = |\mathbf{k}|$$

$$\tau_k^{-1} = \left(\frac{N}{\Omega}\right)\frac{mW^2 k}{2\pi} \quad \text{for } k < a$$

$$= 0 \quad \text{for } k > a$$

What is the effective mass in the limit $k \ll a$?

3.8 Write (3.36') with lines labelled in $(\mathbf{k}, t)$-space. (Answer: see (10.15).) Translate into functions and show that you get an integral equation of form (3.56).

3.9 Consider the fourth-order diagram (drawn on its side to save space):

$$D \equiv \quad \overset{k}{\longrightarrow}\!\!\Big(i\Big)\!\!\overset{p}{\longrightarrow}\!\!\Big(j\Big)\!\!\overset{q}{\longrightarrow}\!\!\Big(j\Big)\!\!\overset{r}{\longrightarrow}\!\!\Big(i\Big)\!\!\overset{s}{\longrightarrow}$$

where $j \neq i$. Calculate its average value, using the same technique as, e.g., in (3.63) or (3.65). Show that this is the same result as you get from applying Table 3.2 to the last diagram in (3.66).

Chapter 4

Quasi Particles in Fermi Systems

4.1 Propagator method in many-body systems

We have thus far defined the quantum Green's function propagator for $t_2 > t_1$, shown what it looks like for free and quasi particles, and evaluated it by partial summation for the case of a single particle in an external potential. In this chapter, the technique will be generalized to many-body systems.

The starting point will be a system consisting of N non-interacting fermions in an external field. This is really a fake many-body system, since, as pointed out in chapter 0, if there are no mutual interactions between particles the problem is actually only a one-body problem. Nevertheless, such a 'trivial' system paves the way for the bona fide many-body case. First, it shows us how to describe Fermi systems very simply in terms of a few particles above the Fermi level, and a few removed particles, or 'holes' below. Second, it allows us to introduce the language of the many-body problem, i.e., 'occupation number formalism' or 'second quantization'. We won't really start talking this language until the second half of the book, but it helps to learn some of the easier words in it now. Finally, it shows us how to extend the definition of the propagator to the case where $t_2 < t_1$. This is the time domain where we have the apparent paradox that the particle is observed in the system before it is put in! In this case, the Green's function turns out to describe the propagation of removed particles, or 'holes', which are represented diagrammatically by a downward-going arrow Υ.

As an illustration of a real many-body system, we will take a Fermi system with interaction between each pair of particles (no external potential). Examples of such systems are N electrons or nucleons in a macroscopic box. By introducing a special diagram: $\succ\!\!\wedge\!\!\wedge\!\!\prec$ for the two-body interaction, it is again possible to represent the propagator for this case as an infinite series of diagrams, which may be evaluated approximately by partial summation. Some of these partial sums are listed in Table 4.1.

The Hartree and Hartree–Fock are the crudest of the approximations and yield quasi particles with infinite lifetimes. The RPA yields the energy and lifetime of quasi particles in a high-density electron gas, while the ladder approximation is good for low-density systems like nuclear matter. Only the Hartree and Hartree–Fock will be discussed in detail in this chapter; the latter two are in chapter 10.

Table 4.1. *Some important partial sum approximations*

Types of diagrams summed over	Name of approximation
Bubbles	Hartree
Bubbles and open oysters	Hartree–Fock
Rings	Random phase approx. (RPA)
Ladders	Ladder approximation

4.2 Non-interacting Fermi system in external potential: particle–hole picture

Let us first talk about the particle–hole way of describing Fermi systems. Suppose we have a single particle in a potential $U(\mathbf{r})$, with energy eigenstates $\phi_k(\mathbf{r})$ ($\equiv \phi_{p_1}, \phi_{p_2}, \ldots$) and energies given by (3.4) (see note on notation after (3.5)!). The energy levels may be represented as in Fig. 4.1, where for simplicity the system is assumed non-degenerate.

The ground state of the single particle has energy ϵ_{p_1}. If we now put $N-1$ other particles into the system (with no mutual interaction), as for example when filling up atomic energy levels with electrons, we find that by the Pauli principle there can be no more than one particle in each state. The lowest energy for the whole system will occur when each state is filled in turn, starting from the bottom, as shown in Fig. 4.1(a) for the case $N=5$. The highest filled single-particle level is called the *Fermi level*, and has energy ϵ_F.

In the case where $U(\mathbf{r})=0$, the particles are free, and the k-subscript means momentum, or, more precisely, wavenumber. Then, in the ground state, the free particles fill a sphere in $\mathbf{k}$-space having radius $k_F=\sqrt{(2m\epsilon_F)}$, where k_F is called the *Fermi momentum*. The filled sphere is called the *Fermi sea*. The surface of this sphere is the *Fermi surface*. If $U(\mathbf{r})\neq0$, then k is just a set of three indices (we are neglecting spin for simplicity) which in general can no longer be interpreted as momentum components. The Fermi surface is then no longer spherical and k_F becomes the vector $\mathbf{k}_F$. (Any reader unfamiliar with the above should see Raimes (1961), chap. 7.)

The various excited states of the system are formed by removing a particle from a state below the Fermi level and placing it in a state above, as shown for example in Fig. 4.1(b). The empty state, e.g., the state p_3 in Fig. 4.1(b), is called a 'hole'. This is just the hole defined in connection with Fig. 0.10, except that here it is in 'p'-space instead of real space.

To avoid the strain of drawing all the particles which were not transferred in forming the excited state, it is convenient to refer everything to the ground state, Fig. 4.1(a), and just record *changes* from the ground state. To draw this, we remove the filled Fermi sea from the picture, yielding Figs. 4.1(c) and 4.1(d). This is called '*particle–hole* description'. Note that the Fermi sea is *physically* still present—it has only been removed from the *drawing* of the system.

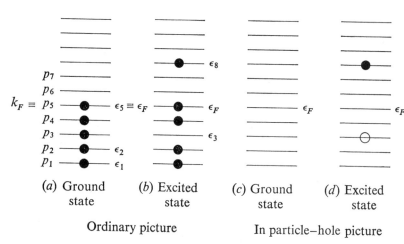

(a) Ground   (b) Excited   (c) Ground   (d) Excited
   state   state   state   state

Ordinary picture   In particle–hole picture

Fig. 4.1 *Non-interacting Fermi System*

Observe that the 'hole' in Fig. 4.1(d) is not the same as that in Fig. 4.1(b), since a Fermi sea particle has been removed from the empty state in p_3 in order to produce this new type of hole. That is, the hole in Fig. 4.1(d) is a 'minus particle' or 'anti-particle' rather than just an 'empty place'. Thus it is analogous to a 'positron' in Dirac's electron theory. This new type of hole can also be defined in position space instead of 'p'-space if we imagine that in Fig. 0.10, the undisturbed electron gas (coloured grey) is removed from the entire figure, *including* the empty places (coloured white) where the old holes are. Thus, each of the empty places will now be coloured 'minus grey' (evidently a job for the surrealist painter Salvador Dali!) which indicates the presence of an anti-particle.

Note that 'particles' in the new sense exist only above the Fermi surface. In cases where there is a possibility of confusion, we will distinguish between 'particle-hole' type particles and ordinary particles by writing the 'p' in italics, thus:

  *p*article: particle in particle–hole sense. Exists only above Fermi surface.
  particle: ordinary particle. Exists above and below Fermi surface.   (4.1)

Since a hole in state ϕ_k is actually removal of a particle from the system, the hole represents energy ϵ_k removed. Hence the hole energy is negative and we have

$$\epsilon_k^{\text{hole}} = -\epsilon_k. \tag{4.2}$$

The shape of the hole wave function in space will be exactly the same as the shape of the wave function of the removed particle. This is analogous to removing a single piece from a completed jig-saw puzzle: the 'hole' thus created in the puzzle has exactly the same shape as the removed piece. Thus the time-dependent wave function for a hole in state ϕ_k is (see §7.5, just after (7.77) for rigorous proof):

$$\psi(t)^{\text{hole}} = \phi_k e^{-i(-\epsilon_k)t}, \quad \epsilon_k < \epsilon_F. \tag{4.3}$$

If we now associate the sign change in the $\epsilon_k t$ term with the t instead of the ϵ_k, the hole may be viewed as a particle moving backward in time. This should not be regarded as theoretical grounds for constructing a time machine, but simply as a convenient mode of description. It was originated by Feynman in his theory of positrons.

4.3 ⟦A primer of occupation number formalism (second quantization)⟧

(This section can be skipped on first reading!)

Although we shall not make any essential use of it until after chapter 7, it is a good idea for orientational purposes to inject a few words here on the occupation number formalism or 'second quantization' as it is often called. This formalism is a sort of 'census-taking' notation which is extremely convenient for keeping track of what is going on in a many-particle system. The details are in chapter 7.

The total wave function for the ground and excited states of a system of non-interacting particles is, (see appendix ($\mathscr{A}$.3)), the product of single-particle wave functions. However, because we are dealing with identical fermions, this product must be antisymmetrized and the proper wave function is the Slater determinant

$$\Phi_{k_1,\ldots,k_N}(\mathbf{r}_1,\ldots,\mathbf{r}_N) = \frac{1}{\sqrt{(N!)}} \begin{vmatrix} \phi_{k_1}(\mathbf{r}_1) \ldots \phi_{k_1}(\mathbf{r}_N) \\ \vdots \qquad\qquad \vdots \\ \phi_{k_N}(\mathbf{r}_1) \ldots \phi_{k_N}(\mathbf{r}_N) \end{vmatrix} \tag{4.4}$$

where the ϕ_k's are the single-particle states of (3.4). If the particles are allowed to interact with each other, or with an external perturbing potential, then the exact wave functions of the system are no longer (4.4) but a linear combination of Φ's thus:

$$\Psi(\mathbf{r}_1,\ldots,\mathbf{r}_N) = \sum_{k_1,\ldots,k_N} A_{k_1,\ldots,k_N} \Phi_{k_1,\ldots,k_N}(\mathbf{r}_1,\ldots,\mathbf{r}_N). \tag{4.5}$$

That is, the $\Phi_{k_1,\ldots,k_N}$ for the non-interacting system are the basis states used to describe the interacting system.

Now these are rather clumsy expressions to carry around, so it would be desirable to have a more compact way of writing them. This may be gotten by noting that since all particles are indistinguishable, the essential information in (4.4) is just how many particles there are in each single-particle state. Therefore, we could equally well specify the state of the non-interacting system by writing Φ as

$$\Phi_{k_1,\ldots,k_N}(\mathbf{r}_1,\ldots,\mathbf{r}_N) = \Phi_{n_{p_1},n_{p_2},\ldots,n_{p_i},\ldots}(\mathbf{r}_1,\ldots,\mathbf{r}_N) \tag{4.6}$$

For short, we shall represent this as

$$\Phi_{n_{p_1},n_{p_2},\ldots,n_{p_i},\ldots} \equiv |n_{p_1},n_{p_2},\ldots,n_{p_i},\ldots\rangle \tag{4.6'}$$

meaning: n_{p_1} particles in state ϕ_{p_1}, n_{p_2} in ϕ_{p_2}, etc., where $n_k=0$ or 1 by the Pauli principle. This is called 'occupation number notation'. It is similar to the shell notation for atoms, where $(1s)^2(2s)^2(2p)^1$ means two electrons in the $1s$ state (one in the spin up and one in the spin down state), two in the $2s$ state, etc.

For the ground state in Fig. 4.1(a) we have in occupation number notation

$$\Phi_{k_1=p_1,\,k_2=p_2,\,k_3=p_3,\,k_4=p_4,\,k_5=p_5} = |1_{p_1},1_{p_2},1_{p_3},1_{p_4},1_{p_5},0_{p_6},0_{p_7},0_{p_8},\ldots,0,0,\ldots\rangle. \tag{4.7}$$

The excited state in Fig. 4.1(b) is

$$\Phi_{k_1=p_1,\,k_2=p_2,\,k_3=p_4,\,k_4=p_5,\,k_5=p_8} = |1_{p_1},1_{p_2},0_{p_3},1_{p_4},1_{p_5},0_{p_6},0_{p_7},1_{p_8},0_{p_9},\ldots\rangle. \tag{4.8}$$

For brevity, from now on we will drop the p's and just use the numerical subscripts. Then

$$\Phi = |n_1,n_2,n_3,\ldots,n_i,\ldots\rangle. \tag{4.9}$$

For example, (4.7) becomes

$$\Phi_0 = |1_1,1_2,1_3,1_4,1_5,0_6,0_7,0_8,\ldots\rangle.$$

It is important to note that just as the original Slater determinants form a complete orthogonal set of basis functions, so do the states in occupation number notation and we have

$$\langle n_1',\ldots n_i',\ldots |n_1,\ldots,n_i,\ldots\rangle = \int d^3\mathbf{r}_1\ldots d^3\mathbf{r}_N \times$$
$$\times \Phi^*_{n_1',\ldots,n_{i'},\ldots}(\mathbf{r}_1,\ldots,\mathbf{r}_N) \times$$
$$\times \Phi_{n_1,\ldots,n_i,\ldots}(\mathbf{r}_1,\ldots,\mathbf{r}_N)$$
$$= \delta_{n_1' n_1}\ldots\delta_{n_{i'} n_i}\ldots. \tag{4.10}$$

Just as in (4.5), the $|n_1, \ldots, n_i, \ldots\rangle$ may be used as the basis states for describing the interacting system's wave function thus:

$$\Psi = \sum_{n_1, \ldots, n_i, \ldots} A_{n_1, \ldots, n_i, \ldots} |n_1, \ldots, n_i, \ldots\rangle. \tag{4.11}$$

Now in most cases of interest, only a few of the particles change their position from that in the ground state, since we deal primarily with only weakly excited states. Hence, carrying along all the unchanged 1's in a wave function like (4.8) is about as useful as taking along every piece of clothing one owns, on a two-day trip. The excess baggage may be avoided by regarding the ground state (4.7) as the 'zero' or so-called '*Fermi vacuum*' of our description, and recording in the $|\ldots\rangle$ only *changes* from the ground state. Thus, the ground state is written as though it has no particles in it:

$$\Phi_0 = |0\rangle \quad (\text{'Fermi vacuum'}) \tag{4.12}$$

corresponding to Fig. 4.1(*c*). The excited state of Fig. 4.1(*b*) according to this viewpoint is a *particle* (see (4.1)!) above ϵ_F and a hole below, as shown in Fig. 4.1(*d*), with the corresponding state vector

$$\Phi = |1_3^h, 1_8^p\rangle \tag{4.13}$$

where h, p, stand for hole, *particle*. This is called 'particle–hole' notation.

Quantum mechanical operators have a new form in the occupation number formalism. Imagine that we have initially a single-particle system in its lowest energy eigenstate $\phi_1(\mathbf{r})$ ($\equiv \phi_{p_1}(\mathbf{r})$). In occupation number notation this is

$$\Phi_{\text{initial}} = \phi_1 = |1000\ldots\rangle. \tag{4.14}$$

If the system is now acted on by some perturbing operator $V(\mathbf{r}, \mathbf{p})$, it may undergo a transition, say, to state ϕ_3, so that

$$\Phi_{\text{final}} = \phi_3 = |001000\ldots\rangle. \tag{4.15}$$

Thus, when written in this formalism, the effect of the operator V appears as the *destruction* of a particle in ϕ_1 and the *creation* of a particle in ϕ_3. This suggests that if we define two primitive operators—c_i (which is short for c_{p_i}), which destroys a particle in $\phi_i (\equiv \phi_{p_i})$ and $c_i^\dagger$ which creates a particle in ϕ_i— it may be possible to write all operators as various combinations of these primitive ones.

This is indeed the case. Look first at the detailed expression for the effect of the c's:

$$c_i |n_1, n_2, \ldots, n_i, \ldots\rangle = n_i |n_1, n_2, \ldots, n_i - 1, \ldots\rangle$$

$$c_i^\dagger |n_1, n_2, \ldots, n_i, \ldots\rangle = (1 - n_i) |n_1, n_2, \ldots, n_i + 1, \ldots\rangle \tag{4.16}$$

where the factors in front mean that c_i cannot destroy a particle in ϕ_i if there is no particle there to start with, and $c_i^\dagger$ cannot create another particle in an already occupied state. (A factor of ± 1 has been left out for simplicity—see chapter 7.) For example:

$$c_3 |11111000...\rangle = |1101100...\rangle$$
$$c_2 |0000...\rangle = 0$$
$$c_n |00...1_j...\rangle = \delta_{nj} |00...\rangle$$
$$c_3^\dagger |11111000...\rangle = 0$$
$$c_2^\dagger |00...\rangle = |0100...\rangle$$
$$c_m^\dagger |00...0_m...\rangle = |00...1_m...\rangle. \qquad (4.17)$$

In the particle–hole notation, it is necessary to introduce hole creation and destruction operators, $b_i^\dagger$, b_i, and similarly particle operators $a_i^\dagger$, a_i, as follows: if $k_i < k_F$, then c_i destroys a particle under the Fermi level, thus creating a hole. Hence

$$\text{for } k_i > k_F, \quad c_i = a_i \quad \text{(particle destruction operator)}$$
$$k_i < k_F, \quad c_i = b_i^\dagger \quad \text{(hole creation operator)}$$

and

$$\text{for } k_i > k_F, \quad c_i^\dagger = a_i^\dagger \quad \text{(particle creation operator)}$$
$$k_i < k_F, \quad c_i^\dagger = b_i \quad \text{(hole destruction operator)}. \qquad (4.18)$$

This change to particle–hole operators may be expressed compactly as the transformation

$$c_i = \theta_{k_i - k_F} a_i + \theta_{k_F - k_i} b_i^\dagger$$
$$c_i^\dagger = \theta_{k_i - k_F} a_i^\dagger + \theta_{k_F - k_i} b_i \qquad (4.19)$$

where

$$\theta_x = 1 \text{ for } x > 0; \quad \theta_x = 0 \text{ for } x < 0.$$

Simple examples of how the particle–hole operators work are:

$$a_i^\dagger |0\rangle = |1_i^p\rangle, \qquad a_i |1_i^p\rangle = \delta_{il} |0\rangle, \qquad b_j^\dagger a_i^\dagger |1_m^p\rangle = |1_m^p, 1_i^p, 1_j^h\rangle,$$
$$b_j^\dagger |0\rangle = |1_j^h\rangle, \qquad b_j |1_m^h\rangle = \delta_{jm} |0\rangle, \qquad a_i |0\rangle = b_i |0\rangle = 0,$$
$$a_i^\dagger |1_i^p\rangle = 0, \qquad b_j^\dagger |1_j^h\rangle = 0. \qquad (4.20)$$

In order to express other operators in terms of the c's (or a's and b's), it is simply required that: if $\mathcal{O}^{\mathrm{old}}$ is the operator in the old notation, and $\mathcal{O}^{\mathrm{occ}}$ is the same operator in the occupation number (or particle–hole) formalism, then $\mathcal{O}^{\mathrm{occ}}$ must give the same matrix elements when sandwiched between states in occupation number (or particle–hole) formalism that $\mathcal{O}^{\mathrm{old}}$ gave when sandwiched between Slater determinants. Consider first a one-particle case, where the Slater determinant is just ϕ_i. Then we must have

$$\langle 0 0 \ldots 1_i \ldots | \mathcal{O}^{\mathrm{occ}} | \ldots 1_j \ldots \rangle = \langle \phi_i | \mathcal{O}^{\mathrm{old}} | \phi_j \rangle$$

$$\left(\equiv \int \phi_i^*(\mathbf{r}) \mathcal{O}(\mathbf{r}) \phi_j(\mathbf{r}) \, d^3\mathbf{r} \equiv \mathcal{O}_{ij} \right). \qquad (4.21)$$

A bit of Buddhistic contemplation shows that

$$\mathcal{O}^{\mathrm{occ}} = \sum_{mn} \mathcal{O}_{mn} c_m^\dagger c_n \qquad (4.22)$$

does the trick, since

$$\langle \ldots 1_i \ldots | \mathcal{O}^{\mathrm{occ}} | \ldots 1_j \ldots \rangle = \sum_{mn} \mathcal{O}_{mn} \langle \ldots 1_i \ldots | c_m^\dagger c_n | \ldots 1_j \ldots \rangle$$

$$= \sum_{mn} \mathcal{O}_{mn} \delta_{im} \delta_{nj} = \mathcal{O}_{ij} \qquad (4.23)$$

where (4.17) and (4.10) have been used. Equation (4.22) can be converted to particle–hole formalism by (4.18). This result turns out to hold also for systems with an arbitrary number of particles.

The Hamiltonian for an arbitrary system may be expressed in occupation number or particle–hole formalism. Suppose the system Hamiltonian in old Neanderthal notation describes a system in an external perturbing potential:

$$H_{\mathrm{Neand.}} = \underbrace{\sum_i \left[\frac{p_i^2}{2m} + U(\mathbf{r}_i) \right]}_{H_0} + \underbrace{\sum_i V(\mathbf{r}_i)}_{H_1 \text{ (perturbation)}} \qquad (4.24)$$

and the single-particle states ϕ_k satisfy

$$\left[\frac{p^2}{2m} + U(\mathbf{r}) \right] \phi_k = \epsilon_k \phi_k. \qquad (4.25)$$

Then it is found that (see chapter 7):

$$H_0 = \sum_k \epsilon_k c_k^\dagger c_k = \sum_{k > k_F} \epsilon_k a_k^\dagger a_k + \sum_{k < k_F} \epsilon_k b_k b_k^\dagger$$

$$H_1 = \sum_{m,\, n > k_F} V_{mn} a_m^\dagger a_n + \sum_{\substack{m > k_F \\ n < k_F}} V_{mn} a_m^\dagger b_n^\dagger + \sum_{\substack{m < k_F \\ n > k_F}} V_{mn} b_m a_n +$$

$$+ \sum_{m,\, n < k_F} V_{mn} b_m b_n^\dagger. \qquad (4.26)$$

For a system of mutually interacting particles with old Hamiltonian

$$H_{old} = \underbrace{\sum_i \frac{p_i^2}{2m}}_{H_0} + \underbrace{\tfrac{1}{2} \sum_{i,j} V(\mathbf{r}_i - \mathbf{r}_j)}_{H_1 \text{ (perturbation)}} \qquad (4.27)$$

we find

$$H_0 = \sum_{k>k_F} \epsilon_k a_k^\dagger a_k + \sum_{k<k_F} \epsilon_k b_k b_k^\dagger, \quad \text{with} \quad \epsilon_k = k^2/2m$$

$$H_1 = \tfrac{1}{2} \sum_{k,l,m,n>k_F} V_{klmn} a_l^\dagger a_k^\dagger a_m a_n + \tfrac{1}{2} \sum_{\substack{k,l,m>k_F \\ n<k_F}} V_{klmn} a_l^\dagger a_k^\dagger a_m b_n^\dagger +$$

$$+ \cdots + \tfrac{1}{2} \sum_{k,l,m,n<k_F} V_{klmn} b_l b_k b_m^\dagger b_n^\dagger \qquad (4.28)$$

with V_{klmn} as defined in (4.42).

It should be carefully remembered that in the case of systems with inter-action, the wave functions are given by the linear combination (4.11).

4.4 Propagator for non-interacting Fermi system in external perturbing potential

Up to now we have worked with a propagator defined only for positive time differences, i.e., for $t_2 > t_1$. This was adequate for solving the super-simple quantum pinball problem, but fails when we try to use it on more complicated cases. To treat the general situation, it is necessary to extend the definition to times $t_2 < t_1$. This of course sounds peculiar, since it seems to describe a particle propagating backward in time. However, as explained in connection with (4.3), such 'time-machine' particles are not science fiction but simply removed particles or 'holes'. That is, a particle moving backward in time from t_1 to t_2 $(t_2 < t_1)$ is just a hole moving forward in time from t_2 to t_1.

This leads us to the definition

$$iG(k_2, k_1, t_2 - t_1)_{t_2 \leqslant t_1} \equiv iG^-(k_2, k_1, t_2 - t_1)$$

$= (-1) \times$ probability amplitude that if at time t_2 we remove a particle in state ϕ_{k_2} from (i.e., if we add a hole in ϕ_{k_2} to) the interacting system in its ground state, then at time t_1 the system will be in its ground state with a particle removed from (i.e., an added hole in) ϕ_{k_1}. \qquad (4.29)

Analogous to (3.3), for $t_2 > t_1$ (but not for $t_2 = t_1$!), G^- is defined so that

$$iG^-(k_2, k_1, t_2 - t_1)_{t_2 > t_1} = 0. \qquad (4.30)$$

Thus, G^- is just the hole propagator. (The factor of (-1) here compared with (3.1) comes because we have fermions—see chapter 9. Note that G^- is called an 'advanced' propagator or Green's function.)

The use of the word hole in the sense of 'removed particle' is more general than the way it was used in §4.2. There we dealt with the non-interacting system, so a particle could be removed only from $k < k_F$ and therefore all holes had $k < k_F$. However, in the ground state of the interacting system, by (4.11) there is a finite probability of finding particles above k_F. Hence we can remove a particle or create a hole (in this more general sense) above $k = k_F$. If the 'hole propagator' G^- is, as above, defined as being the propagator for $t_2 < t_1$, then the free G_0^- has $k < k_F$ but the exact G^- can have any k.

In the case of a free hole, an argument like that in (3.8) applied to the single hole state in (4.3) yields (note that (3.8) and G_0^+ in (3.9) describe particle propagation, since $\epsilon_k > \epsilon_F$)

$$G_0^-(k, t_2 - t_1) = \begin{cases} i\theta_{t_1 - t_2} e^{-i\epsilon_k(t_2 - t_1)}, & \text{for } t_2 \neq t_1, \quad \epsilon_k < \epsilon_F \\ i, & \text{for } t_2 = t_1 \quad \text{(see (9.2), (9.4), end of appendix F)} \end{cases} \quad (4.31)$$

with Fourier transform

$$G_0^-(k, \omega) = \frac{1}{\omega - \epsilon_k - i\delta}, \quad \epsilon_k < \epsilon_F. \quad (4.32)$$

Suppose now that we turn on an external perturbing potential $V(\mathbf{r})$ (this is distinct from $U(\mathbf{r})$ which is part of the unperturbed Hamiltonian), and wish to find, say, the single-particle propagator $G^+(k_2, k_1, t_2 - t_1)$ or $G^+(k_2, k_1, \omega)$. This will be the sum of the amplitudes for all the ways the particle can move through the system interacting zero or more times with $V(\mathbf{r})$. Previously, we wrote down the series for the propagator and translated it into diagrams. Now we turn the trick and pull the hat out of the rabbit, i.e., write down the diagrams first, then translate them into the numerical series. To do this, we need a modified dictionary, analogous to Table 3.1 with downward directed lines for the hole propagators, as shown in Table 4.2. Observe the reversed time order for the hole propagator diagrams! This is of course due simply to the fact that $t_2 < t_1$ for holes. The reason why these diagrams are labelled 'Goldstone method' is discussed in §9.5.

The interaction amplitude, V_{kl}, merits some discussion. It is given by

$$V_{kl} = \int d^3\mathbf{r} \phi_k^*(\mathbf{r}) V(\mathbf{r}, \mathbf{p}) \phi_l(\mathbf{r}). \quad (4.33)$$

The four possibilities shown in Table 4.2 mean: (a) scattering of a particle (remember (4.1)!) from state ϕ_l to ϕ_k, (b) the potential scatters a particle out of state ϕ_l, where $\epsilon_l < \epsilon_F$, into state ϕ_k, $\epsilon_k > \epsilon_F$, thus simultaneously creating a particle in ϕ_k and a hole in ϕ_l, (c), etc. [Note that these four possibilities correspond to the four interaction terms in the particle–hole Hamiltonian for this case (4.26).]

Of course, the particle which emerges in state $\mathbf{k}$ after interaction, is not necessarily the same particle which entered in state $\mathbf{l}$, since this has no meaning in a system of indistinguishable particles. Nevertheless, for the sake of verbal simplicity, it is customary to describe interactions *as if* particles were distinguishable; the reader should always bear in mind that this is just a manner of talking.

With the aid of Table 4.2, the diagrammatic series for G^+ may be drawn as the sum of all possible different diagrams which can be built up out of sequences of interaction dots connected by particle and hole lines, beginning in state k_1 and ending in state k_2:

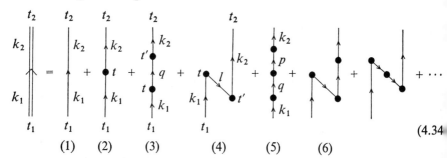

$$(4.34)$$

$$(1) \quad (2) \quad (3) \quad (4) \quad (5) \quad (6)$$

(Note that the first diagram disappears if $k_2 \neq k_1$, by (3.9).) The physical significance of the hole lines in the diagrams may be understood by looking at the fourth diagram. A *particle* enters the system in state k_1 ($\equiv \phi_{k_1}$) at time t_1. At time t', the potential knocks a particle out of the state l into state k_2 thus creating a *particle* in k_2 and a *hole* in l. At time t, the *particle* in k_1 is knocked into the hole in l causing mutual annihilation; the *particle* in k_2 continues propagating until t_2.

It should be pointed out that many diagrams in this series violate the Pauli exclusion principle. For example, when $k_1 = k_2$, in diagram 4 we have two particles in the same state, k_1. The reason why such diagrams must be included is discussed at the end of Appendix G (see also §4.6).

⟦It is amusing to do the 'book-keeping' on these processes by means of the particle–hole notation, with H_1 as in (4.26). We have the sequence:

(1) Put in *particle* in state k_1 at time t_1:

$$a^\dagger_{k_1} |0\rangle = |1^p_{k_1}\rangle$$

(2) At t', one of the terms in H_1 acts on system creating *particle* in k_2, hole in l:

$$V_{k_2 l} a^\dagger_{k_2} b^\dagger_l |1^p_{k_1}\rangle = V_{k_2 l} |1^p_{k_1}, 1^h_l, 1^p_{k_2}\rangle$$

(Note that if $k_2 = k_1$, we have $a^\dagger_{k_1} |1^p_{k_1}, 1^h_l\rangle$ which equals zero by (4.20). Nevertheless, the diagram which includes this process (number (4) in (4.34)) does not have the value zero! It is, as just mentioned, an exclusion-principle-violating diagram, and it must be kept (see §4.6).)

Table 4.2 *Diagram dictionary for many-fermion system in external perturbing potential (Goldstone method)*

(k,t)-space		(k,ω)-space	
Word	Diagram	Word	Diagram
$iG^+(k_2,k_1,t_2-t_1)$ (particle), see (3.2)	k_2,t_2 / k_1,t_1	$iG^+(k_2,k_1,\omega)$	ω k_2 / k_1
$iG^-(k_2,k_1,t_2-t_1)$ (hole), see (4.29)	k_1,t_1 / k_2,t_2	$iG^-(k_2,k_1,\omega)$	ω k_1 / k_2
$iG_0^+(k,t_2-t_1)=\theta_{t_2-t_1}\,\theta_{\epsilon_k-\epsilon_F}\exp[-i\epsilon_k(t_2-t_1)]$ (free particle) ($=0$ for $t_2=t_1$)	k $\uparrow$ t_2 / t_1	$iG_0^+(k,\omega)=\dfrac{i\theta_{\epsilon_k-\epsilon_F}}{\omega-\epsilon_k+i\delta}$	k,ω
$iG_0^-(k,t_2-t_1)=-\theta_{t_1-t_2}\,\theta_{\epsilon_F-\epsilon_k}\exp[-i\epsilon_k(t_2-t_1)]$ (free hole) ($=-1$ for $t_2=t_1$)	k $\downarrow$ t_1 / t_2	$iG_0^-(k,\omega)=\dfrac{i\theta_{\epsilon_F-\epsilon_k}}{\omega-\epsilon_k-i\delta}$	k,ω
$-iV_{kl}$ (interaction with external perturbation), see (4.33)	$\bullet$ or $\circ$ 4 possibilities k l (a) (b) (c) (d)	$-iV_{kl}$	$\bullet$ or $\circ$ 4 possibilities

(3) At t, H_1 acts again, destroying hole in l, particle in k_1:

$$V_{lk_1} b_l a_{k_1} [V_{k_2 l} | 1^p_{k_1}, 1^h_l, 1^p_{k_2}\rangle] = V_{k_2 l} V_{lk_1} | 1^p_{k_2}\rangle$$

(4) At t_2, take particle out:

$$a_{k_2} [V_{k_2 l} V_{lk_1} | 1^p_{k_2}\rangle] = V_{k_2 l} V_{lk_1} | 0\rangle.]] \tag{4.35}$$

The above diagram series may be written out in words by means of the dictionary Table 4.2. This gives in (k, t)-space (cancel the i's):

$$G^+(k_2, k_1, t_2 - t_1) = G_0^+(k_1, t_2 - t_1) \delta_{k_1 k_2}$$

$$+ \int_{-\infty}^{+\infty} dt \, G_0^+(k_2, t_2 - t) V_{k_2 k_1} G_0^+(k_1, t - t_1) +$$

$$+ \sum_{q > k_F} \int_{-\infty}^{\infty} dt \int_{-\infty}^{\infty} dt' \cdots + \cdots \tag{4.36}$$

or in (k, ω)-space (leave out ω's for brevity):

$$G^+(k_2, k_1) = \delta_{k_1 k_2} G_0^+(k_1) + G_0^+(k_1) V_{k_2 k_1} G_0^+(k_2)$$

$$+ \sum_{q > k_F} G_0^+(k_1) V_{qk_1} G_0^+(q) V_{k_2 q} G_0^+(k_2) +$$

$$+ \sum_{l < k_F} G_0^+(k_1) V_{lk_1} G_0^-(l) V_{k_2 l} G_0^+(k_2) + \cdots \tag{4.37}$$

where we have remembered to sum over all possible intermediate states, q, l, etc., since, for example, the single diagram with q on it actually stands for an infinite number of diagrams, each one with a different value of q. (The notation $q > k_F$ is short for $\epsilon_q > \epsilon_F$, etc.)

[The time integrations in (4.36) are automatically restricted by the θ-functions found in G^+ and G^-. Thus, in the third diagram in (4.34) since all lines are particle propagators, we see that $t_1 < t < t' < t_2$. In the fourth diagram, since the l-line is a hole propagator, t must be $> t'$, and we find: if $t_1 < t < t_2$ then $-\infty < t' < t$, while if $t_2 < t < \infty$, then $-\infty < t' < t_2$. (Strictly speaking, in the Goldstone method, diagrams are 'time-ordered' (see §9.5) so that for diagram (4), $t_1 < t' < t$, $t_1 < t < t_2$. There will be other diagrams like (4), but with $-\infty < t' < t_1$ and/or $t_2 < t < \infty$, which may be added to (4) to obtain the stated region of integration.)

However, when the time integrations here are actually performed, one is dismayed to discover the page jumping with exponentials oscillating at ∞ just as in (3.12). The remedy is to change the integration limits from $\pm \infty$ to $\pm \infty(1 - i\eta)$ where η is a positive infinitesimal like the δ in (3.12'): it is such that $\eta \times \infty = \infty$. The justification for these new limits lies in the rigorous derivation of the propagator expansion (see Appendix E, especially (E.11)).

One might imagine that these modified limits would cause trouble in Fourier transforming from (4.36) to (4.37). That is, we would expect that the limits $\pm \infty$ were required in order to get the δ-functions, like $\delta(\omega - \omega')$ in (2.23). However, one finds that the sort of integral which arises, i.e., of form

$$\int_{T_1}^{T_2} dt \exp(i\Delta\epsilon t),$$

where $T_1 = -\infty(1 - i\eta)$ and $T_2 = +\infty(1 - i\eta)$, is also a legitimate δ-function, so this causes no difficulty.]]

And now an easy example showing how to evaluate G^+ by partial summation. Suppose $k_1 = k_2 = k$ ($k > k_F$), and the potential is such that V_{km} and V_{mk} ($m < k_F$—remember this is short for $\epsilon_m < \epsilon_F$) are large, and all the other V's are small. Then the propagator in (4.34) may be approximated by the sum of the following diagrams:

$$(4.38)$$

or

$$G^+(k, \omega) = \frac{1}{[G_0^+(k, \omega)]^{-1} - V_{km} V_{mk} G_0^-(m, \omega)}$$

$$= \frac{1}{(\omega - \epsilon_k + i\delta) - \dfrac{|V_{km}|^2}{(\omega - \epsilon_m - i\delta)}}. \qquad (4.39)$$

This result is evidently not of the quasi particle form (3.16). However, by (3.14), the poles of G^+ give the excited state energies of the perturbed system. Thus, dropping the $i\delta$'s (they have no significance in this simple calculation) yields

$$\omega - \epsilon_k - \frac{|V_{km}|^2}{\omega - \epsilon_m} = 0 \qquad (4.40)$$

which gives

$$\omega = \epsilon_k' = \frac{\epsilon_k + \epsilon_m}{2} + \frac{1}{2}\sqrt{\{(\epsilon_k - \epsilon_m)^2 + 4|V_{km}|^2\}}$$

$$= \epsilon_m' = \frac{\epsilon_k + \epsilon_m}{2} - \frac{1}{2}\sqrt{\{(\epsilon_k - \epsilon_m)^2 + 4|V_{km}|^2\}}. \qquad (4.41)$$

These reduce respectively to ϵ_k, ϵ_m in the weak interaction case, when $V_{km} \to 0$, and are also valid in the strong interaction case when V_{km} is of the order of or greater than the separation between the levels, $\epsilon_k - \epsilon_m$. Note that it is necessary to go to infinite order to get (4.41). If we just go to *any finite order*, the poles are still at the unperturbed energies, ϵ_k, ϵ_m!

The result (4.41) will be recognized by experts as just the formula for the new energies of a single particle two-level system placed in a perturbing field. Of course, we could have predicted this result from the beginning since we've really got a single particle system here, because by assumption the particles don't interact with each other. Again, as mentioned in connection with the quantum pinball game, this should not be regarded as a demonstration that the 'powerful' diagram technique merely provides a complicated method for getting trivial results, but rather as a super-simple illustration of the general principles.

In the next section we go on to the real many-body problem.

4.5 Interacting Fermi system

Imagine now that we've got a genuine many-body system consisting of N fermions interacting by means of two-body forces $V(|\mathbf{r}_i - \mathbf{r}_j|)$, depending just on the interparticle distance $|\mathbf{r}_i - \mathbf{r}_j|$. For simplicity, assume there are no external fields, so that the single particle states are just $\phi_k = \Omega^{-\frac{1}{2}} \exp(i\mathbf{k} \cdot \mathbf{r})$ with $\epsilon_k = k^2/2m$ as in (3.4) and (3.5). Our object is to construct diagrammatically the perturbation expansion of the propagator for this system, evaluate it by partial summation and examine the result for quasi particle behaviour.

The first thing we need is the transition probability amplitude for a process in which two particles, one in state ϕ_m, the other in state ϕ_n collide with each other and are scattered into states ϕ_k, ϕ_l respectively. Analogous to the interaction amplitude V_{kl} in (4.33), this is just the matrix element

$$V_{klmn} = \int d^3\mathbf{r} \int d^3\mathbf{r}' \, \phi_k^*(\mathbf{r}) \phi_l^*(\mathbf{r}') V(|\mathbf{r} - \mathbf{r}'|) \phi_m(\mathbf{r}) \phi_n(\mathbf{r}') = V_{lknm}. \quad (4.42)$$

As we saw in (1.8) in $(\mathbf{r}, t)$-space, such an interaction may be represented diagrammatically by a wiggly line:

$$\equiv (-i)\tfrac{1}{2} V_{klmn}, \quad (4.43)$$

where the left intersection or 'vertex' shows the scattering of one particle from **m** to **k**, and the right vertex shows the scattering of the other from **n** to **l**. (Note: the majority of writers draw the above interaction with a dashed line:

However, we shall always use the wiggle (4.43).) [The $\frac{1}{2}$ comes from H_1 in (4.28). It is eliminated by (4.60).] Using the particle–hole description, this may be drawn in more detail, thus:

$$(4.44)$$

$$(a) \qquad\qquad (b) \qquad\qquad (c)$$

Diagram (a) just pictures ordinary scattering of two particles from states m,n to k,l. In (b) a particle in ϕ_m collides with a particle below the Fermi surface in state ϕ_n. It knocks the particle out of ϕ_n, thus creating a hole in ϕ_n and a particle above the Fermi surface in state ϕ_l. At the same time the original particle undergoes a transition to state ϕ_k. And so on. [Note that the diagrams (4.44) correspond precisely to the interaction terms in the Hamiltonian for this case, (4.28).]

☞ It is extremely important to note the labelling convention used in V_{klmn}: k = line out of left vertex, l = line out of right vertex, m = line into left vertex, n = line into right vertex. A mnemonic aid is to remember the tango dance step: left out, right out, left in, right in.

The interaction $V(|\mathbf{r}-\mathbf{r}'|)$ conserves linear and spin momentum since it depends only on $|\mathbf{r}-\mathbf{r}'|$, therefore cannot move the centre of mass of the two colliding particles or flip their spins. Thus

$$\mathbf{k}+\mathbf{l} = \mathbf{m}+\mathbf{n}; \quad \sigma_k + \sigma_l = \sigma_m + \sigma_n. \qquad (4.45)$$

If the arrows in (4.44) are interpreted as giving the direction of 'momentum flow', then (4.45) shows that the momentum flowing into the interaction equals the momentum flow out. It is convenient to incorporate this into the labelling as follows:

$$\equiv -i \times \tfrac{1}{2} V_{m-q, n+q, m, n}$$

$$= -i \times \tfrac{1}{2} V_q \qquad (4.46)$$

where the form V_q is justified in (7.70). (Observe that the momentum transfer, q, in (4.46) is defined as momentum into left vertex minus momentum out of left vertex. That is, for matrix element V_{klmn}, the momentum transfer is $\mathbf{q}=\mathbf{m}-\mathbf{k}(=\mathbf{l}-\mathbf{n}$ by momentum conservation). The element V_{lknm}, corresponding to diagram (4.43) twisted through 180°, has momentum transfer $\mathbf{q}'=\mathbf{n}-\mathbf{l}=-\mathbf{q}$. Hence, since by (4.42) $V_{klmn}=V_{lknm}$, we have $V_q=V_{-q}$.) All this implies that no matter how complicated the chain of collision processes is, the momentum at the beginning of the chain is equal to the momentum at the end. This can be seen, for instance, in the second-order diagram:

$$(4.47)$$

This is analogous to the flow of current in a network without sources or sinks, so that the flow of current into the network = current out. Hence it is only necessary to deal with propagators $G(\mathbf{k}, \mathbf{l}, t_2 - t_1)$ such that $\mathbf{k} = \mathbf{l}$.

It is important to note here that although the collisions conserve momentum, they do not conserve energy. For example, at the lower interaction of (4.47) we see that the energy flow into the interaction (in units of $\hbar^2/2m$) is $k^2 + l^2$, while the energy flow out is $(k-q)^2 + (l+q)^2$. Hence we are dealing here with *virtual* scattering processes, not real ones.

We may now construct the perturbation series for the single-particle propagator, G^+, as the sum of all possible different diagrams which can be built up out of sequences of interactions (4.43), connected by particle and hole lines, with a particle entering the system in state $\mathbf{k}$ and leaving in $\mathbf{k}$. One such sequence is just that in (4.47). It depicts a *particle* in $\mathbf{k}$ being scattered into $\mathbf{k} - \mathbf{q}$ and simultaneously knocking a particle out of $\mathbf{l}$ into $\mathbf{l} + \mathbf{q}$ (i.e., creating a *particle* in $\mathbf{l} + \mathbf{q}$ and a hole in $\mathbf{l}$). At later time t', the *particle* in $\mathbf{k} - \mathbf{q}$ knocks the *particle* in $\mathbf{l} + \mathbf{q}$ back into the hole state $\mathbf{l}$ (thus annihilating the *particle*–hole pair) and is itself scattered into state $\mathbf{k}$. This is a second-order process, because it involves two interactions.

There are also several first-order sequences which can occur. Although these are simpler than (4.47) because they involve only one interaction, they are more difficult to interpret physically. Let us see what first-order processes can be constructed using the interaction in (4.44). Since one particle enters in $\mathbf{k}$ and one leaves in $\mathbf{k}$, by conservation of momentum the only possibilities are

$$(4.48)$$

Such diagrams as

$$\text{(diagrams)} \qquad , \qquad \text{etc.} \qquad (4.49)$$

are not allowed since they have a particle and a hole in the same state, l, which is impossible—by definition, particles exist only above k_F and holes only below. It can also be shown that diagrams (1), (3), (5) and (7) above do not occur. [The argument for this requires use of the interaction Hamiltonian, H_1, as in (4.28). The term in H_1 corresponding to diagram (1) is

$$V_{klkl} a_l^\dagger a_k^\dagger a_k a_l. \qquad (4.50)$$

When this acts on the state with one incoming particle in ϕ_k, we find

$$V_{klkl} a_l^\dagger a_k^\dagger a_k a_l |1_k^p\rangle = 0 \qquad (4.51)$$

by (4.20). Diagrams (3), (5) and (7) are similarly eliminated. Note that the term in H_1 corresponding to diagram (2), for example, is

$$V_{klkl} b_l a_k^\dagger a_k b_l^\dagger \qquad (4.52)$$

which gives, by (4.20):

$$V_{klkl} b_l a_k^\dagger a_k b_l^\dagger |1_k^p\rangle = V_{klkl} |1_k^p\rangle \neq 0.] \qquad (4.53)$$

(Note: Let us not make the mistake of thinking that in diagrams (1)–(4) in (4.48), 'nothing has happened' just because the particles and holes emerge in exactly the same momentum state in which they entered. This would only be true if we were dealing with classical particles. In the present quantum case something has indeed happened, i.e., two particles in states k,l enter and interact with each other, but instead of being scattered into new states different from k,l, they are simply scattered into the same states, k,l. This is the same as what occurred in the quantum pinball game, where the potential $V_{M_{kk}}$ scattered the particle from the state k into the same state k.)

The possible first-order processes may then be drawn using (4.48)—(2), (4), (6) and (8). This can be done in only one way, e.g., by in each case attaching the outgoing l line to the incoming one (otherwise we would have a particle and a hole entering and leaving the diagram, which would violate the definition of the single-particle propagator, or we would have to introduce more interaction lines, making it a higher-order process). Thus we find:

$$\text{(diagrams)} \qquad (4.54)$$

(a)   Bubble diagrams   (b)

$$ t \quad \text{(diagram)} \quad \left(\text{or} \quad \text{(diagram)} \right) \quad \text{(diagram)} \quad \left(\text{or} \quad \text{(diagram)} \right) . \quad (4.55) $$

(c)   Open oyster diagrams   (d)

(closed oyster appears in (0.23))

The bubble processes can be physically interpreted as follows: a particle
enters in **k**, knocks a particle out of state **l** ($|\mathbf{l}| < k_F$) at time t, then knocks the
particle instantaneously back into **l** at time t, then continues freely in state **k**.
Thus the hole which is created in **l** lasts only zero seconds, and there is no
accompanying particle. Of course it is impossible to draw instantaneous
processes like this, and the bubble picture is purely schematic. This process
is also called '*forward scattering*', since the particle emerges in the same
direction (i.e., momentum state) as it entered. (Note again that by the argument after (4.53), something has really 'happened' in these forward scattering
processes!)

This bubble process undoubtedly sounds so bizarre that it may seem far-
fetched to consider it physical. The fact is that, while in the classical pinball
case, each diagram described a real physical process, the quantum diagrams
describe only what might be called '*quasi-physical*' processes. This will be
discussed further in the next section, §4.6. At the end of Appendix G, it is
proved rigorously that the bubble is a legitimate diagram.

The open-oyster processes are just like the bubbles, except that a quick-
change act occurs in which at time t the incoming particle simultaneously
(*a*) strikes the particle in **l**, (*b*) creates an instantaneous hole in **l** and (*c*) is
exchanged for the particle in **l**. Diagrams (4.55) are often called 'first-order
exchange diagrams', and the process is referred to as an '*exchange scattering*'.
The instantaneous hole lines in the bubble and open oyster are called '*non-
propagating*' lines.

Note that the situation shown in (4.54, 55) is general, i.e., whenever the
interaction (4.43) occurs in a diagram, there is also another diagram possible
in which the two outgoing (or incoming) particles have exchanged momentum.
This is usually drawn thus:

For example, diagram (5) in (4.63) is the exchange of diagram (4).

Let us see how one evaluates these diagrams. Consider first the bubbles. Using dictionary Table 4.2 and (4.43):

$$\begin{array}{l} t_2 \\ \uparrow k \quad 1 \\ t \, \text{\Large www} \, \bigcirc \end{array} = (-1) \sum_{l<k_F} \int_{-\infty}^{+\infty} dt [iG_0^+(\mathbf{k}, t-t_1)] \times \left[-\frac{i}{2} V_{klkl}\right] \times$$

$$\times [iG_0^-(\mathbf{l}, t-t)] \times [iG_0^+(\mathbf{k}, t_2-t)], \qquad (4.56)$$

where we have integrated over the 'intermediate' time, t, and summed over the 'intermediate' momentum, $\mathbf{l}$, as in (4.36). (The extra factor of (-1) in front comes from the fact that the diagram contains one 'fermion loop', namely $\bigcirc$. [A fermion loop is any set of directed lines, in a diagram, which can be traversed in the direction of the arrow, returning to the starting point without lifting pencil from the paper. For example, the $\mathbf{l}, \mathbf{l+q}$ lines in (4.47) form a fermion loop.] This is one of the annoying 'phase factors' which comes out of the rigorous mathematical development of the theory (see end of Appendix G).) Note that an additional factor of (-1) appears because the propagator line for the bubble is:

$$iG_0^-(\mathbf{l}, t-t) = i \times i e^{-i\epsilon_l \times 0} = -1. \qquad (4.57)$$

The Fourier transform of (4.56) may be taken just as was done in the pinball case (2.23). This yields

$$\begin{array}{l} \mathbf{k}, \omega \\ \\ \mathbf{k}, \omega \end{array} \text{\Large www} \, \bigcirc = (-1) [iG_0^+(\mathbf{k}, \omega)]^2 \sum_{l<k_F} \left[-\frac{i}{2} V_{klkl}\right](-1). \qquad (4.58)$$

The (-1) after V_{klkl} comes from (4.57) and is the value of the 'non-propagating' bubble line in $(\mathbf{k}, \omega)$-space as well as $(\mathbf{k}, t)$-space. Note that we cannot get (4.58) just by using the (k, ω) side of dictionary Table 4.2! This is because the bubble (and open oyster) diagrams are special cases.

It should be remarked here that if spin is included, then $\mathbf{k}$ is short for $\mathbf{k}, \sigma$ where σ is the spin quantum number (see p. 106), and $\mathbf{l} \equiv \mathbf{l}, \sigma'$. For a spin-independent interaction, (7.70) holds, and the sum over σ' then produces a factor 2 which multiplies (4.58).

In a similar fashion, the reversed bubble gives

$$\bigcirc \text{\Large www} \begin{array}{l} \mathbf{k}, \omega \\ \\ \mathbf{k}, \omega \end{array} = (-1) [iG_0(\mathbf{k}, \omega)]^2 \sum_{l<k_F} \left(-\frac{i}{2}\right) V_{lklk}(-1). \qquad (4.59)$$

But by (4.42), $V_{klkl} = V_{lklk}$ so these two diagrams are equal. This is quite general and we may write:

> If we are given a diagram, and form a new diagram from it by twisting one or more of its interaction wiggles through 180 degrees, then the new diagram has the same value as the original one. Hence all twisted diagrams may be omitted if we just multiply (4.43) by a factor of 2. (4.60)

Thus, for instance, of the diagrams

$$\text{} \qquad (4.61)$$

it is only necessary to keep the first.

In a manner similar to (4.59), the open oyster gives

$$\text{} = [iG_0^+(\mathbf{k}, \omega)]^2 \sum_{l<k_F} (-i) V_{lkkl}(-1). \qquad (4.62)$$

The factor of 2 recommended in (4.60) has been included. If spin is included, so $\mathbf{k} \equiv \mathbf{k}, \sigma$, and $\mathbf{l} \equiv \mathbf{l}, \sigma'$, then for a spin-independent interaction like (7.70), we find that $\sigma' = \sigma$. Hence there is *no* factor 2 from a spin sum, in contrast to the case of the bubble (4.58).

Observe that the *frequency* (or 'energy parameter'), ω, associated with the propagator line coming out of the interaction in (4.59, 62) is the same as that entering. This is a special case illustrating the general rule called '*conservation of frequency*'. It is the same thing we saw in the pinball model (2.23), (2.25), and results from the fact that the Hamiltonian is time-independent, so the propagators depend only on time differences. This gives rise to δ-functions similar to the $2\pi\delta(\omega' - \omega)$ in (2.23). Conservation of frequency may be incorporated into the labelling of diagrams in $\mathbf{k},\omega$-space, as shown in (4.62')

$$\text{} \qquad (4.62')$$

All momenta and frequencies in this diagram, aside from those entering and leaving, are called 'intermediate'. Thus $\mathbf{q},\mathbf{l}$ and β,ϵ are the intermediate momentum and frequencies. Note that it is convenient to associate a frequency with the wiggly line, even though the interaction itself is independent of ω.

☞   Do not make the mistake of confusing the frequency of a line with the *particle energy*! For example, in the line $\mathbf{k}, \omega$ the frequency is ω while the particle energy is $\epsilon_k = k^2/2m$. Also, the frequency is conserved while the particle energy is not.

Now we can collect the information in Table 4.2, equations (4.43, 57, 60) to produce an unabridged dictionary for the interacting many-body fermion system. This is shown in Table 4.3. The whole series for G^+ is then just the sum of all possible diagrams such as (4.47, 54), etc. Chapter 9 will show how to draw all the possibilities systematically, but here we will simply draw a few representative diagrams, written in $(\mathbf{k}, \omega)$-space for simplicity:

Such diagrams are often called '*self-energy diagrams*' since they show the particle interacting with the many-body medium, which in turn acts back on the particle, altering its energy (see just after (0.5)). It should be noted that many writers draw these diagrams lying down, thus:

Table 4.3 *Diagram dictionary for interacting many-fermion system with no external potential (Goldstone method)*

(k, t)-space		(k, ω)-space	
Word	Diagram	Word	Diagram
$iG^{\pm}(\mathbf{k}, t_2-t_1)$ (see (3.2), (4.29))		$iG^{\pm}(\mathbf{k}, \omega)$	
$iG_0^{+}(\mathbf{k}, t_2-t_1) = \theta_{t_2-t_1} e^{-i\epsilon_k(t_2-t_1)}$		$iG_0^{+}(\mathbf{k}, \omega) = \dfrac{i}{\omega-\epsilon_k+i\delta}$	
$iG_0^{-}(\mathbf{k}, t_2-t_1) = -\theta_{t_1-t_2} e^{-i\epsilon_k(t_2-t_1)}$		$iG_0^{-}(\mathbf{k}, \omega) = \dfrac{i}{\omega-\epsilon_k-i\delta}$	
Non-propagating: $iG_0(\mathbf{k}, t_2-t_2) = -1,\ |\mathbf{k}|<k_F$ $= 0,\ |\mathbf{k}|>k_F$		Non-propagating $iG_0^{-}(\mathbf{k}) = -1,\ |\mathbf{k}|<k_F$ $= 0,\ |\mathbf{k}|>k_F$	
Factor of -1	Each fermion loop. Example:	Factor of -1	Each fermion loop. Example:
$-iV_{klmn}$ or $-iV_q$ (see (4.42), (4.46))		$-iV_{klmn}$ or $-iV_q$	
Each intermediate $\mathbf{k}, t$: $\displaystyle\sum_k, \int dt$		Each intermediate $\mathbf{k}, \omega$: $\displaystyle\sum_k, \int \frac{d\omega}{2\pi}$	

The diagrams may also be interpreted physically from another point of view. Look at the diagrams in $(\mathbf{k}, t)$-space at a particular time t_0:

$$(4.64)$$

At t_0 we see that besides the bare particle, there may exist in the many-body system two 'virtual' particles plus one hole created by second-order process d, or two particles and a hole created by second-order sequence e, and so on, with the particle plus three particle–hole pairs created during the eighth-order poodle process illustrating a typical higher-order case. That is, the diagrams show all the particles and holes which may be kicked up by the bare particle as it churns through the Fermi sea. Now, since the propagator given by (4.64) describes quasi particles (as will be proved in chapter 11) it follows that the diagrams reveal the content of the cloud of particles and holes surrounding the bare particle and converting it into a quasi particle.

Equation (4.63) may be translated into functions by Table 4.3, giving

$$G^+(\mathbf{k}, \omega) = G_0^+(\mathbf{k}, \omega) + (-1) G_0^+(\mathbf{k}, \omega)^2 \sum_{p < k_F} V_{kpkp}(-1) +$$

$$+ G_0^+(\mathbf{k}, \omega)^2 \sum_{m < k_F} V_{kmmk}(-1) + \cdots. \qquad (4.65)$$

4.6 The 'quasi-physical' nature of Feynman diagrams

In the classical pinball game, each individual diagram in the perturbation expansion of the propagator described a real physical process. Using 'physical intuition' based on the analogy to the classical case, we developed the diagrammatic perturbation expansion for the quantum propagator in a one-particle system (3.32), and in non-interacting and interacting many-particle systems, (4.34) and (4.63). Our intuitive methods are, in fact, similar to those used by Feynman when he first introduced diagrams into quantum electrodynamics (Feynman (1962), p. 167 ff.).

However, by now the reader is doubtless aware that Feynman diagrams describe processes which are considerably less 'real' than those described by the classical pinball diagrams. For example, in the case of a single particle in an external potential, it was difficult to see how a particle could be scattered several times by a potential, despite the fact that it was in the potential the whole time. Later on, in the non-interacting fermion system, graphs appeared which violated the Pauli exclusion principle. And then, in the interacting case, we met the simple bubbles which seemed to elude any common-sense physical interpretation. Finally, we found that higher-order diagrams involved 'virtual', rather than real, processes.

Nevertheless, the situation is not as bad as it might seem at first sight. For, although the individual diagrams in quantum propagator expansions have unphysical properties, the sum as a whole does not. In fact, the full propagator, G^+, describes an actual physical experiment—for instance, the elastic scattering of a single nucleon by a nucleus in its ground state (Thouless (1961), p. 69). This means that the unphysical aspects arise because of the manner in which we have decomposed the propagator into a perturbation series. This is roughly analogous to breaking a sentence up into words: the individual words, even though they are meaningful, are not thoughts in themselves. It is only when they are put together to form the sentence that a thought emerges.

Because of the unphysical properties of Feynman diagrams, many writers do not give them any physical interpretation at all, but simply regard them as a mnemonic device for writing down any term in the perturbation expansion. However, the diagrams are so vividly 'physical-looking', that it seems a bit extreme to completely reject any sort of physical interpretation whatsoever. As Kaempffer (1965, p. 209) points out, one has to go back in the history of physics to Faraday's 'lines of force' if one wants to find a mnemonic device which matches Feynman's graphs in intuitive appeal. Therefore, we shall here adopt a compromise attitude, i.e., we will 'talk about' the diagrams as if they were physical, but remember that in reality they are only 'apparently physical' or '*quasi-physical*'.

There is still an important question left: the quantum propagator diagrams describe only quasi-physical processes, whereas the classical pinball diagrams describe real physical processes. How, then, can we justify obtaining the quantum series by analogy to the classical case? Evidently, the only satis- factory answer to this question would be to derive the diagram expansion directly from the Schrödinger equation. This was done at the end of chapter 3 in the single-particle case. It can also be done in the many-body case, but unfortunately the argument there is so long and labyrinthine that the average non-specialist tends to get completely lost in it. It is for this reason that we prefer to use the intuitive approach in the body of the text, and have postponed the rigorous derivation to the appendix. However, for those who feel ex- tremely uncomfortable with intuitive arguments, we offer the following alternative:

IMPORTANT

Those readers who wish to see the rigorous derivation of the many-body diagrams before going any further, should leave the direct path through the book, and instead propagate along the following detour:

§4.6→—chapter 7 (second quantization)→—§9.1, 9.2 (mathematical definition of propagator)→—Appendices B through G (derivation of diagrams. Note: skip Appendix C, and all sections referring to 'vacuum amplitude' or 'finite temperature')→—§4.7→—etc.→—

All others should go on from here directly to §4.7.

4.7 Hartree and Hartree–Fock quasi particles

We will now consider the simplest of all partial sum approximations for the propagator, i.e., the Hartree and the Hartree–Fock. Imagine we have a hypothetical system with no external potential and with an interaction between particles which is dominated by forward-scattering processes (i.e., both particles emerge from the interaction with the same momentum they had when they entered). We ask for the energy dispersion law of the elementary excitations (quasi particles) in this case. The procedure will be to calculate the propagator approximately by picking out the most important set of diagrams in (4.63) for this system, and sum over this set to infinite order.

Let us first write down the interaction, V_{klmn}, in (4.43, 44). This will be dominated by a large forward-scattering term, so we have

$$V_{klmn} = \delta_{mk}\delta_{nl}\underbrace{V_{klkl}}_{large}+\underbrace{W_{klmn}}_{small}(\mathbf{m}\neq\mathbf{k},\mathbf{n}\neq\mathbf{l}) \qquad (4.66)$$

Thus the most important interaction diagrams are the forward-scattering ones shown in (4.48 (1), ..., (4)). The diagrams which will dominate the series (4.63) will therefore be just those in which every interaction is of the forward-scattering type. A few trials reveal that the only diagrams of this sort are

$$\left(1\pm x\right)^{-1}=1\mp x+x^2\mp x^3+\dots \qquad x^2<1$$

those containing just bubbles, so that the propagator may be approximated by a partial summation over repeated bubbles (see (4.67)).

Using Table 4.3 and substituting for the propagators, this becomes

$$iG^+(\mathbf{k}, \omega) = \frac{1}{[iG_0^+(\mathbf{k}, \omega)]^{-1} - (-1) \sum\limits_{l < k_F} (-iV_{klkl})(-1)} \qquad (4.68)$$

or

$$G^+(\mathbf{k}, \omega) = \frac{1}{\omega - \epsilon_k - \sum\limits_{l < k_F} V_{klkl} + i\delta}. \qquad (4.69)$$

Comparing with (3.16) reveals that we have here a live many-body quasi particle with energy dispersion law and lifetime:

$$\epsilon_k' = \epsilon_k + \sum_{l < k_F} V_{klkl}; \quad \tau_k = 1/\delta = \infty. \qquad (4.70)$$

The quantity $\sum\limits_{l < k_F} V_{klkl}$ is the 'self-energy' of the particle as described just after (0.5). If spin is included (see after (4.58)) there is a factor 2 multiplying V_{klkl}.

This result has a simple physical meaning. First we note that (4.67) has exactly the same form as the diagram series (3.33) for a single particle moving through an external potential, with

$$\sim\!\!\!\!\circ \equiv (-1) \sum_{l < k_F} (-i) V_{klkl}(-1) \qquad (4.71)$$

playing the same role as

$$\overset{|}{\textcircled{M}} \equiv -iV_{M\mathbf{k}\mathbf{k}'}. \qquad (4.72)$$

Thus, (4.71) can be interpreted as a transition probability for $\phi_k \to \phi_k$ scattering caused by an 'effective external potential', v_{eff}. We can find v_{eff} by writing out (4.71) in detail, using (4.42):

$$\sum_{l < k_F} V_{klkl} = \int d^3 \mathbf{r} \phi_k^*(\mathbf{r}) \underbrace{\left\{ \sum_{l < k_F} \int |\phi_l(\mathbf{r}')|^2 V(\mathbf{r} - \mathbf{r}') d^3 \mathbf{r}' \right\}}_{v_{\text{eff}}} \phi_k(\mathbf{r}). \qquad (4.73)$$

Comparing with (4.33) shows that the quantity in brackets is just v_{eff}. Since $|\phi_l(\mathbf{r}')|^2$ is the density at point $\mathbf{r}'$ of a particle in ϕ_l, v_{eff} is evidently the average potential at point $\mathbf{r}$ due to all the particles in the Fermi sea. (In the present case, since the ϕ_l are plane waves, v_{eff} is independent of $\mathbf{r}$.)

We now recall that for the quantum pinball propagator, the quasi particle energy (3.38) could be obtained both by the diagram method and directly from the Schrödinger equation using Hamiltonian (3.20). In the present case, it's easy to write a Schrödinger equation with energy eigenvalues ϵ'_k by just using v_{eff} as external potential:

$$\left[\frac{p^2}{2m} + v_{\text{eff}}(\mathbf{r})\right]\phi_k(\mathbf{r}) = \epsilon'_k \phi_k(\mathbf{r}). \tag{4.74}$$

It is easily checked that this is correct by multiplying both sides by $\phi_k^*(\mathbf{r})$ and integrating—the result is just (4.70).

In our intuitive argument for (4.74), the ϕ_k were given (plane waves). However, if we regard them as eigenfunctions to be solved for, then (4.74) is just the famous Hartree equation. Remember that by (4.73), v_{eff} is a function of all the ϕ_k's. This means that we must calculate ϕ_k *self-consistently*, i.e., put an assumed ϕ_k in v_{eff}, find a new ϕ_k from (4.74), put the new ϕ_k in v_{eff}, calculate a newer ϕ_k, etc., until ϕ_k stops changing appreciably. In the present case with no external potential, we find immediately that the correct ϕ_k is just a plane wave. However, in a system with an external potential, like an atom, or a molecule the whole self-consistent procedure must be carried out. In such cases, the ϕ_k may correspond to atomic or molecular orbitals, and V may be interpreted as scattering between orbitals. (See §11.1 for further discussion.)

From here, it is only a baby step away to the quasi particle in Hartree–Fock (HF) approximation. Imagine that exchange scattering is just as important as forward scattering in our hypothetical system, i.e., that

$$V_{klmn} = \delta_{mk}\delta_{nl}V_{klkl} + \delta_{ml}\delta_{nk}V_{kllk} + \text{small terms}. \tag{4.75}$$

Then the open oysters must also be included in the approximation for the propagator, and the partial sum carried out as in (3.39):

$$= \frac{\uparrow}{1 - \uparrow \times (\text{⌒O} + \text{⌣⌣})} = \frac{1}{\uparrow^{-1} - (\text{⌒O} + \text{⌣⌣})}. \tag{4.76}$$

Translating by means of dictionary Table 4.3 yields

$$G^+(\mathbf{k}, \omega) = \frac{1}{\omega - \epsilon_k - \sum_{l<k_F} (V_{klkl} - V_{lkkl}) + i\delta}. \quad (4.77)$$

(If spin is included, multiply V_{klkl} by 2.) This also has quasi particle form with

$$\epsilon_k' = \epsilon_k + \sum_{l<k_F} (V_{klkl} - V_{lkkl})$$

$$\tau_k = \infty. \quad (4.78)$$

This is the quasi particle energy and lifetime in HF approximation. The V_{lkkl} is the well-known 'exchange term'. Analogous to what was done in the Hartree case, we can here construct a Schrödinger equation including the effective external 'exchange' potential; this turns out to be the Hartree–Fock equation. (Note that plane waves are the self-consistent solution of the HF equation in the present case with no real external potential, just as with the Hartree equation (4.74).) It should be mentioned that the lifetime here is infinite because of the crudeness of the HF approximation. Better approximations, which include sums over diagrams like (4.47), produce finite lifetimes.

4.8 Hartree–Fock quasi particles in nuclear matter

Real-life physical systems have interactions considerably more complicated than the hypothetical 'forward plus exchange scattering' model in the previous section. Nevertheless, the HF can be used as a very crude 'first approximation' to the propagator, as we show now for the case of nuclear matter.

Nuclear matter is *not* matter in a nucleus! It is a *hypothetical* stuff concocted in the following way (see Thouless (1961), p. 20, for details): On the basis of the 'liquid drop' model of the nucleus, Weizsäcker constructed the famous 'semi-empirical mass formula' for nuclear binding energy:

$$E(N,Z) = \underbrace{-a_1 A}_{\substack{nuclear \\ forces}} + \underbrace{a_2 A^{\frac{2}{3}}}_{\substack{surface \\ correction}} + \underbrace{a_3 Z^2 A^{-\frac{1}{3}}}_{\substack{Coulomb \\ forces}} + \underbrace{\tfrac{1}{4}a_4(N-Z)^2/A}_{\substack{Pauli\ principle \\ correction}} \quad (4.79)$$

where N and Z = number of neutrons and protons respectively, $A = N + Z$ and the a_i's are constants determined by fitting (4.79) to known nuclear masses. In the first term, $-a_1$ is the binding energy of a single nucleon, well inside the nucleus (i.e., not near the surface), due to the attractive nuclear forces—it is about -15.9 MeV. The second, third and fourth terms are respectively corrections due to the presence of the nuclear surface, Coulomb forces between protons, and the effect of the exclusion principle.

If there were no Coulomb forces, and if the number of nucleons was so large that the nucleus was the size of, say, a coconut (making the surface term negligible in comparison with the first term) and if $N=Z$, then we would have a simple system with binding energy proportional to the number of nucleons, A. This hypothetical system consisting of a huge number of protons and an equal number of neutrons interacting by purely nuclear forces (no Coulomb forces) is called *nuclear matter*. It is of great interest because a calculation of the binding energy of nuclear matter, using some model of the nuclear force, is evidently a calculation of $-a_1$, and can be compared with the experimental value of -15.9 MeV.

We will assume that the nuclear interaction has the form of a simple Yukawa potential $(V_0 < 0)$

$$V = + aV_0 \frac{e^{-|\mathbf{r}-\mathbf{r}'|/a}}{|\mathbf{r}-\mathbf{r}'|}. \tag{4.80}$$

(Such a purely attractive interaction is clearly science-fiction, since it would cause the nuclear matter to collapse to a point. This can be prevented by adding a 'hard core' to the potential, as described in §12.4.) The quantity a ($\sim 10^{-13}$ cm) is called the 'range' of the interaction, since the exponential becomes very small for $|\mathbf{r}-\mathbf{r}'| > a$.

The quasi particle energy in HF approximation may be calculated using (4.78, 80) (Brown (1972)). Noting that the density of points in $\mathbf{k}$-space is $\Omega/(2\pi)^3$, where Ω is the normalization volume, (see after (3.64)), we may convert from a sum to an integral using

$$\sum_l \rightarrow \Omega \int \frac{d^3\mathbf{l}}{(2\pi)^3} \tag{4.81}$$

so that (multiply V_{klkl} by 2 if spin is included)

$$\epsilon'_k = \frac{k^2}{2m} + \Omega \int\limits_{|\mathbf{l}| < k_F} \frac{d^3\mathbf{l}}{(2\pi)^3} (V_{klkl} - V_{lkkl}). \tag{4.82}$$

The transition matrix element V_{klmn} is (using (3.5), (4.42, 45, 80))

$$V_{klmn} = + \frac{V_0}{\Omega^2} \int\int d^3\mathbf{r}\, d^3\mathbf{r}'\, e^{-i(\mathbf{k}\cdot\mathbf{r}+\mathbf{l}\cdot\mathbf{r}'-\mathbf{m}\cdot\mathbf{r}-\mathbf{n}\cdot\mathbf{r}')} \frac{e^{-|\mathbf{r}-\mathbf{r}'|/a}}{|\mathbf{r}-\mathbf{r}'|/a}$$

$$= \frac{1}{\Omega} \frac{4\pi V_0 a^3 \delta_{k+l,\, m+n}}{[1+(\mathbf{k}-\mathbf{m})^2 a^2]} = \frac{1}{\Omega} \frac{4\pi V_0 a^3 \delta_{k+l,\, m+n}}{[1+(k^2+m^2-2km\cos\theta)a^2]} \tag{4.83}$$

where θ is the angle between $\mathbf{k}$ and $\mathbf{m}$ and we have used that $\Omega \gg a^3$. Hence

$$V_{klkl} = + \frac{4\pi V_0 a^3}{\Omega}, \quad V_{lkkl} = + \frac{1}{\Omega} \frac{4\pi V_0 a^3}{[1+(l^2+k^2-2kl\cos\theta)a^2]}. \tag{4.84}$$

Substituting these expressions in (4.82), we find that the V_{klkl} integral is trivial and yields $2V_0 a^3 k_F^3/3\pi$. The V_{lkkl} integral is first integrated over ϕ and θ, which yields terms involving $(k+l)a$ and $(k-l)a$. The remaining l-integration is easily carried out with the aid of the substitutions $y = (k+l)a$, and $z = (k-l)a$, and we obtain for the quasi particle energy

$$\epsilon_k' = \frac{k^2}{2m} + \frac{2V_0 a^3 k_F^3}{3\pi} - \frac{V_0}{2\pi}[F(ka+k_F a) - F(ka-k_F a)], \qquad (4.85)$$

where

$$F(z) = \frac{1}{2ka}[1+z^2][\ln(1+z^2)-1] - [z\ln(1+z^2)-2z+2\tan^{-1}z] \qquad (4.86)$$

This expression can be evaluated to find the effective mass in the limit when ka and $k_F a$ are both $\ll 1$, so that $z \ll 1$. In order to get a non-vanishing contribution from $[F(ka+k_F a) - F(ka-k_F a)]$, it is necessary to expand the logarithm and $\tan^{-1}$ functions up through order z^6. Keeping only terms up through order k^2 we find:

$$\epsilon_k' \approx \frac{2V_0 a^5 k_F^5}{5\pi} + \left[\frac{1}{2m} + \frac{2V_0 a^5 k_F^3}{3\pi}\right]k^2, \qquad (4.87)$$

from which we see that the effective mass is

$$m^* = \frac{m}{1 + \dfrac{4mV_0 a^5 k_F^3}{3\pi}}. \qquad (4.88)$$

4.9 Quasi particles in the electron gas, and the random phase approximation

A real metal consists of $\sim 10^{23}$ positively charged ions arranged in the form of a regular lattice, with $\sim 10^{23}$ electrons moving more or less freely among these ions. The ions execute oscillations about their equilibrium positions ('lattice vibrations'). Such a complicated system poses a nasty problem for the many-body physicist. To make life easier, he often postulates a utopian metal in which the ions are motionless, and the positive ion charge is smeared out to form a fixed uniform positive background against which the electrons move. The electrons are assumed to interact by purely Coulomb forces. This theoretician's pipe dream is called the 'electron gas'. (See Fig. 0.7).

Let us first examine the electron gas in the HF approximation. The Coulomb interaction and its transition matrix element are just the Yukawa interaction (4.80) (with $V_0 > 0$) and its matrix element (4.83) with $V_0 a = e^2$, $a \to \infty$, i.e.:

$$(a)\ V(\mathbf{r}, \mathbf{r}') = \frac{e^2}{|\mathbf{r} - \mathbf{r}'|}, \qquad (b)\ V_{klmn} = \frac{4\pi e^2}{|\mathbf{k} - \mathbf{m}|^2}, \qquad (4.89)$$

where spins are left out for simplicity, and we take $\Omega = 1$ cm^3. That is, the Coulomb interaction has the form of a Yukawa interaction with 'infinite range'. Alternatively, one often says that the Yukawa potential has the form of a 'shielded' Coulomb potential, the $\exp(-r/a)$ in (4.80) being the 'shielding factor'. Note also that (4.89) becomes infinite for $\mathbf{k} = \mathbf{m}$ whereas (4.83) remains finite.

The quasi particle energy may be evaluated in exactly the same way as for the nuclear matter case. There is a slight simplification because of the fact that the bubble term in (4.76) is cancelled by the positive charge background (see §10.4), so that

$$\left.\begin{array}{c} \| \\ \uparrow \\ \| \end{array}\right|_{\text{HF (electron gas)}} = \frac{1}{\Big|^{-1} - \smile} \cdot \qquad (4.90)$$

The expression for the quasi particle energy turns out to be (take limit of (4.85), (4.86) when $V_0 a = e^2$ and $a \to \infty$):

$$\epsilon_k' = \frac{k^2}{2m} - \frac{e^2 k_F}{2\pi}\left[2 + \frac{(k_F^2 - k^2)}{kk_F}\ln\left|\frac{k+k_F}{k-k_F}\right|\right]. \qquad (4.91)$$

We are mainly interested in quasi particles near k_F, since it is primarily these which take part in physical processes. For $|\mathbf{k}|$ near k_F, the effective mass may be found by expanding ϵ_k' about k_F:

$$\epsilon_k' = \epsilon_{k_F}' + \left(\frac{\partial \epsilon_k'}{\partial k}\right)_{k_F}(k - k_F) + \cdots, \qquad (4.92)$$

where $k = |\mathbf{k}|$.

For the non-interacting system

$$\epsilon_k = \frac{k_F^2}{2m} + \frac{k_F}{m}(k - k_F) + \cdots. \qquad (4.93)$$

Comparing (4.92) and (4.93), we may regard the effective mass as given by

$$\frac{k_F}{m^*} = \left(\frac{\partial \epsilon_k'}{\partial k}\right)_{k_F} \quad \text{or} \quad m^* = k_F \Big/ \left(\frac{\partial \epsilon_k'}{\partial k}\right)_{k_F}. \qquad (4.94)$$

For ϵ_k' as in (4.91) we obtain

$$m^*_{\text{HF (electron gas)}} = 0! \qquad (4.95)$$

This is of course an absurd result, and it disagrees with experiments, all of which show m^* to be of the same order of magnitude as m.

The reason why the Coulomb interaction produces zero effective mass at the Fermi surface, whereas the Yukawa interaction does not, can be traced back to the fact that the Coulomb interaction is infinite for zero momentum transfer, as was pointed out after (4.89). The HF approximation is not adequate to handle such singular interactions.

The physical reason for the inadequacy of HF lies in the fact that it treats the effect of all the other particles on the test particle by means of a time-independent average potential. But we know from §0.2 that the quasi particle is a bare particle plus a cloud which in a sense 'follows' the bare particle. The HF approximation thus gives us what might be called the 'static' part of this cloud, but misses out on the 'moving' part.

The usual way of putting this is to say that the HF neglects 'correlations', which means that it neglects that movement of the other particles which 'is correlated with' (i.e., 'follows') the movement of the bare particle. As mentioned in §0.2, we would expect that these correlations would have the effect of 'shielding' the interaction between particles, making it much weaker. The diagram method which we discuss now (very briefly) gives us the way to calculate this shielding effect. (It should be observed that although we consider only the electron gas here, this is the same sort of problem one has to deal with when trying to improve on HF calculations of atoms and molecules.)

How is it possible to take account of correlations diagrammatically? Evidently the correlation effects must lie in those diagrams which were omitted in the HF approximation. Of course it is impossible to take account of all the omitted diagrams, but we can at least sum over the most important ones.

It turns out (as will be shown in §10.4) that in the limit of a high density electron gas, the most important diagrams are those occurring in the following approximation for G:

$$(4.96)$$

These diagrams may be summed to infinity and it is found that (see chapter 10)

$$\| = \frac{1}{\Big|^{-1} - \bigcirc\!\!\!\!\sum}, \qquad (4.97)$$

where $\sum$ (the self energy) is given by

$$\bigcirc\!\!\!\!\sum = \underset{q,\,\omega}{\curvearrowright} + \;\;+\;\;+\;\;+\cdots. \qquad (4.98)$$

The diagrams in (4.98) are called 'ring' diagrams because of their ring-like structure. For historical reasons, this approximation for G is called the 'Random Phase Approximation' or 'RPA'.

In order to interpret (4.98), we twist the top interaction wiggle in each diagram through 180° (this has no effect on the value—see (4.61)), and factor out a free propagator:

$$\bigcirc\!\!\!\!\sum = \Big| \times \Big[\;\;+\;\;+\;\;+\;\;+\cdots \Big]. \qquad (4.99)$$

The series in brackets

$$\underset{q,\,\omega}{\nwarrow} = \;\;+\;\;+\;\;+ \qquad$$

$$(a) \qquad\qquad (b) \qquad\qquad (c)$$

$$+ \;\;+\cdots \qquad (4.100)$$

$$(d)$$

is called the 'effective interaction'. The reason for this is as follows: Diagram (a) shows the direct or 'bare' interaction between two particles. Diagram (b) shows the first particle creating a particle–hole pair in the system, and the second particle interacting with this pair. There is thus an indirect or 'effective' interaction between two particles via the many-body system. The higher order diagrams describe interactions which are more and more indirect. We may thus write $\sum$ in terms of the effective interaction:

$$\textcircled{\textstyle\sum} = \text{(diagram)} \qquad (4.101)$$

which shows an electron interacting with itself via the effective interaction.

We may obtain further insight into the nature of the effective interaction by carrying out the sum (4.99). In the limit when $\omega = 0$ and q is small, this yields (see chapter 10):

$$V_{\text{eff}}(q) = \frac{4\pi e^2}{q^2 + \lambda^2} \qquad (4.102)$$

This has the same form as (4.83), so that, assuming it is true for all q, it must correspond to an effective interaction having the same r-dependence as the Yukawa potential in (4.80):

$$V_{\text{eff}}(\mathbf{r}) = 4\pi e^2 \frac{e^{-\lambda r}}{r}. \qquad (4.103)$$

In contrast to the Coulomb interaction, which is 'long range', dropping off as $1/r$, this drops off exponentially for $r \gg \lambda^{-1}$ so it has only a short range $\sim \lambda^{-1}$ cm. It is referred to as a 'shielded' or 'screened' interaction.

Such a screened interaction is just what we would expect physically on the basis of the argument in §0.2. The real electron repels other electrons from it; this exposes the positive charge background so that the electron is effectively 'followed' by a positive charge cloud of width λ^{-1}. This turns it into a quasi electron because the positive cloud 'screens' the electron's own charge, thus drastically reducing its interaction with the other particles of the system at distances greater than λ^{-1}.

Since (4.101) with V_{eff} as in (4.102) has the same form as the HF self-energy in (4.90), the quasi particle energy is easily calculated by placing $V_{\text{eff}}(\mathbf{k} - \mathbf{l})$ into (4.82), with the V_{klkl} term equal to zero:

$$\epsilon_k' = \frac{k^2}{2m} - \int\limits_{|\mathbf{l}| < k_F} \frac{d^3\mathbf{l}}{(2\pi)^3} \frac{4\pi e^2}{[(\mathbf{k} - \mathbf{l})^2 + \lambda^2]}. \qquad (4.104)$$

The calculation of the effective mass then goes just as for the Yukawa potential. In the simple case of large λ (i.e., $k_F\lambda^{-1}\ll1$, a condition which is not actually satisfied in the electron gas), it is found that parallel to (4.88),

$$m^* = \frac{m}{1+\dfrac{16\,m\pi\,e^2}{3}\dfrac{e^2}{\lambda^5}\,k_F^3} \qquad (4.105)$$

which is evidently finite. Thus the inclusion of the correlation (screening) effects represented by the ring diagrams has produced a physically reasonable result. (The result for m^* in RPA when $\omega\neq0$ and $\mathbf{q}$ is not small appears in §10.4.)

Exercises

4.1 In a system of free particles, a hole is created in the single-particle state $\phi_k(\mathbf{r})=\Omega^{-1/2}e^{i\mathbf{k}\cdot\mathbf{r}}$. What is the momentum of the hole?

4.2 For the five-particle system in Fig. 4.1:

    (a) Evaluate $c_3^\dagger\,c_6^\dagger\,c_4\,c_6^\dagger\,c_3|11111000...\rangle$.

    (b) Write $|1101100100...\rangle$ in particle–hole notation.

    (c) Find $\sum_k \epsilon_k\,c_k^\dagger\,c_k|1111100...\rangle$.

4.3 Suppose we have a non-interacting system with external perturbing potential such that V_{km}, V_{mk}, V_{kl}, V_{lk} $(m<k_F,\ l>k_F,\ k>k_F)$ are large, and all other V's are small. Find $G^+(k_1=k,\ k_2=k,\ \omega)$.

4.4 Show that for a system of fermions with a momentum-conserving interaction, the following diagrams are not allowed:

4.5 Consider diagram 5 on the right-hand side of (4.63).
    (a) Label it, showing momentum conservation explicitly in the labelling.
    (b) Show that the scattering processes at each interaction are virtual.

4.6 Translate (4.62') into functions. What variables does it depend on?

4.7 Show that:

4.8 Suppose we have a hypothetical system in which the most important scattering processes are the forward and exchange scattering of (4.58), (4.62) and the double scattering in (4.62′). Find an approximate expression for the propagator by partial summation. Do not attempt to evaluate the integrals!

4.9 Verify in detail equations (4.83) through (4.88).

4.10 Verify (4.91).

4.11 We have a system of N non-interacting Fermi particles. (a) They are acted upon by an external perturbing potential such that $V_{kl} = A$ for all $k,\, l$. Find the propagator $G^+(\mathbf{q}, \mathbf{p}, \omega)$, $p > k_F$, $q > k_F$, by summing exactly over all diagrams. (b) Generalize your result to the case where V_{kl} has the form: $V_{kl} = A f_k f_l$ (factorizable potential).

Chapter 5

Ground State Energy and the Vacuum Amplitude or 'No-particle Propagator'

5.1 Meaning of the vacuum amplitude

One of the first many-body problems to be tackled by the field theoretical diagram techniques was that of finding the ground state energy, E_0, of a system of interacting fermions. This quantity is directly related to experimentally measured properties—such as for example the cohesive energy in a metal or the binding energy in nuclear matter. Calculating it theoretically is a tough job. The interactions are large and hard to handle, and naïve approaches simply drown one in a deluge of infinities. Thus in the nuclear case, because of the hard core interaction, one gets $V_{klmn} = \infty$ making the interaction Hamiltonian infinite. The electron gas is equally psychotic, yielding ∞ for every order of perturbation theory higher than first.

The diagrammatic methods to be discussed in this chapter provide a neat way of handling such delinquents as the above nuclear and electron interactions. In both cases, we can perform a partial sum over an infinite series of infinite terms and get a finite result! In order to do this, it is necessary to have a general way of writing down the nth-order term in the ordinary perturbation series for E_0, i.e., in

$$E_0 = W_0 + \langle \Phi_0 | H_1 | \Phi_0 \rangle + \sum_{m \neq 0} \frac{\langle \Phi_0 | H_1 | \Phi_m \rangle \langle \Phi_m | H_1 | \Phi_0 \rangle}{W_0 - W_m} + \cdots \quad (5.1)$$

where W_0, W_m are the ground and excited state energies of the unperturbed Hamiltonian, and Φ_0, Φ_m are the corresponding wave functions. The general term is hard to obtain from the time-independent theory usually used to get (5.1). However, there is a time-dependent technique which gives a pictorial recipe for finding the desired nth-order term; this is the method of the vacuum amplitude expansion.

The *vacuum amplitude*, $R(t)$, is defined as follows: Let Φ_0 be the ground state of the unperturbed system as defined in (4.12) (i.e., Φ_0 is the 'Fermi vacuum'). Then $R(t)$ is the probability amplitude that if the system is in Φ_0 at time 0, and the external potential and/or interactions between particles are allowed to act, then the system will be in Φ_0 at time t. That is, $R(t)$ is the 'Fermi vacuum to Fermi vacuum transition amplitude'.

Since no particles are added to the system, and none emerges, $R(t)$ may be called a 'no-particle propagator'.

Let us define the vacuum amplitude in more detail. Suppose that at $t=0$ the system is in state Φ_0. If there is no interaction, then the wave function at time t will be $\Phi_0 e^{-iW_0 t}$ where W_0 is the ground state energy. If the interaction is now switched on at time $t=0$, the system will start to make transitions from Φ_0 to all possible N-particle states. We ask for the probability amplitude, $R(t)$, that at the end of time t the system is in the (time-developed) ground state, $\Phi_0 e^{-iW_0 t}$. Let the state after time t be $\Psi(t)$; this must be obtainable from the ground state Φ_0, by some sort of operation, thus:

$$\Psi(t) = U(t)\Phi_0 \tag{5.2}$$

which may be regarded as the equation defining the 'time development operator', $U(t)$ (see appendix B). The probability amplitude $R(t)$ is just the scalar product of $\Phi_0 e^{-iW_0 t}$ and $\Psi(t)$ or:

$$R(t) = (\Phi_0 e^{-iW_0 t}, \Psi(t)) = \int \Phi_0^* e^{+iW_0 t} U(t)\Phi_0 \, d\mathbf{r}_1 \ldots d\mathbf{r}_N$$

$$\equiv \langle \Phi_0 | U(t) | \Phi_0 \rangle \, e^{+iW_0 t} = \text{vacuum amplitude.} \tag{5.3}$$

The importance of the vacuum amplitude lies in the fact that the ground state energy, E_0, may be obtained from it with the aid of the theorem

$$E_0 = W_0 + \lim_{t \to \infty(1-i\eta)} i\frac{d}{dt}\ln R(t), \tag{5.4}$$

where η is an infinitesimal. This is proved in appendix C. Thus, if we can get a diagrammatic expansion of $R(t)$, then the diagram series for E_0 follows from (5.4).

The diagrammatic perturbation expansion of $R(t)$ is rather similar to the corresponding expansion of the propagator. However, it is considerably more complicated, because in addition to diagrams which consist of only one piece ('linked' diagrams) there are also diagrams consisting of two or more pieces ('unlinked' diagrams). Luckily, the logarithm of R, which appears in (5.4), turns out to be the sum over just linked diagrams. This is the famous 'linked cluster theorem'. Placing this result into (5.4) yields the ordinary Rayleigh–Schrödinger perturbation series together with a general rule for writing out the nth-order term.

In this chapter, we will first investigate the analogue of the vacuum amplitude in the classical pinball machine case. Then, we will show how to calculate the vacuum amplitude and ground state energy diagrammatically for the case of a single quantum particle in an external perturbing potential. Finally, the method will be generalized to many-body systems. The application to the electron gas and nuclear matter is in chapter 12, which may be read directly after this chapter, for those who wish.

5.2 The pinball machine vacuum amplitude

The pinball accelerator, Fig. 2.3, used to illustrate the calculation of the single-particle propagator may also be employed as the classical analogue for the vacuum amplitude. Before firing, the particle is in its 'ground state' or 'vacuum' at the point O. The accelerating mechanism propels it through the collimator, after which it undergoes a series of interactions with various scattering centres, winding up at one of the points r_2, r_3, ..., r_6 or possibly back at the original point O itself. (Note: In what follows, the position of the point O will be designated by r_O. The vectors r_1, r_2, will be variables denoting the initial and final points of the particle, i.e., they will no longer be the fixed points r_1, r_2, labelled on Fig. 2.3.)

The classical analogue of the vacuum amplitude here is the probability $P(r_2 = r_O, t_2 = t; r_1 = r_O, t_1 = 0)$ that if the test particle is in its 'ground state' at $r_1 = r_O$ at time $t_1 = 0$, then it returns to the 'ground state' at $r_2 = r_O$ at time $t_2 = t$. For the sake of simplicity, let us leave time out of the argument to begin with, and consider just $P(r_2 = r_O, r_1 = r_O)$, or $P(r_O, r_O)$ for short; this is the probability that if the particle begins at the point $r_1 = r_O$, then it will finish at $r_2 = r_O$, regardless of the time. This $P(r_O, r_O)$ may be broken up into the sum of the probabilities for all the different ways the particle can go through the machine and still wind up back at the point O. Thus, the first possibility is that the trigger is not pulled at all so the particle just continues lying at O. The next possibility is that the trigger is pulled, but not hard enough to propel the particle out of the collimator, so that it just falls back to O. Next, the particle could come out of the collimator and be scattered from the giraffe's tail back into the collimator, rolling down to O, and so on. Let P_u = probability that the trigger is not pulled, $P_0(r_j, r_i)$ = probability that if the particle leaves the point r_i then it travels to r_j without being scattered by any animal, and $P(A)$ = 'interaction' term, giving the probability that if the particle arrives at animal A, then it is scattered. Then, assuming independent probabilities, we find, similar to the propagator case, that $P(r_O, r_O)$ is given by (Note: $P(O)$ is the probability that the particle is scattered away from the point O; this is just $1 - P_u$, i.e., the probability that the trigger is pulled):

$$P(r_O, r_O) = P_u + P(O)P_0(r_O, r_O) + P(O)P_0(r_G, r_O)P(G)P_0(r_O, r_G) +$$
$$+ P(O)P_0(r_L, r_O)P(L)P_0(r_O, r_L) + \cdots$$
$$+ (PO)P_0(r_G, r_O)P(G)P_0(r_G, r_G)P(G)P_0(r_O, r_G) + \cdots. \quad (5.5)$$

Associating diagrams with probabilities, using Table 2.1, the above series becomes

$$P(\mathbf{r}_O, \mathbf{r}_O) = P_u + \bigcirc + \bigcirc + \bigcirc + \cdots + \bigcirc + \cdots + \bigcirc + \cdots \quad (5.6)$$

where $\bigcirc\!\!\!O$ stands for the scattering at O. This series may be evaluated by partial summation parallel to (2.17, 18, 19), but we shall not bother to do it.

When generalized to include time, using Table 2.2, the diagrammatic series becomes (we abandon, in this particular case, the convention that time increases in the positive-y direction!):

$$P(\mathbf{r}_O, \mathbf{r}_O, t) = P_u + \bigcirc + \bigcirc + \cdots \quad (5.7)$$

which is equivalent to

$$P(\mathbf{r}_O, \mathbf{r}_O, t) = P_u + P_0(\mathbf{r}_O, \mathbf{r}_O, t) P(O) +$$
$$+ \int_0^t dt_G P_0(\mathbf{r}_G, \mathbf{r}_O, t_G - 0) P(G) P_0(\mathbf{r}_O, \mathbf{r}_G, t - t_G) P(O) + \cdots. \quad (5.8)$$

With the aid of a simple modification of the time-dependent case, it is possible to obtain a more precise analogue to the quantum vacuum amplitude. The modification is to allow the trigger to be pulled an arbitrary number of times during the time interval $(0, t)$, instead of just no times or once at $t = 0$. Then in the event that the particle has returned to O before time t, there is the possibility that it can be shot out again and return to O again before time t has elapsed. This may happen many times, if t is long enough, and can be represented by diagrams of the following sort

$$(5.9)$$

These are the so-called '*unlinked*' diagrams. When this type is included, the series for $P(\mathbf{r}_O, \mathbf{r}_O, t)$ becomes

$$P(\mathbf{r}_O, \mathbf{r}_O, t) = P_u + \text{(diagrams)} \qquad (5.10)$$

We go on now to the quantum mechanical case.

5.3 Quantum vacuum amplitude for one-particle system

The calculation of the quantum vacuum amplitude proceeds parallel to the classical case just discussed. Consider the simplest situation first: a Fermi system consisting of one particle in an external potential, with non-degenerate energy levels—for example, an electron in a one-dimensional harmonic oscillator potential. Let the unperturbed Hamiltonian be

$$H_0 = \frac{p^2}{2m} + U(r) \qquad (5.11)$$

with eigensolutions $\phi_k(\mathbf{r})$, ϵ_k. In order of increasing energy, we label the single particle states ϕ_1, ϕ_2, ..., with corresponding energies ϵ_1, ϵ_2, ... The ground state of the system consists of one particle in ϕ_1 and no particles in any higher states; in occupation number formalism this is $\Phi_0 = |1_1, 0_2, 0_3, ...\rangle$. The corresponding ground state energy, W_0, is evidently just ϵ_1. The Fermi energy is also ϵ_1. A typical excited state is one particle in ϕ_k and no particle in any other state: $\Phi_{\text{excited}} = |0_1, 0_2, ..., 1_k, ...\rangle$. In particle–hole notation, the ground state is $\Phi_0 = |0\rangle$, while a typical excited state consists of a hole in ϕ_1 and a particle in ϕ_k: $\Phi_{\text{exc}} = |1_1^h, 1_k^p\rangle$. Note that in this one-particle system, there is only one possible hole state, e.g., ϕ_1.

Suppose now a perturbation $V(\mathbf{r})$ is added to H_0. The vacuum amplitude in that case is the probability amplitude that if the system starts in its ground state Φ_0 at $t=0$, and is acted upon zero or more times by $V(\mathbf{r})$, then it will be in $\Phi_0 e^{-iW_0 t}$ at time t. By analogy with the pinball case, $R(t)$ will be the sum of the probability amplitudes for all the different ways the system can start out in Φ_0, interact with $V(\mathbf{r})$ zero, one, or more times and return to Φ_0. In the zeroth-order process, nothing at all happens as illustrated in (5.12a):

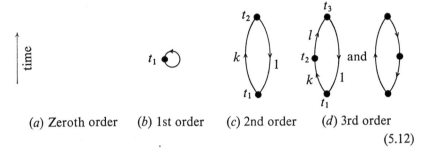

(a) Zeroth order   (b) 1st order   (c) 2nd order   (d) 3rd order

$$(5.12)$$

In first order, $V(\mathbf{r})$ can lift the particle out of ϕ_1 at t_1, thus creating a hole, and instantaneously put it back in, destroying the hole. This is shown by (5.12b) (compare with (4.54)). In second order, at t_1, $V(\mathbf{r})$ can scatter the particle up into the state ϕ_k, thus simultaneously creating a hole in ϕ_1 and a particle in ϕ_k, and at t_2 scatter the particle back into ϕ_1, destroying the hole and particle, and so on. The third-order processes are in (5.12d), and the fourth-order ones are

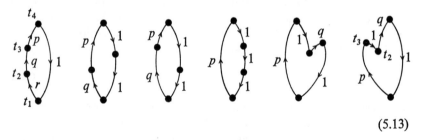

$$(5.13)$$

(Note that in the last two diagrams of (5.13), there are *two* particle lines and *two* hole lines between t_2 and t_3, whereas our one-particle system can have at most one particle and one hole. However, it is easily shown that these diagrams are exactly cancelled by unlinked diagrams of the sort in (5.17) below. For example, because of the (-1) from the extra fermion loop, the last diagram in (5.13) is cancelled by the fourth-order diagram in (5.17), as is easily seen by evaluating diagrams using Table 4.2. Nevertheless, it is necessary to retain such diagrams which violate conservation of particle number, in order to prove the linked cluster theorem described in the next section. The same argument holds for diagrams which violate the Pauli exclusion principle.)

[These processes may be conveniently described in particle–hole formalism by using H_1 as in (4.26). Thus in first order:

$$H_1 \text{ acts at } t_1: V_{11} b_1 b_1^\dagger |0\rangle = V_{11} |0\rangle. \tag{5.14}$$

In second order, H_1 acts at t_1, creating a particle–hole pair from the Fermi vacuum:

$$V_{k1} a_k^\dagger b_1^\dagger |0\rangle = V_{k1} |1_k^p, 1_1^h\rangle. \tag{5.15}$$

At t_2, H_1 acts again, destroying the pair and returning the system to the vacuum state:

$$V_{1k} b_1 a_k V_{k1} |1_k^p, 1_1^h\rangle = V_{1k} V_{k1} |0\rangle.] \tag{5.16}$$

The diagrams in (5.12), (5.13) are often called '*vacuum fluctuation*' or '*vacuum polarization*' diagrams, since they show all the virtual processes (see paragraph after (4.47)) taking place in the Fermi vacuum as a result of the perturbing potential.

It is also necessary to include higher order processes which are composed of several complete lower order ones, like for example

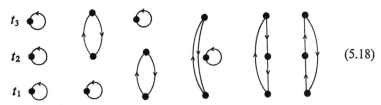

2nd order   4th order   5th order

$$\tag{5.17}$$

These disconnected or 'unlinked' diagrams are the analogues of the classical ones appearing in (5.9).

In order to draw all diagrams in nth order, draw n dots in a vertical row, label them $t_1, t_2, ..., t_n$ and connect them up in all possible 'topologically distinct' (see below) ways with one line entering and one leaving each dot. For example in third order we find the six diagrams

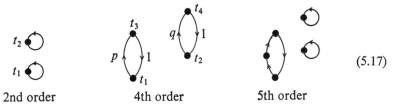

$$\tag{5.18}$$

Two diagrams are '*topologically equivalent*' if one can be distorted into the other without changing the vertical ordering of the dots; otherwise they are distinct. This is illustrated by the fourth-order diagrams (note significance of the direction of the arrows! It helps to visualize the distortions if we imagine the dots and lines to be buttons connected by rubber bands, carrying attached arrow-heads; all three spatial dimensions may be used in making the deformations):

$$(5.19)$$

Finally, as in the pinball case, the diagrammatic expansion for the vacuum amplitude will just be the sum of all diagrams such as the above:

$$(5.20)$$

where the 1 expresses the fact that in the unperturbed case, the probability amplitude for the system staying in its ground state is 1. This is the analogue of the P_u in the classical case. Compare the above expansion with the corresponding classical one in (5.10). Using Table 4.2 these diagrams may be translated, leading to the series

$$R(t) = 1 - V_{11} \int_0^t dt_1\, G_0^-(1, t_1 - t_1) -$$

$$- \sum_{p>1} V_{1p} V_{p1} \int_0^t dt_1 \int_{\substack{0 \\ t_2 > t_1}}^t dt_2 \quad G_0^+(p, t_2 - t_1) G_0^-(1, t_1 - t_2) + \cdots$$

$$+ V_{11} V_{11} \int_0^t dt_1 \int_{\substack{0 \\ t_2 > t_1}}^t dt_2 \quad G_0^-(1, t_1 - t_1) G_0^-(1, t_2 - t_2) + \cdots.$$

$$(5.21)$$

We have summed over all state indices and integrated over all intermediate times from 0 to t, since each diagram actually stands for an infinite number of diagrams having different state indices and different $t_1, t_2, \ldots$, etc.

5.4 Linked cluster theorem for one-particle system

Examination of the many-headed monster (5.20), shows two types of heads: the linked graphs, which are all in one connected piece and the 'unlinked' ones which have two or more disconnected, internally linked, parts. We now discuss a famous theorem which enables us to perform the Herculean feat of cutting off all the unlinked heads—the so-called 'linked cluster theorem'. The theorem states that

$$\ln R(t) = \sum \text{ all linked graphs}$$

$$(5.22)$$

The proof is based on the fact that (stated roughly) the contribution from an unlinked diagram is proportional to the product of the contribution of its various parts. Consider for example the $V_{11} V_{11}$ term in (5.21):

$$\begin{aligned}
t_2 \;\; \text{⬡} \\
t_1 \;\; \text{⬡}
\end{aligned} \equiv V_{11} V_{11} \int_0^t dt_1 \int_0^t {dt_2 \atop t_2 > t_1} G_0^- G_0^- = V_{11}^2 G_0^{-2} \int_0^t dt_1 \int_0^t {dt_2 \atop t_2 > t_1}$$

$$= V_{11}^2 G_0^{-2} \int_0^t dt_1 \int_0^t dt_2 \times \tfrac{1}{2} \quad (\text{where } t_2 > \text{ or } < t_1)$$

$$= \tfrac{1}{2} \left[V_{11} \int_0^t dt_1 \, G_0^-(1, t_1 - t_1) \right] \times \left[V_{11} \int_0^t dt_2 \, G_0^-(1, t_2 - t_2) \right]$$

$$= \tfrac{1}{2} \times \text{⬡} \times \text{⬡} = \frac{1}{2!} \times \text{⬡}^2 \qquad (5.23)$$

(we have used the fact that the G_0^- are time independent in this case). In general, it turns out that the value of an unlinked diagram with n identical links L, is just $(1/n!) \times L^n$.

A similar factorization occurs for non-identical links if we first sum over all possible time orders of the links. For instance, three of the diagrams in (5.18) may be summed thus (drop i's for brevity):

$$= \sum_k V_{k1} V_{1k} V_{11} \int \int \int dt_a \, dt_b \, dt_c \times$$
$$\times G_0^+(k, t_c - t_b) \, G_0^-(1, t_b - t_c) \, G_0^-(1, t_a - t_a) \times$$
$$\times [\theta_{t_c - t_b} \theta_{t_b - t_a} + \theta_{t_c - t_a} \theta_{t_a - t_b} + \theta_{t_a - t_c} \theta_{t_c - t_b}]$$

$$= \qquad \times \qquad \qquad (5.24)$$

where

$$\theta_\chi = 1 \quad \text{if } \chi > 0$$
$$= 0 \quad \text{if } \chi < 0. \qquad (5.25)$$

The θ's are used as a convenient way of writing the time order in the three diagrams. For $t_c > t_b$, some concentration shows that the term in brackets $= 1$ regardless of where t_a lies. This means the integral over t_a is independent of that over t_b and t_c, so the triple integral factorizes into two parts producing the result shown.

Combining these results, one finds that R may be written

$$R = 1 + \left[\quad + \quad + \quad + \cdots \right] + \frac{1}{2!} \left[\quad^2 + 2 \times \quad \times \quad + \cdots \right]$$

$$+ \frac{1}{3!} \left[\quad^3 + 3 \quad^2 \times \quad + \cdots \right] + \cdots$$

$$= 1 + \left[\quad + \quad + \quad + \cdots \right] + \frac{1}{2!} \left[\quad + \quad + \quad + \cdots \right]^2$$

$$+ \frac{1}{3!} \left[\quad + \quad + \quad + \cdots \right]^3 + \cdots$$

$$= e^{\left[\quad + \quad + \quad + \cdots \right]} = e^{\sum \text{all linked diagrams}} \qquad (5.26)$$

from which the linked cluster theorem (5.22) follows immediately.

5.5 Finding the ground state energy in one-particle system

The importance of this remarkable result lies in the fact that by (5.4), the ground state energy depends only on the logarithm of R:

$$E_0 = \epsilon_1 + \lim_{t \to \infty(1-i\eta)} i \frac{d}{dt} \ln R(t) \tag{5.27}$$

so that we may write it in terms of a sum over linked diagrams only, thus

$$\tag{5.28}$$

[In the many-body case, there is a deeper reason for the importance of the linked cluster theorem, e.g., if we do not use it, we find that the perturbation series for the energy appears to diverge badly as the number of particles $N \to \infty$. See Brout (1963), p. 47.]

It is now possible to obtain the expression for the ground state energy by translating the above diagrams using Table 4.2. This converts (5.28) to (remember the (-1) factor for the 'fermion loop'):

$$E_0 = \epsilon_1 - \lim_{t \to \infty(1-i\eta)} i \frac{d}{dt} \left\{ (-i)V_{11} \int_0^t dt_1 \, iG_0^-(1, t_1 - t_1) + \right.$$

$$\left. + \sum_{p \neq 1} (-i)^2 V_{1p} V_{p1} \int_0^t dt_1 \int_{\substack{0 \\ t_2 > t_1}}^t dt_2 \, iG_0^+(p, t_2 - t_1) \, iG_0^-(1, t_1 - t_2) + \cdots \right\}. \tag{5.29}$$

This is evaluated by substituting for the G's from Table 4.2 which yields

$= (-1)(-i) V_{11} \int_0^t dt_1 (-1) = -iV_{11}t$

$$E_0^{(1)} = \lim_{t \to \infty(1-i\eta)} i \frac{d}{dt} \;\;\bigcirc\;\; = V_{11} . \tag{5.30}$$

The second term produces

$$
\bigcirc = (-1)^2(-i)^2 \sum_{p \neq 1} V_{1p} V_{p1} \int_0^t dt_1 \int_0^t dt_2\, \theta_{t_2-t_1}\, e^{-i(\epsilon_p - \epsilon_1)(t_2 - t_1)}
$$

$$
= (-1)^2(-i)^2 \sum_{p \neq 1} V_{1p} V_{p1} \int_0^t dt_1 \int_0^{t-t_1} d(t_2 - t_1)\, e^{-i(\epsilon_p - \epsilon_1)(t_2 - t_1)}.
$$

$$(5.31)$$

Integrating, and taking the derivative and limit yields for $E_0^{(2)}$:

$$
\lim_{t \to \infty(1 - i\eta)} i \frac{d}{dt} \bigcirc = -i \sum_{p \neq 1} V_{1p} V_{p1} \left[-\frac{e^{-i(\epsilon_p - \epsilon_1)\infty(1 - i\eta)}}{i(\epsilon_p - \epsilon_1)} + \frac{1}{i(\epsilon_p - \epsilon_1)} \right]
$$

or

$$
E_0^{(2)} = \sum_{p \neq 1} \frac{V_{1p} V_{p1}}{\epsilon_1 - \epsilon_p} \tag{5.32}
$$

where the oscillating exponential is killed because $(\epsilon_1 - \epsilon_p)$ and the infinitesimal η are both positive. (Note that η is chosen such that $\eta \times \infty = \infty$.) Proceeding in this way yields the third- and fourth-order terms:

$$
E_0^{(3)} = \sum_{p,q \neq 1} \frac{V_{1p} V_{pq} V_{q1}}{(\epsilon_1 - \epsilon_p)(\epsilon_1 - \epsilon_q)} - \sum_{p \neq 1} \frac{V_{1p} V_{p1} V_{11}}{(\epsilon_1 - \epsilon_p)^2} \tag{5.33}
$$

$$
E_0^{(4)} = \sum_{p,q,r \neq 1} \frac{V_{1p} V_{pq} V_{qr} V_{r1}}{(\epsilon_1 - \epsilon_p)(\epsilon_1 - \epsilon_q)(\epsilon_1 - \epsilon_r)} - \sum_{p,q \neq 1} \frac{V_{1p} V_{pq} V_{q1} V_{11}}{(\epsilon_1 - \epsilon_p)^2(\epsilon_1 - \epsilon_q)} -
$$

$$
- \sum_{p,q \neq 1} \frac{V_{1p} V_{pq} V_{q1} V_{11}}{(\epsilon_1 - \epsilon_p)(\epsilon_1 - \epsilon_q)^2} + \sum_{p \neq 1} \frac{V_{1p} V_{p1} V_{11} V_{11}}{(\epsilon_1 - \epsilon_p)^3} -
$$

$$
- \sum_{p,q \neq 1} \frac{V_{1p} V_{p1} V_{1q} V_{q1}}{(\epsilon_1 - \epsilon_p)^2(\epsilon_1 - \epsilon_q)}. \tag{5.34}
$$

This is just the well-known Rayleigh–Schrödinger perturbation series carried out to fourth order!

If this commonplace textbook result is regarded as the end product of the elaborate vacuum amplitude approach, we might justifiably conclude that a rocket launcher has been built to fire a spitball. However, the aim here is not to do textbook-type perturbation theory, which no one has lived long enough to carry beyond twenty-seventh order, but rather to do the more exotic type, which is carried to infinite order. And the diagrams allow us to do this by providing a systematic method for writing out the nth-order term in the expansion of E_0.

To see this, look first at the third-order terms in (5.33) and compare them with the corresponding diagrams in (5.28). The product of the V_{q1}, V_{pq}, and V_{1p} factors associated with the interaction dots in the first third-order diagram, yields the right numerator for the first $E_0^{(3)}$ term (that is, provided we drop the $-i$ factor associated with the V's in evaluating the vacuum amplitude diagrams), and similarly for the second third-order diagram. This is easily shown to be general, leading to the rule: the numerators of the nth-order terms in the perturbation expansion of E_0 are obtained from the corresponding vacuum amplitude diagrams by associating a factor of V_{ij} with each interaction vertex (i.e., dot). To get the denominators, draw light dotted horizontal lines between successive (in time) pairs of vertices, thus

$$
\begin{array}{cccc}
\cdots\bullet\cdots & \dfrac{1}{\epsilon_1-\epsilon_p} & \cdots\bullet 1\cdots & \dfrac{1}{\epsilon_1-\epsilon_p} \\
& & & \\
\cdots\bullet\cdots & \dfrac{1}{\epsilon_1-\epsilon_q} & \cdots\bullet 1\cdots & \dfrac{1}{\epsilon_1-\epsilon_p}
\end{array}
\tag{5.35}
$$

and associate a factor of

$$
\cfrac{1}{\underset{\substack{\text{all hole lines}\\\text{intersected by dotted}\\\text{horizontal line}}}{\sum \epsilon_1} \;-\; \underset{\substack{\text{all particle lines}\\\text{intersected by dotted}\\\text{horizontal line}}}{\sum \epsilon_p}}
\tag{5.36}
$$

as shown in (5.35). The proper sign is obtained by multiplying by the factor $(-1)^{h+1}$ where $h=$ number of hole lines in the diagram. The final rule is to sum over all particle indices. Applying these rules (which can be rigorously proved from the vacuum amplitude expansion) yields both of the third-order terms and it is a simple matter for the reader to show that they also produce the correct result in the other orders. (Note that in fourth order, the last term in (5.34) is obtained by summing the two 'mitten' diagrams of (5.13).)

We are now in a position to show how the perturbation series may be evaluated approximately even if the perturbing term H_1 is strong, by summing to infinite order over certain types of diagrams. Imagine that the perturbing potential is so large that it is impossible to get a decent result by using the usual method of cutting off the series after the first few orders. But suppose, for example, that the potential happens to have big matrix elements only between the ground and first excited states, 1 and 2, i.e., that V_{12} and V_{21} are large and all others—V_{11}, V_{13}, V_{31}, V_{33}, ...—are small. Then the perturbation series may be approximated by a partial sum over just those special diagrams in which all vertices connect '1' lines and '2' lines. This means that E_0 reduces to a sum over just the following diagrams

$$E_0 \approx \epsilon_1 + 2 \qquad + \qquad + \qquad + \qquad \qquad (5.37)$$

The odd-order diagrams do not occur. There are sixteen sixth-order diagrams of which three typical ones have been drawn. Using the rules and assuming $V_{12} = V_{21}^*$ this expansion yields

$$E_0 \approx \epsilon_1 + \frac{|V_{12}|^2}{\epsilon_1 - \epsilon_2} - \frac{|V_{12}|^4}{(\epsilon_1 - \epsilon_2) \times 2(\epsilon_1 - \epsilon_2) \times (\epsilon_1 - \epsilon_2)} \times 2$$

$$+ \frac{|V_{12}|^6}{(\epsilon_1 - \epsilon_2) \times 2(\epsilon_1 - \epsilon_2) \times (\epsilon_1 - \epsilon_2) \times 2(\epsilon_1 - \epsilon_2) \times (\epsilon_1 - \epsilon_2)} \times 4$$

$$+ \frac{|V_{12}|^6}{(\epsilon_1 - \epsilon_2) \times 2(\epsilon_1 - \epsilon_2) \times 3(\epsilon_1 - \epsilon_2) \times 2(\epsilon_1 - \epsilon_2) \times (\epsilon_1 - \epsilon_2)} \times 12$$

$$+ \cdots$$

$$= \epsilon_1 + \frac{|V_{12}|^2}{\epsilon_1 - \epsilon_2} - \frac{|V_{12}|^4}{(\epsilon_1 - \epsilon_2)^3} + \frac{2|V_{12}|^6}{(\epsilon_1 - \epsilon_2)^5} + \cdots. \qquad (5.38)$$

This can be brought into a more recognizable form by adding and subtracting $\epsilon_1/2$ and factoring out $\frac{1}{2}(\epsilon_1 - \epsilon_2)$ yielding

$$E_0 = \frac{\epsilon_1 + \epsilon_2}{2} + \frac{(\epsilon_1 - \epsilon_2)}{2} \left[1 + \frac{2|V_{12}|^2}{(\epsilon_1 - \epsilon_2)^2} - \frac{2|V_{12}|^4}{(\epsilon_1 - \epsilon_2)^4} + \frac{4|V_{12}|^6}{(\epsilon_1 - \epsilon_2)^6} + \cdots \right]. \qquad (5.39)$$

The bracketed term is seen to be just the infinite series for the square root, giving us the final result

$$E_0 \approx \frac{\epsilon_1 + \epsilon_2}{2} + \frac{(\epsilon_1 - \epsilon_2)}{2} \sqrt{\left\{ 1 + \frac{4|V_{12}|^2}{(\epsilon_1 - \epsilon_2)^2} \right\}} \qquad (5.40)$$

Thus we have found the ground state energy in the case of a strong perturbing interaction by means of a partial summation of the perturbation series to all orders.

In the present case, this result is not as remarkable as it might seem at first sight, since it will be recognized as just the same result we can obtain much more easily by directly diagonalizing the Hamiltonian between the two levels 1 and 2 (see (7.64)). This single-particle example, like the others we have used, is obviously much too simple to show the power of the diagrammatic method; it should be regarded only as a transparent illustration of the general technique.

5.6 The many-body case

The single-particle example may be generalized to the many-particle case (no external potential) by using the interaction diagrams of (4.43, 44). The vacuum amplitude may then be built up as the sum of all possible sequences of interactions beginning and ending in the many-body ground state, or vacuum, thus:

$$(5.41)$$

The ground state energy again involves just the sum over linked diagrams and may be written as follows:

$$(5.42)$$

These diagrams may be evaluated by rules essentially the same as those for the one particle case—these are postponed until chapter 12. Here we will content ourselves with briefly mentioning a few popular approximations for the ground state energy which can be made with (5.42).

The simplest approximation is the Hartree–Fock (HF), which is just the sum of the double-bubble and oyster diagrams:

$$E_{0(HF)} = W_0 + \raisebox{-0.5em}{\includegraphics{}} \, {}^l + \raisebox{-0.5em}{\includegraphics{}} \,. \tag{5.43}$$

These diagrams involve only three simple rules, so we will evaluate them here. The rules are: (1) V_{klmn} for each interaction, (2) a factor of (-1) for each hole line and each fermion loop and (3) a factor of $\frac{1}{2}$ because the graphs are symmetric. Remembering that all lines are hole lines here, we find

$$E_{0(HF)} = \sum_{k<k_F} \epsilon_k + \frac{1}{2} \sum_{k,\,l<k_F} V_{klkl} - \frac{1}{2} \sum_{k,\,l<k_F} V_{lkkl}. \tag{5.44}$$

The approximation which is good for the high density electron gas (random phase approximation, or 'RPA') involves a partial sum over all 'ring' diagrams in second and higher order

$$E_{0(RPA)} = W_0 + \cdots \tag{5.45}$$

In the case of nuclear matter, we have the 'ladder approximation' involving a partial sum over all ladders:

$$E_{0(ladder)} = W_0 + \cdots \tag{5.46}$$

The reader who wishes to look now at the Gell-Mann–Brueckner calculation of the ground state energy of an electron gas, and the Brueckner theory
of nuclear matter, will find it possible to go directly to chapter 12 from here.

Exercises

5.1 Translate the last vacuum amplitude diagram in (5.13) into functions.

5.2 Classify each of the following diagram pairs as topologically distinct or equivalent:

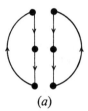 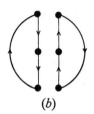 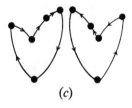

          (*a*)   (*b*)   (*c*)

5.3 Suppose we have a system of N non-interacting fermions in an external potential,
$U(\mathbf{r})$. A perturbation $V(\mathbf{r})$ is added to the Hamiltonian. Draw all diagrams up
through and including fourth order in the expansion of the vacuum amplitude.

5.4 Apply the diagram rules in the paragraphs containing (5.35, 36) to the diagrams
in (5.13), and verify that their energy contribution is given by (5.34). (Combine
the contributions from the two 'mitten' diagrams.)

5.5 For a system with no external potential and a momentum-conserving interaction, show why the fifth and sixth diagrams after the '1' in (5.41) do not
occur.

Chapter 6

Bird's-Eye View of Diagram Methods in the Many-Body Problem

This chapter brings us to the conclusion of the kindergarten part of the book. We have seen in some detail how the diagram methods of quantum field theory work on simple one-body systems, and have obtained a glimpse of the machinery in the many-body case. All steps were 'justified' with the aid of the 'argument by monkey', i.e., by analogy with the animal game in Fig. 2.3. No essential use was made of the occupation number formalism (second quantization).

Now we move on to the elementary part of the book. This will involve a more extensive use of second quantization, more impressive juggling of

Table 6.1 *Quantum field theory in the many-body problem*

Field theoretic ingredient	Significance in many-body theory
(1) Occupation number notation (chaps. 4.3, 7)	Expresses arbitrary state of many-body system
(2) Creation and destruction operators (chaps. 4.3, 7)	Primitive operators out of which all many-body operators are built
(3) Single particle propagator (Green's function) (chaps. 2, 3, 4, 8, 9, 10, 11, 15)	Yields quasi particle energies, particle momentum distribution, particle density, ground energy
(4) Vacuum amplitude (chaps. 5, 12)	Gives ground state energy
(5) Two-particle Green's function propagator (chap. 13)	Yields energies of collective excitations, electrical conductivity, other non-equilibrium properties
(6) Finite temperature vacuum amplitude (chap. 14)	Gives equilibrium thermodynamic properties of system
(7) Finite temperature propagator (chap. 14)	Yields temperature dependence of properties in (3)

diagrams, more real many-body systems, and less monkeys. But before continuing, it's a good idea to pause for a moment and take a look at the field as a whole. In Table 6.1 is a list of basic ingredients from quantum field theory, and their significance in the many-body problem.

The various field theoretical quantities, (3) to (7), may be calculated either by partial summation of Feynman diagrams, or, equivalently, by decoupling the differential equations which they obey (see appendix M). We consider only the former method in this book. The diagram method may be outlined as follows:

(1) The field theoretic quantity—call it M, standing for the propagator or vacuum amplitude, etc.—is expanded in a time-dependent perturbation series

$$M = M^{(0)} + M^{(1)} + M^{(2)} + \cdots$$

where $M^{(0)}$ is the value of M when the interaction is zero, and $M^{(1)}$, $M^{(2)}$, ... give the effect of the interaction to first order, second order, etc. The expansion can be carried out by the 'intuitive' methods we have used until now, or by a rigorous technique requiring the use of second quantization (items (1) and (2) in the above table). It is not assumed that the higher orders here are small compared to the zeroth order; in fact, because of the large interactions in many-body systems, they are large, sometimes infinite!

(2) Each term in the perturbation expansion is represented by a Feynman diagram. For example:

(a) Single-particle propagator:

(b) Vacuum amplitude

(c) Polarization propagator (special case of 2-particle propagator—see chapter 13):

Each one of the lines and wiggles in the diagrams has a definite factor associated with it; by writing down these factors, the original perturbation series in (1) can be reproduced completely. It is in principle possible to do many-body perturbation theory without diagrams, just as it is possible to go through the jungles of the Amazon without a map. However, the probability of survival is much greater if we use them (see Fig. 6.1).

Fig. 6.1 *Feynman Diagrams in the Amazon Jungle*

(Reproduced with the
kind permission of *Physics Today*)

(3) The perturbation series is evaluated approximately by summing to infinite order over certain types of diagrams (that is, by summing over the terms in the perturbation expansion corresponding to certain types of diagrams). This is the famous 'partial' or 'selective' summation. The partial sum is usually evaluated by showing that it involves a summable infinite series, or that it is equivalent to an integral equation already solved by someone else. For example:

(*a*) Single-particle propagator: a partial sum over repeated bubble and open oyster diagrams,

$$ G \approx \quad | \quad + \quad \multimap \quad + \quad \multimap \quad + \quad \multimap \quad + \quad \multimap \quad + \quad \multimap \quad + \quad \cdots $$

yields Hartree–Fock quasi particles.

(*b*) Vacuum amplitude: a partial sum over just 'ring' graphs,

$\ln R \approx A +$ 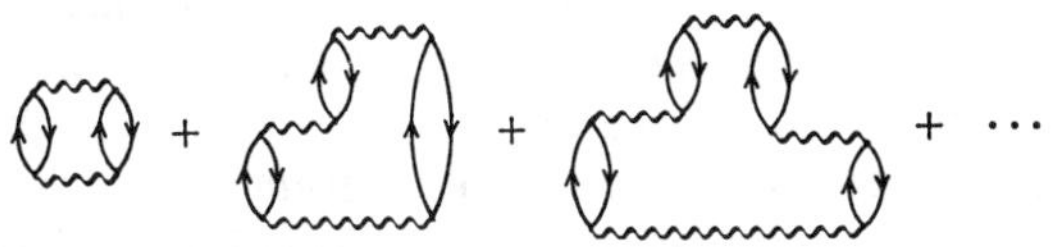

gives the ground state energy of a high density electron gas.

(*c*) Polarization propagator: a partial sum over repeated 'pair bubbles',

$F \approx$ 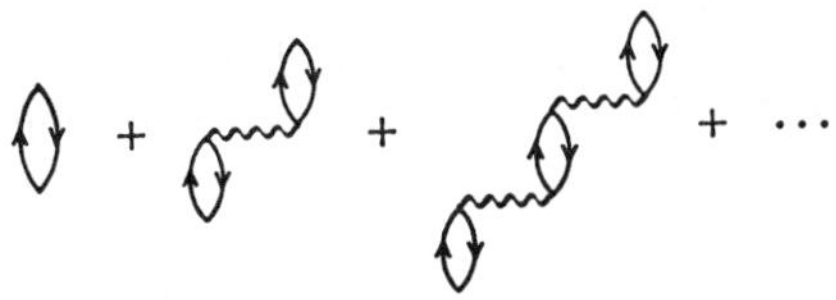

produces the energy dispersion law for plasmons.

The crucial point here is that although it employs a perturbation series, this technique is radically different from ordinary perturbation theory and is generally called a 'non-perturbative technique'. In the ordinary theory it is assumed that the perturbation is small so that the terms in the expansion obey the size ordering

$$M^{(0)} \gg M^{(1)} \gg M^{(2)} \gg M^{(3)} \gg \cdots.$$

In other words, the terms are arranged according to order, and we can estimate the accuracy of the approximation by looking at the order to which it is carried out.

In many-body theory, on the other hand, the perturbation is generally large, and the above size ordering does not hold. As was mentioned, in some cases the terms are individually infinite. So what we do is to re-arrange the terms according to type, pick out the type that gives the largest contribution and sum over terms of this type to all orders.

Of course selecting and summing the most important diagrams is generally very difficult. In the early days of many-body theory, the terms to be summed over were sometimes chosen simply on the basis of mathematical feasibility—on one occasion this was justified by the statement: 'Well, it's certainly better than not summing over anything!'. Today, the requirements for a good many-body calculation are (or should be!) more stringent, since there are now several model cases known in which the partial sum has been carried out very accurately. In these cases, the result is correct to lowest order in a well-defined small parameter (generally, this parameter is not directly related to the size of the two-body interaction V_{klmn} in the Hamiltonian). Two examples of this will be discussed in detail in this book: (1) RPA (random phase approximation) in the electron gas. The result here is correct to lowest order in r_s, the average distance between two electrons in the gas, and it becomes exact in the high-density limit ($r_s \to 0$)—see chapters 10, 12, 13. (2) Ladder approximation in a system with short-range repulsive interaction. The results are valid to lowest order in ρ, the density of the system, and are exact in the low density limit ($\rho \to 0$)—see chapters 10, 12. Two other examples which we will only refer to are: (3) Self-consistent renormalization approximation in the X-ray problem. The result is good to lowest order in g the coupling constant between the conduction electron and the 'deep hole' created by the X-ray. (See Roulet, et al. (1969), Nozières et al. (1969), Nozières and de Dominicis (1969).) (4) Renormalized parquet approximation in the theory of critical phenomena, The results are valid to lowest order in $\epsilon = 4 - d$ ($d =$ number of spatial dimensions in the system). (See Wilson (1971a, b, c), Wilson (1972) and Tsuneto and Abrahams (1972).)

If we use the above examples as models for calculations in the years to come, then there are good grounds to be optimistic about the future of the many-body problem!

Chapter 7

Occupation Number Formalism
(Second Quantization)

7.1 The advantages of occupation number formalism

The occupation number language in which modern many-body physics is written, was presented in primer form in §4.3 (which the reader should go through now if he hasn't already done so). We saw that its basic elements are the system wave function $|n_1, n_2, \ldots, n_i, \ldots\rangle$ telling how many particles n_i are in each single-particle state ϕ_i, and the $c_i^\dagger$, c_i operators which create and destroy particles in this state. Just enough details will be added now to make it possible to understand how the formalism is actually used. However, anyone who wishes to see the subject with all proofs in full regalia is advised to look at Raimes (1972), p. 21, or Dirac (1947, 58), p. 225 ff., p. 248 ff. It will help the reader to understand the shorthand notation in this chapter if he first goes through appendix A.

Since simple things can sometimes get to look pretty formidable in second quantization it's a good idea to understand why many-body physicists all use it. The first reason is that it enables us to deal with systems containing a variable number of particles. Since most systems we have to deal with at zero temperature have fixed particle number, this may sound as useful as a pair of trousers with five legs. However, it turns out to give an enormous flexibility in the formalism if N is allowed to vary in intermediate stages of a calculation and becomes fixed only at the end. For example, we can put in and remove test particles at will, as in the case of the propagator. Or we can introduce the particle–hole formalism in which the number of particles and holes is variable. (Of course, in statistical mechanics (see chapter 14) variable N systems are commonplace.)

The second reason for the occupation number formalism has to do with the symmetry properties of Fermi and Bose systems. Doing things the old way, we always have to worry about the complicated business of keeping the wave function properly symmetrized. But it turns out that in second quantization, the creation and destruction operators obey certain commutation rules which have built into them all the symmetry properties of the system. By just using these rules we are automatically free from symmetrization headaches.

7.2 Many-body wave function in occupation number formalism

Imagine that we are given a system of N identical fermions (bosons are in §7.7), which are in general interacting with each other and with an external potential. In §4.3 we saw that such a system may be described in terms of a set of basis states, $|n_1,\ldots,n_i,\ldots\rangle$ in which the n_i meant n_i particles in the unperturbed single-particle energy eigenstate, ϕ_i. Actually, the single-particle states used can be any orthonormal set—like for example position eigenstates (see §7.6) in which case n_i means n_i particles at the point r_i. This means that in general, $|n_1,\ldots,n_i,\ldots\rangle$ are not energy eigenstates of either the interacting or the non-interacting system of particles and their choice is determined by convenience. For the moment, we will use the same single-particle states as in §4.3, that is, the ϕ's which satisfy the Schrödinger equation

$$H\phi_{k\sigma}(\mathbf{r},\gamma) = \epsilon_{k\sigma}\phi_{k\sigma}(\mathbf{r},\gamma)$$

where

$$H = \frac{p^2}{2m} + U(\mathbf{r}) = -\frac{1}{2m}\nabla^2 + U(\mathbf{r}) \tag{7.1}$$

$$(\hbar = 1)$$

and γ, σ are the spin co-ordinate and quantum number respectively. In the case $U(\mathbf{r})=0$, this has the solutions (see appendix K):

$$\phi_{k\sigma}(\mathbf{r},\gamma) = \frac{1}{\sqrt{\Omega}}e^{+i\mathbf{k}\cdot\mathbf{r}}\eta_\sigma(\gamma)$$

$$\epsilon_k = \frac{k^2}{2m} \quad (\hbar = 1) \tag{7.2}$$

where η is the spin eigenfunction. In general, σ, γ will be suppressed for brevity, and $\mathbf{k}$ will be short for $\mathbf{k}$, σ, and $\mathbf{r}\equiv\mathbf{r}$, γ. The energy levels for (7.1) or (7.2) were shown schematically in Fig. 4.1(a).

If there are now N identical non-interacting fermions having Hamiltonian (7.1), the Hamiltonian and Schrödinger equation become

$$H_0 = \sum_{i=1}^{N} H_i, \quad H_0\Phi(\mathbf{r}_1,\ldots,\mathbf{r}_N) = E\Phi(\mathbf{r}_1,\ldots,\mathbf{r}_N) \tag{7.3}$$

where

$$H_i = \frac{p_i^2}{2m} + U(\mathbf{r}_i); \quad H_i\phi_{k_i} = \epsilon_{k_i}\phi_{k_i}. \tag{7.4}$$

Equation (7.3) has the solution

$$\Phi_{k_1,\ldots,k_N}(\mathbf{r}_1,\ldots,\mathbf{r}_N) = \prod_{i=1}^{N}\phi_{k_i}(\mathbf{r}_i), \quad E = \sum\epsilon_{k_i}. \tag{7.5}$$

However, since the system consists of identical fermions, the wave function must be antisymmetric, i.e., change sign when any two particle co-ordinates are interchanged. This is accomplished by forming a new Φ given by the sum

$$\Phi_{k_1,\ldots,k_N}(\mathbf{r}_1,\ldots,\mathbf{r}_N) = \frac{1}{(N!)^{\frac{1}{2}}} \sum_P \gamma_P P[\phi_{k_1}(\mathbf{r}_1)\phi_{k_2}(\mathbf{r}_2)\ldots\phi_{k_N}(\mathbf{r}_N)] \quad (7.6)$$

or

$$\Phi_{k_1,\ldots,k_N}(\mathbf{r}_1,\ldots,\mathbf{r}_N) = \frac{1}{(N!)^{\frac{1}{2}}} \begin{vmatrix} \phi_{k_1}(\mathbf{r}_1)\phi_{k_1}(\mathbf{r}_2)\ldots\phi_{k_1}(\mathbf{r}_N) \\ \vdots \qquad\qquad \vdots \\ \phi_{k_N}(\mathbf{r}_1)\phi_{k_N}(\mathbf{r}_2)\ldots\phi_{k_N}(\mathbf{r}_N) \end{vmatrix}. \quad (7.7)$$

In the first form, P is the permutation operator which interchanges the $\mathbf{r}_i$'s in all possible ways (starting from some standard order), and $\gamma_P = -1$ for an odd number of interchanges, and $+1$ for an even number. The last form is the well-known Slater determinant; the fact that $\Phi = 0$ when any two k_i's are equal means that there can't be more than one particle in any state (exclusion principle). The filling of states proceeds as in §4.2, and the energy of any set of filled states is just given by (7.5).

A tricky thing about (7.7) is its sign. For example, in a two-particle system with one particle in state ϕ_{p_1} ($\equiv \phi_1$; see note on notation after (3.5)!), and the other in ϕ_3, the wave function is $\Phi_{k_1=1, k_2=3}$ ($\equiv \Phi_{13}$) or $\Phi_{k_1=3, k_2=1}$ ($\equiv \Phi_{31}$). Since the particles are identical, these obviously represent the same state, but by (7.7) they differ by a minus sign. To remove this ambiguity, we always write Φ with the k's in standard order given by

$$\Phi_{k_1 < k_2 < \ldots < k_N}, \quad (7.8)$$

with associated $+$ sign. Thus, Φ_{31} will never appear again.

As pointed out in §4.3, a compact way of writing Φ is

$$\Phi_{k_1, k_2, \ldots, k_N}(\mathbf{r}_1, \mathbf{r}_2, \ldots, \mathbf{r}_N) = \Phi_{n_1, \ldots, n_i, \ldots}(\mathbf{r}_1, \mathbf{r}_2, \ldots, \mathbf{r}_N)$$
$$= \langle \mathbf{r}_1, \mathbf{r}_2, \ldots, \mathbf{r}_N | n_1, \ldots, n_i, \ldots \rangle, \quad (7.9)$$

where the last expression is just another way of writing the middle one. The n_i give the number of particles in the single-particle state $\phi_{k_i}(\mathbf{r})$ of (7.2). We assume for the moment that the number of particles is fixed so that

$$\sum_i n_i = N. \quad (7.10)$$

For brevity, the $\mathbf{r}$'s will generally be left out so that (7.9) becomes:

$$\Phi_{k_1, k_2, \ldots, k_N} = \Phi_{n_1, n_2, \ldots, n_i, \ldots} = |n_1, n_2, \ldots, n_i, \ldots \rangle. \quad (7.11)$$

Those familiar with the Dirac formalism (see appendix A) will recognize (7.11) as a state vector in an abstract space while the uninitiated may regard it as simply shorthand for (7.9). Simple examples of (7.11) appeared in (4.7, 8).

There is nothing occult about (7.9) or (7.11). It may be regarded either as a new notation for the same old wave function or as a simple change of variable from the k_j's to the occupation numbers, n_i. In fact, this change may be written out explicitly; it is

$$n_i = \sum_{j=1}^{N} \delta_{k_j, i}. \tag{7.12}$$

For example, if the wave function for a two-particle system is

$$\Phi_{k_1=1, k_2=3} = |1_1 0_2 1_3 0_4 0_5 0_6 \ldots\rangle \tag{7.13}$$

then

$$n_1 = \sum_{j=1}^{2} \delta_{k_j, 1} = \delta_{1,1} + \delta_{3,1} = 1 \tag{7.14}$$

and so on. The change from $\Phi_{k_1, \ldots, k_N}$ to $|n_1, n_2, \ldots, n_i, \ldots\rangle$ is a trivial unitary transformation in which the transformation matrix is a product of unit matrices. To see this, consider the $N=2$ case with one particle in state ϕ_i and the other in ϕ_j. Then by (7.11):

$$|00 \ldots 1_i \ldots 1_j \ldots\rangle = \Phi_{k_1=i, k_2=j}. \tag{7.15}$$

This means the 'transformation' is just

$$|00 \ldots 1_i \ldots 1_j \ldots\rangle = \sum_{k_1 < k_2} \delta_{k_1, i} \delta_{k_2, j} \Phi_{k_1 k_2}, \tag{7.16}$$

showing the transformation coefficients are just the unity matrix. In particular, it should be noticed that the change to occupation numbers is *not* in any sense a unitary transformation from 'position space' to 'occupation number space'. This may be emphasized by just including the r's in (7.16):

$$\Phi_{0, 0, \ldots, 1_i, \ldots, 1_j, \ldots}(\mathbf{r}_1, \mathbf{r}_2) = \sum_{k_1 < k_2} \delta_{k_1, i} \delta_{k_2, j} \Phi_{k_1 k_2}(\mathbf{r}_1, \mathbf{r}_2). \tag{7.17}$$

It is important to remember that the $|n_1, \ldots, n_i, \ldots\rangle$ are orthogonal and normal because the $\Phi_{k_1, \ldots, k_N}$ are, and we may write this in the various equivalent ways

$$\langle n_1', n_2', \ldots, n_i', \ldots | n_1, n_2, \ldots, n_i, \ldots \rangle$$

$$\equiv (\Phi_{n_1', n_2', \ldots, n_i', \ldots}, \Phi_{n_1, n_2, \ldots, n_i, \ldots})$$

$$\equiv \int d^3 \mathbf{r}_1 \ldots d^3 \mathbf{r}_N \Phi^*_{n_1', n_2', \ldots}(\mathbf{r}_1, \ldots, \mathbf{r}_N) \Phi_{n_1, n_2, \ldots}(\mathbf{r}_1, \ldots, \mathbf{r}_N)$$

$$= \delta_{n_1', n_1} \delta_{n_2', n_2} \ldots \delta_{n_i', n_i} \ldots . \tag{7.18}$$

Up to this point we have been dealing with systems containing a fixed number of particles and have made a relatively trivial change of variables in the wave function. Now we take the important step, and, even though the particle number in the real system is fixed, allow N to be variable, running from 0 to ∞. This generates the set of basis functions in Table 7.1.

Table 7.1 *Complete set of basis functions used in second quantization*

N	$\Phi_{k_1, k_2, \ldots, k_N}$	$= \lvert n_1, n_2, \ldots, n_i, \ldots \rangle$		
0	Φ_0	$\lvert 000\ldots\rangle$		
1	$\Phi_1, \Phi_2, \Phi_3, \ldots$	$\lvert 100\ldots\rangle,$	$\lvert 0100\ldots\rangle,$	$\lvert 00100\ldots\rangle, \ldots$
2	$\Phi_{12}, \Phi_{13}, \Phi_{23}, \ldots$	$\lvert 1100\ldots\rangle,$	$\lvert 101000\ldots\rangle,$	$\lvert 01100\ldots\rangle, \ldots$
$\vdots$	$\vdots$	$\vdots$		

The state Φ_0 or $\lvert 000\ldots\rangle$ with no particles at all in it is called the 'true vacuum'. The set of all $\lvert n_1, \ldots, n_i, \ldots\rangle$ (or $\Phi_{k_1, \ldots, k_N}$) in Table 7.1 is a complete orthogonal set of basis functions in an extended Hilbert space in which the number of particles is variable. This Hilbert space may be pictured as follows:

$$
\text{Extended Hilbert space} = \xrightarrow{\ \lvert 000\ldots\rangle\ } + \quad + \quad + \cdots . \tag{7.19}
$$

This set is often called 'occupation number basis', and the whole formalism is sometimes referred to as 'occupation number representation'. Note carefully that we did *not* get this new basis by unitary transformation (like, for example, is done in going from position to momentum basis). We got it instead by (1) a trivial (although convenient) change of variable from the k_i's to the n_i's, (2) extending the Hilbert space to an arbitrary number of particles (this is non-trivial). Furthermore, because of the fact that any convenient set of single-particle states may be used, the occupation number basis is not unique. For this reason, it is a good idea to refer to 'energy occupation number basis', meaning the single-particle states are energy eigenstates, or 'position occupation number basis' (position eigenstates—see §7.6), etc. In each case, however, every basis vector has definite values of the n_i—i.e., n_i is a good quantum number.

Only systems of independent fermions without perturbing interactions of any sort have been considered thus far. In the presence of such interactions, the $|n_1,\ldots,n_i,\ldots\rangle$ are no longer eigenstates of the total Hamiltonian for the system and the correct eigenstates must be obtained as the linear combination

$$\Psi = \Phi_0 + \sum_{k_1} A_{k_1}\Phi_{k_1} + \sum_{k_1<k_2} A_{k_1,k_2}\Phi_{k_1,k_2} + \cdots$$

$$= \sum_{n_1,\ldots,n_i,\ldots} A_{n_1,\ldots,n_i,\ldots}|n_1,\ldots,n_i,\ldots\rangle. \qquad (7.20)$$

7.3 Operators in occupation number formalism

It was pointed out in §4.3 that all operators in this new formalism may be expressed in terms of the creation and destruction operators $c_i^\dagger$, c_i defined in (4.16). Actually, these definitions left out a factor of ± 1 which is necessary because of antisymmetry. That is, $c_i^\dagger$, c_i must have the property that, if they act in such a sequence on the wave function that their net effect is to exchange two particles, then the wave function must change sign. (An example of such a sequence is:

$$|1100\ldots\rangle \to |0110\ldots\rangle \to |1010\ldots\rangle \to |1100\ldots\rangle.)$$

Some thought shows that the proper definition is

$$c_i^\dagger|n_1,\ldots,n_i,\ldots\rangle = (-1)^{\Sigma_i}(1-n_i)|n_1,\ldots,n_i+1,\ldots\rangle$$

$$c_i|n_1,\ldots,n_i,\ldots\rangle = (-1)^{\Sigma_i}n_i|n_1,\ldots,n_i-1,\ldots\rangle \qquad (7.21)$$

where

$$(-1)^{\Sigma_i} = (-1)^{[n_1+n_2+\ldots+n_{i-1}]}. \qquad (7.22)$$

That is, we get a factor of (-1) for each particle (i.e., each occupied state) standing to the left of the state i in the wave function. For example,

$$c_i|\ldots,0_i,\ldots\rangle = 0, \quad c_i^\dagger|\ldots,1_i,\ldots\rangle = 0$$

$$c_3|11111000\ldots\rangle = +|11011000\ldots\rangle$$

$$c_4^\dagger|1110100\ldots\rangle = -|11111000\ldots\rangle$$

$$c_2^\dagger c_3 c_1^\dagger c_2 c_3^\dagger c_1|1100\ldots\rangle = c_2^\dagger c_3 c_1^\dagger c_2 c_3^\dagger|0100\ldots\rangle$$

$$= c_2^\dagger c_3 c_1^\dagger c_2(-1)|01100\ldots\rangle = \cdots$$

$$= -|1100\ldots\rangle \quad \text{(particle exchange)}. \qquad (7.23)$$

One of the nice properties of the $c_i^\dagger$ operators is that by applying them repeatedly to the 'true vacuum' state (state with no particles in it), it is possible to generate all other states, thus:

$$\Phi_{k_1,k_2,\ldots,k_N} = c_{k_1}^\dagger c_{k_2}^\dagger \ldots c_{k_N}^\dagger|000\ldots\rangle$$

or

$$|n_1,n_2,\ldots\rangle = (c_1^\dagger)^{n_1}(c_2^\dagger)^{n_2}\ldots|0000\ldots\rangle. \qquad (7.24)$$

For example

$$|0\,1\,1\,0\,0\,0\ldots\rangle = (c_1^\dagger)^0 (c_2^\dagger)^1 (c_3^\dagger)^1 (c_4^\dagger)^0 \ldots |0\,0\ldots\rangle$$

$$= c_2^\dagger c_3^\dagger |0\,0\ldots\rangle. \tag{7.25}$$

Another important property of the $c_i^\dagger$, c_i operators is that they are 'hermitian adjoints' of each other. This can be seen by constructing matrices for them, using the $|n_1 n_2,\ldots,n_i,\ldots\rangle$'s as basis states

$$\langle n_1, n_2, \ldots, n_i', \ldots | c_i | n_1, n_2, \ldots, n_i, \ldots \rangle = (-1)^{\Sigma_i} \times \begin{array}{c|c|c} & |\ldots 0_i \ldots\rangle & |\ldots 1_i \ldots\rangle \\ \hline \langle\ldots 0_i \ldots| & 0 & 1 \\ \hline \langle\ldots 1_i \ldots| & 0 & 0 \end{array}$$

$$\tag{7.26}$$

$$\langle n_1, n_2, \ldots, n_i', \ldots | c_i^\dagger | n_1, n_2, \ldots, n_i, \ldots \rangle = (-1)^{\Sigma_i} \times \begin{array}{c|c|c} & |\ldots 0_i \ldots\rangle & |\ldots 1_i \ldots\rangle \\ \hline \langle\ldots 0_i \ldots| & 0 & 0 \\ \hline \langle\ldots 1_i \ldots| & 1 & 0 \end{array}$$

$$\tag{7.27}$$

so that

$$c_i^\dagger = (c_i)^\dagger \tag{7.28}$$

where $\dagger$ means hermitian adjoint. This further shows that $c_i^\dagger$, c_i are non-hermitian and are therefore not observables.

It is, however, easy to construct a hermitian operator from $c_i^\dagger$, c_i. Multiplying the above matrices shows that

$$(c_i^\dagger c_i)^\dagger = c_i^\dagger c_i \tag{7.29}$$

so that $c_i^\dagger c_i$ is hermitian. This combination:

$$\hat{n}_i = c_i^\dagger c_i; \qquad (\hat{N} = \sum_i c_i^\dagger c_i) \tag{7.30}$$

is an extremely important observable called the number operator ($\hat{N}$ = total number operator). To understand its properties, let it operate on some typical state vectors:

$$c_i^\dagger c_i |n_1, n_2, \ldots, 1_i, \ldots\rangle = (-1)^{\Sigma_i} c_i^\dagger |n_1, n_2, \ldots, 0_i, \ldots\rangle$$

$$= (-1)^{\Sigma_i + \Sigma_i} |n_1, n_2, \ldots, 1_i, \ldots\rangle$$

$$= (+1) |n_1, n_2, \ldots, 1_i, \ldots\rangle.$$

Similarly

$$c_i^\dagger c_i |n_1, n_2, \ldots, 0, \ldots\rangle = 0 |n_1, n_2, \ldots, 0_i, \ldots\rangle,$$

so that in general

$$c_i^\dagger c_i |n_1, \ldots, n_i, \ldots\rangle = n_i |n_1, \ldots, n_i, \ldots\rangle. \tag{7.31}$$

Thus, the eigenvalue of the number operator for the state ϕ_i is just the occupation number for that state. Hence, in the occupation number basis, all number operators are diagonal and the total system wave functions $|n_1, n_2, \ldots, n_i, \ldots\rangle$ are just the simultaneous eigenfunctions of the number operators, $\hat{n}_1, \hat{n}_2, \ldots, \hat{n}_i, \ldots$.

The $c_i^\dagger$, c_i operators obey the following important 'fermion commutation rules':

(1) $[c_l, c_k^\dagger]_+ = c_l c_k^\dagger + c_k^\dagger c_l = \delta_{lk}$

(2) $[c_l, c_k]_+ = 0,$ (3) $[c_l^\dagger, c_k^\dagger]_+ = 0. \tag{7.32}$

These can be easily proved from the definitions in (7.21):

$$c_l c_k |n_1, \ldots, n_l, \ldots, n_k, \ldots\rangle = (-1)^{\Sigma_k} n_k c_l |n_1, \ldots, n_l, \ldots, n_k - 1, \ldots\rangle$$

$$= (-1)^{\Sigma_k + \Sigma_l} n_k n_l |n_1, \ldots, n_l - 1, \ldots, n_k - 1, \ldots\rangle$$

$$c_k c_l |n_1, \ldots, n_l, \ldots, n_k, \ldots\rangle = (-1)^{\Sigma_l} n_l c_k |n_1, \ldots, n_l - 1, \ldots, n_k, \ldots\rangle$$

$$= (-1)(-1)^{\Sigma_k + \Sigma_l} n_k n_l |n_1, \ldots, n_l - 1, \ldots, n_k - 1, \ldots\rangle$$

$$\tag{7.33}$$

where the extra (-1) on line four comes from the fact that there is one less particle to the left of state k. Adding the two equations yields the second rule in (7.32); the other rules may be established in a similar fashion.

The importance of the above sets of 'anti-commutation' relations lies in the fact that all the antisymmetry properties are built into them. Therefore, by using them in the right places, we don't have to worry either about the symmetry of the wave functions themselves, or even about the awkward $(-1)^\Sigma$ factors. A simple example of this is to evaluate the matrix element $\langle \Phi_0 | c_k c_l^\dagger | \Phi_0 \rangle$ where Φ_0 is the ground state of the non-interacting Fermi system (see (4.7)), and it is assumed that $k, l > k_F$. Calculating in the ordinary way we have

$$\langle \Phi_0 | c_k c_l^\dagger | \Phi_0 \rangle = (-1)^N \langle \Phi_0 | c_k | \Phi_0, 1_l \rangle$$
$$= (-1)^{N+N} \delta_{kl} = \delta_{kl}, \qquad (7.34)$$

where N is the number of particles in the system. Using the commutation relations

$$\langle \Phi_0 | c_k c_l^\dagger | \Phi_0 \rangle = \langle \Phi_0 | \delta_{kl} - c_l^\dagger c_k | \Phi_0 \rangle = \delta_{kl} - \langle \Phi_0 | c_l^\dagger c_k | \Phi_0 \rangle$$
$$= \delta_{kl} \qquad (7.35)$$

where the last term vanishes since $c_k | \Phi_0 \rangle = 0$ (because $k > k_F$ but $| \Phi_0 \rangle$ has no particles above k_F). In this way the $(-1)^\Sigma$ factor is avoided.

Let us now consider how to express the usual quantum operators in terms of $c_i^\dagger$, c_i. It is a good idea to review the argument used in §4.3. We require equality between the matrix elements of the operator as computed in occupation number formalism and in the old cave-man formalism. For example, in a one-particle system, the operator $\mathcal{O}(\mathbf{r}, \mathbf{p})$ with matrix elements

$$\mathcal{O}_{ij} = \langle \phi_i | \mathcal{O} | \phi_j \rangle = \int \phi_i^*(\mathbf{r}) \mathcal{O}(\mathbf{r}, \mathbf{p}) \phi_j(\mathbf{r}) d^3\mathbf{r} \qquad (7.36)$$

has the occupation number form

$$\mathcal{O}^{occ} = \sum_{k,l} \mathcal{O}_{kl} c_k^\dagger c_l. \qquad (7.37)$$

This is easily checked, as in (4.23):

$$\langle 00 \ldots 1_i \ldots | \mathcal{O}^{occ} | 00 \ldots 1_j \ldots \rangle = \sum_{k,l} \mathcal{O}_{kl} \langle 00 \ldots 1_i \ldots | c_k^\dagger c_l | 00 \ldots 1_j \ldots \rangle$$
$$= \sum_{k,l} \mathcal{O}_{kl} \delta_{lj} \delta_{ik}$$
$$= \mathcal{O}_{ij}. \qquad (7.38)$$

Equation (7.38) may be generalized to the N-particle case. Suppose we have an operator

$$\mathcal{O} = \sum_{i=1}^{N} \mathcal{O}(\mathbf{r}_i, \mathbf{p}_i) \qquad (7.39)$$

like for example the external potential:

$$V(\mathbf{r}_1, \ldots, \mathbf{r}_N) = \sum_{i=1}^{N} V(\mathbf{r}_i). \qquad (7.40)$$

Such operators are called '*one-body*' operators since they are a sum of operators each of which acts separately on one particle. Then it can be shown (the proof is difficult—see Dirac (1947), pp. 230, 251) that (7.37) still holds, with $\mathcal{O}_{kl}$ still given by (7.36). (See exercise (7.10).) Thus we have the valuable result that in occupation number formalism, the single-particle operators have a form independent of N. (Compare with (7.39) which involves N explicitly.)

In a similar way, it can be shown that the 'two-body' operator

$$\mathcal{O} = \tfrac{1}{2} \sum_{\substack{i,j=1 \\ (i \neq j)}}^{N} \mathcal{O}(\mathbf{r}_i, \mathbf{p}_i, \mathbf{r}_j, \mathbf{p}_j), \tag{7.41}$$

like for instance the interaction potential

$$V(\mathbf{r}_1, \ldots, \mathbf{r}_N) = \tfrac{1}{2} \sum_{\substack{i,j=1 \\ (i \neq j)}} V(\mathbf{r}_i - \mathbf{r}_j), \tag{7.42}$$

becomes

$$\mathcal{O}^{\text{occ}} = \tfrac{1}{2} \sum_{klmn} \mathcal{O}_{klmn} c_l^\dagger c_k^\dagger c_m c_n \tag{7.43}$$

where

$$\mathcal{O}_{klmn} = \int d^3\mathbf{r} \int d^3\mathbf{r}' \, \phi_k^*(\mathbf{r}) \, \phi_l^*(\mathbf{r}') \, \mathcal{O}(\mathbf{r}, \mathbf{r}'; \mathbf{p}, \mathbf{p}') \, \phi_m(\mathbf{r}) \, \phi_n(\mathbf{r}') \tag{7.44}$$

(note reversal of order in k, l indices of the c_k's!). We remark here that the results (7.37) and (7.43) also hold true in the case of bosons.

7.4 Hamiltonian and Schrödinger equation in occupation number formalism

Let us translate into second quantized form the Hamiltonian for a system of N identical fermions in an external potential $U(\mathbf{r})$ (bosons give the same result). Assume that the particles interact by means of a two-body force of form (7.42), and that there is in addition an external perturbing potential $V(\mathbf{r})$. In the stone-age notation this is

$$H = \sum_i \underbrace{\left[\frac{p_i^2}{2m} + U(\mathbf{r}_i) \right]}_{H_0} + \underbrace{\tfrac{1}{2} \sum_{i,j} V(\mathbf{r}_i - \mathbf{r}_j)}_{H_1} + \underbrace{\sum_i V(\mathbf{r}_i, \mathbf{p}_i)}_{H_2}. \tag{7.45}$$

The first term has the form of the one-body operator, $\mathcal{O}$ in (7.39). Hence by (7.36, 37):

$$H_0 = \sum_{k,l} \langle \phi_k | \frac{p^2}{2m} + U(\mathbf{r}) | \phi_l \rangle c_k^\dagger c_l. \tag{7.46}$$

If the ϕ_k are chosen to be eigenstates of $p^2/2m + U(\mathbf{r})$, with eigenvalues ϵ_k, then this becomes (see appendix A.22):

$$H_0 = \sum_{k,l} \epsilon_k \delta_{kl} c_k^\dagger c_l = \sum_k \epsilon_k c_k^\dagger c_k. \tag{7.47}$$

Similarly, H_1 is translated by (7.43) into

$$H_1 = \tfrac{1}{2} \sum_{k,l,m,n} V_{klmn} c_l^\dagger c_k^\dagger c_m c_n \tag{7.48}$$

where

$$V_{klmn} = \int d^3\mathbf{r} \int d^3\mathbf{r}' \, \phi_k^*(\mathbf{r}) \, \phi_l^*(\mathbf{r}') \, V(\mathbf{r}-\mathbf{r}') \, \phi_m(\mathbf{r}) \, \phi_n(\mathbf{r}') \qquad (7.49)$$

and H_2 becomes by (7.37)

$$H_2 = \sum_{k,l} V_{kl} c_k^\dagger c_l; \quad V_{kl} = \int d^3\mathbf{r} \phi_k^*(\mathbf{r}) \, V(\mathbf{r}, \mathbf{p}) \, \phi_l(\mathbf{r}). \qquad (7.50)$$

(Note that in (7.49, 50), the $\int d^3\mathbf{r}$ is short for integration over $\mathbf{r}$ and sum over spins.) Hence in the occupation number formalism H is:

$$H = \sum_k \epsilon_k c_k^\dagger c_k + \tfrac{1}{2} \sum_{k,l,m,n} V_{klmn} c_l^\dagger c_k^\dagger c_m c_n$$

$$+ \sum_{k,l} V_{kl} c_k^\dagger c_l. \qquad (7.51)$$

For practice, let us solve the Schrödinger equation, $H\Psi = E\Psi$, in occupation number formalism in some trivial cases. Suppose first that both $V_{klmn}=0$ and $V_{kl}=0$. Then we have

$$H|\Psi\rangle = \sum_k \epsilon_k c_k^\dagger c_k |\Psi\rangle = E|\Psi\rangle. \qquad (7.52)$$

The ordinary form of this is (7.3). It is easy to see that the solution is

$$|\Psi\rangle = |n_1, n_2, \ldots, n_i, \ldots\rangle \qquad (7.53)$$

since by (7.31):

$$\sum_k \epsilon_k c_k^\dagger c_k |n_1, \ldots, n_k, \ldots\rangle = \sum_k \epsilon_k n_k |n_1, \ldots, n_k, \ldots\rangle. \qquad (7.54)$$

The energy eigenvalues are evidently

$$E = \sum_k \epsilon_k n_k \qquad (7.55)$$

which is the sum of the individual particle energies and is exactly the same result as in (7.5).

Another extremely easy example involves N free particles in the perturbing potential

$$V(\mathbf{p}) = Mp^2 + Lp^4 = -M\nabla_r^2 + L\nabla_r^4 \qquad (7.56)$$

of (3.19). We have from (3.27) and (3.29) that

$$V_{kl} = [Mk^2 + Lk^4]\delta_{kl}. \qquad (7.57)$$

Hence, using (7.51):

$$H = \sum_k \frac{k^2}{2m} c_k^\dagger c_k + \sum_{k,l} (Mk^2 + Lk^4)\delta_{kl} c_k^\dagger c_l$$

$$= \sum_k \left(\frac{k^2}{2m}\right) c_k^\dagger c_k + \sum_k (Mk^2 + Lk^4) c_k^\dagger c_k, \qquad (7.58)$$

which evidently is just the Hamiltonian of a set of particles with single-particle energy

$$\epsilon_k'' = \frac{k^2}{2m} + Mk^2 + Lk^4 \qquad (7.59)$$

exactly as in (3.24). The eigenfunctions of (7.58) are just (7.53) and the eigenenergies are

$$E = \sum_k \epsilon_k'' n_k. \qquad (7.60)$$

A third example is for a one-particle system subjected to an external perturbing potential $V(\mathbf{r})$. The Schrödinger equation is

$$H\Psi = (H_0 + V)\,\Psi = E\Psi$$

with

$$H_0 \phi_k = \epsilon_k \phi_k, \qquad (7.61)$$

which yields the secular equation for the energy:

$$\det\left[(\epsilon_i - E)\,\delta_{ij} + V_{ij}\right] = 0. \qquad (7.62)$$

In the ultra-simple case where all $V_{ij} = 0$ except V_{pq} and V_{qp}, this becomes

$$(\epsilon_1 - E)(\epsilon_2 - E)\ldots \begin{vmatrix} \epsilon_p - E & V_{pq} \\ V_{qp} & \epsilon_q - E \end{vmatrix} \ldots = 0$$

which has the solutions

$$E = \epsilon_i \quad (i \neq p, q) \qquad (7.63)$$

otherwise:

$$E = \frac{\epsilon_p + \epsilon_q}{2} \pm \tfrac{1}{2}\sqrt{\{(\epsilon_p - \epsilon_q)^2 + 4|V_{pq}|^2\}}. \qquad (7.64)$$

Let us see how this goes in the occupation number formalism. We have

$$H\,|\Psi\rangle = \left[\sum_k \epsilon_k c_k^\dagger c_k + \sum_{kl} V_{kl} c_k^\dagger c_l\right]|\Psi\rangle = E\,|\Psi\rangle. \qquad (7.65)$$

Since there is only one particle

$$|\Psi\rangle = \sum_j A_j |00\ldots 1_j \ldots\rangle. \qquad (7.66)$$

Putting this in (7.65) and multiplying on the left by $\langle 00\ldots 1_i \ldots|$ gives

$$\sum A_j [H_{ij} - E\delta_{ij}] = 0 \qquad (7.67)$$

where

$$H_{ij} = \langle 00\ldots 1_i\ldots | H | \ldots 1_j\ldots \rangle$$

$$= \sum_k \epsilon_k \langle \ldots 1_i\ldots | c_k^\dagger c_k | \ldots 1_j\ldots \rangle + \sum_{k,l} V_{kl} \langle \ldots 1_i\ldots | c_k^\dagger c_l | \ldots 1_j\ldots \rangle$$

$$= \epsilon_i \delta_{ij} + V_{ij} \tag{7.68}$$

which evidently yields just (7.62).

The real many-body case we shall deal with most often is one in which the external potential is zero and the interaction potential depends only on the distance between pairs of particles. In this case, H has the form in (7.51) with $V_{kl}=0$, and ϵ_k, ϕ_k given by (7.2). Let us work out the form of V_{klmn} in this case. Using (7.49) and (7.2), remembering $\mathbf{k} \equiv \mathbf{k}, \sigma$, and summing over spin variables (see appendix K) yields

$$V_{k\sigma_1,\, l\sigma_2,\, m\sigma_3,\, n\sigma_4} = \frac{\delta_{\sigma_1\sigma_3}\delta_{\sigma_2\sigma_4}}{\Omega^2} \int d^3\mathbf{r} \int d^3\mathbf{r}'\, V(\mathbf{r}-\mathbf{r}')\, e^{-i[(\mathbf{k}-\mathbf{m})\cdot\mathbf{r}+(\mathbf{l}-\mathbf{n})\cdot\mathbf{r}']}$$

$$= \frac{\delta_{\sigma_1\sigma_3}\delta_{\sigma_2\sigma_4}}{\Omega^2} \int d^3\rho\, V(\rho)\, e^{-i(\mathbf{k}-\mathbf{m})\cdot\rho} \int d^3\mathbf{r}'\, e^{-i(\mathbf{k}-\mathbf{m}+\mathbf{l}-\mathbf{n})\cdot\mathbf{r}'}$$

$$= \delta_{\sigma_1\sigma_3}\delta_{\sigma_2\sigma_4}\, \Omega^{-2}\, V_{k-m}\, \Omega\, \delta_{k+l,\, m+n} \tag{7.69}$$

where V_{k-m} is the Fourier transform of $V(\rho)$ given by:

$$V_{k-m} = \int d^3\rho\, e^{-i(\mathbf{k}-\mathbf{m})\cdot\rho}\, V(\rho). \tag{7.69'}$$

(Note that the inverse transform is

$$V(\rho) = \frac{1}{(2\pi)^3} \int d^3\mathbf{k}\, e^{+i(\mathbf{k}-\mathbf{m})\cdot\rho}\, V_{k-m}.) \tag{7.69''}$$

Observe that in (7.69) we have used the fact that (see after (3.62)):

$$I = \int d^3\mathbf{r}\, e^{-i(p-q)\cdot\mathbf{r}} = \Omega\delta_{pq} \quad \text{(Kronecker } \delta\text{)}. \tag{7.69'''}$$

Equation (7.69) may be written

$$V_{m+q,\,\sigma_1,\, n-q,\,\sigma_2;\, m\sigma_3,\, n\sigma_4} = \Omega^{-1}\, \delta_{\sigma_1\sigma_3}\delta_{\sigma_2\sigma_4}\, V_q. \tag{7.70}$$

The δ-functions express conservation of spin angular momentum (since V does not involve spin) and linear momentum (since V depends only on $\mathbf{r}-\mathbf{r}'$ and therefore cannot move the centre of mass). Note that the interaction terms in (7.51) also conserve particle number since they involve equal numbers of creation and destruction operators.

The Coulomb case where $V(\mathbf{r} - \mathbf{r}') = e^2/|\mathbf{r}' - \mathbf{r}|$ yields (see argument leading to (4.89)):

$$V_{k\sigma_1, \ldots, n\sigma_4} = \frac{\delta_{\sigma_1 \sigma_3} \delta_{\sigma_2 \sigma_4}}{\Omega} \frac{4\pi e^2}{|\mathbf{k} - \mathbf{m}|^2} \delta_{k+l, m+n}$$

or

$$V_{q\sigma_1 \sigma_2 \sigma_3 \sigma_4} = \delta_{\sigma_1 \sigma_3} \delta_{\sigma_2 \sigma_4} \frac{4\pi e^2}{q^2} \times \frac{1}{\Omega}. \tag{7.71}$$

This will be used in chapter 12 in the discussion of the electron gas.

7.5 Particle–hole formalism

In §4.3 it was shown how a lot of excess baggage in the occupation number scheme could be avoided by taking the ground state of the non-interacting Fermi system as the 'vacuum', and recording changes from this in terms of particles and holes. Since this was done in some detail, only a few comments will be added here.

First of all, since 'particles' in the particle–hole formalism exist only above the Fermi level ϵ_F and holes exist only below, we may write:

$$\mathbf{k}_{\text{hole}} \neq \mathbf{k}_{\text{particle}}. \tag{7.72}$$

Second, in the simple examples worked out in (4.20) we neglected the proper sign, which must come from the $(-1)^{\Sigma_i}$ factor in (7.21). In practice, one can avoid this problem by just enlarging the commutation rules for the $c_i^\dagger$, c_i in (7.32) to take care of holes. Using (7.32) together with the definition of the a's and b's in (4.18, 19), one can write out a complete set of rules for all these operators:

$a^\dagger, b$ create
$a, b^\dagger$ annihilate
$[a, b]_+ = ab + ba$

$$[a_k, a_l^\dagger]_+ = \delta_{kl}, \quad [a_k, a_l]_+ = [a_k^\dagger, a_l^\dagger]_+ = 0$$

$$[b_m, b_p^\dagger]_+ = \delta_{mp}, \quad [b_m, b_p]_+ = [b_m^\dagger, b_p^\dagger]_+ = 0$$

$$[a_k, b_m]_+ = [a_k, b_m^\dagger]_+ = [a_k^\dagger, b_m]_+ = [a_k^\dagger, b_m^\dagger]_+ = 0 \tag{7.73}$$

where (7.72) is used to get the last line. For example, suppose we want to evaluate $\langle 0 | b_l b_k^\dagger | 0 \rangle$, where $|0\rangle$ is the Fermi vacuum. To try to include the $(-1)^{\Sigma_i}$ factor here would be a confusing business. But (7.73) makes things easy:

$$\langle 0 | b_l b_k^\dagger | 0 \rangle = \langle 0 | \delta_{kl} - b_k^\dagger b_l | 0 \rangle = \delta_{kl} \tag{7.74}$$

where, analogous to (7.35), the last term $= 0$ because there are no holes in the Fermi vacuum. (The general rule in evaluating matrix elements like (7.74) is to employ the commutation relations to bring systematically all destruction operators a_k, b_k to the right where they operate on the Fermi vacuum and produce zero. This can become very tedious—a much simpler method using *Wick's theorem* is discussed in appendix F.)

Re-writing the many-body Hamiltonian (7.47, 48, 50) in particle–hole scheme is straightforward. Making use of the definition (4.18, 19) and rules for the b's in (7.73) yields

$$H_0 = \sum_{k<k_F} \epsilon_k - \sum_{k<k_F} \epsilon_k b_k^\dagger b_k + \sum_{k>k_F} \epsilon_k a_k^\dagger a_k. \tag{7.75}$$

The operator $b_k^\dagger b_k$ is just the hole number operator, so the second term shows explicitly that the holes have negative energy. The first term is just the energy of the Fermi vacuum. Similarly

$$H_1 = \tfrac{1}{2} \sum_{k,l,m,n>k_F} V_{klmn} a_l^\dagger a_k^\dagger a_m a_n + \tfrac{1}{2} \sum_{\substack{k,l,m>k_F \\ n<k_F}} V_{klmn} a_l^\dagger a_k^\dagger a_m b_n^\dagger$$

$$+ \cdots \tfrac{1}{2} \sum_{k,l,m,n<k_F} V_{klmn} b_l b_k b_m^\dagger b_n^\dagger \tag{7.76}$$

and

$$H_2 = \sum_{m,n>k_F} V_{mn} a_m^\dagger a_n + \sum_{\substack{m>k_F \\ n<k_F}} V_{mn} a_m^\dagger b_n^\dagger$$

$$+ \sum_{\substack{m<k_F \\ n>k_F}} V_{mn} b_m a_n + \sum_{m,n<k_F} V_{mn} b_m b_n^\dagger. \tag{7.77}$$

With the aid of (7.75), we can deduce an equation for the hole wave function. Suppose we have a system of N non-interacting particles filling the Fermi sea up to ϵ_F, and remove N_h of them, thus creating N_h holes (but no particles). Then (7.75) becomes

$$H_0^{\text{hole}} = W_0 - \sum_{k<k_F} \epsilon_k b_k^\dagger b_k$$

where

$$W_0 = \sum_{k<k_F} \epsilon_k \quad \text{and} \quad \sum_k b_k^\dagger b_k = N_h$$

Aside from the constant term, W_0, and the minus sign before ϵ_k, this has just the form of (7.47), if we imagine that (7.47) describes N_h particles, all in states below k_F. Hence we can reason backwards (Heisenberg (1931)) and conclude that H_0^{hole} in ordinary notation must be

$$H_0^{\text{hole}} = W_0 + \sum_{i=1}^{N_h} \left[\frac{-p_i^2}{2m} - U(\mathbf{r}_i) \right],$$

i.e., just like H_0 in (7.45) except for the W_0 (which just shifts the energy zero) and the minus sign. The corresponding wave equation for the single hole is

$$\left[\frac{p_i^2}{2m} + U(\mathbf{r}) \right] \phi_k(\mathbf{r}) = - \epsilon_k \phi_k(\mathbf{r}), \tag{7.78a}$$

showing that the hole wave function is just that of the particle, while its energy is the negative of the particle energy. (See also Kittel (1967), p. 274 ff.)

Finally, it may be remarked that for the same reason that the hole energy is negative we find

$$\text{hole momentum} = -\mathbf{k} \ (\text{for } U = 0)$$

$$\text{hole spin} = -\sigma.$$

$$\text{hole charge} = - \text{ particle charge} \tag{7.78b}$$

7.6 Occupation number formalism based on single-particle position eigenstates

The treatment up to now has been based on single-particle states which are eigenstates of the single-particle Hamiltonian, i.e., the energy operator. However, there is no law against using any convenient set of single-particle states—like for example, eigenstates of the single-particle momentum or position operator. Because of its utility, we will discuss the case of a scheme based on the position operator, $\hat{\mathbf{r}}$.

The single-particle position operator has the eigenvalue equation

$$\hat{\mathbf{r}}\delta(\mathbf{r}-\mathbf{R}) = \mathbf{R}\delta(\mathbf{r}-\mathbf{R}) \tag{7.79}$$

where the eigenvalue $\mathbf{R}$ is at any point in space and δ is the Dirac δ-function describing a particle precisely at the point $\mathbf{R}$. In (7.3) we find the total energy operator, H_0 for N particles, and in (7.7) we find the antisymmetrized eigenfunction for this operator. In a similar way, we have here the total position (i.e., centre of mass) operator for N particles

$$\mathscr{R} = \sum_{i=1}^{N} \hat{\mathbf{r}}_i \tag{7.80}$$

with eigenfunction

$$\Phi_{R_1, R_2, \ldots, R_N}(\mathbf{r}_1, \ldots, \mathbf{r}_N) = \frac{1}{(N!)^{\frac{1}{2}}} \begin{vmatrix} \delta(\mathbf{r}_1-\mathbf{R}_1) \ldots \delta(\mathbf{r}_N-\mathbf{R}_1) \\ \vdots \qquad\qquad \vdots \\ \delta(\mathbf{r}_1-\mathbf{R}_N) \ldots \delta(\mathbf{r}_N-\mathbf{R}_N) \end{vmatrix}. \tag{7.81}$$

(This is of course not an eigenstate of H_0!) For simplicity spin has been omitted.

The transition to occupation number scheme is made by (cf. (4.6) or (7.9)):

$$\Phi_{R_1, R_2, \ldots, R_N} = \Phi_{n_{x_1}, n_{x_2}, \ldots, n_{x_i}}$$

$$\equiv |n_{x_1}, n_{x_2}, \ldots, n_{x_i} \ldots\rangle \tag{7.82}$$

where n_{x_i} means: n_{x_i} particles in the single-particle state $\delta(\mathbf{r}-\mathbf{x}_i)$, that is, n_{x_i} particles at the point $\mathbf{x}$ (again $n_{x_i}=0$ or 1 for Fermi system). Thus, $|n_{x_1},...,n_{x_i},...\rangle$ describes the distribution of particles in space.

Similarly, the creation and destruction operators here are: $c^\dagger_{x_i}$, c_{x_i} which respectively create and destroy a particle at the point $\mathbf{x}_i$. These operators are usually written in an unfortunate form which makes them look as though they were ordinary wave functions:

$$\psi^\dagger(\mathbf{x}_i) \equiv c^\dagger_{x_i}: \text{ creates particle at point } \mathbf{x}_i$$

$$\psi(\mathbf{x}_i) \equiv c_{x_i}: \text{ destroys particle at point } \mathbf{x}_i. \qquad (7.83)$$

The $\psi^\dagger(\mathbf{x})$, $\psi(\mathbf{x})$ are the basic field operators of quantum field theory. The combination

$$\rho(\mathbf{x}) = \psi^\dagger(\mathbf{x})\psi(\mathbf{x}) \qquad (7.84)$$

is the number operator for this case, and has eigenfunctions (7.82). Since its eigenvalues are the number of particles at the point $\mathbf{x}$, it is evidently just a density operator.

It is easy to show that the $\psi^\dagger(\mathbf{x}_i)$, $\psi(\mathbf{x}_i)$, are related to the $c^\dagger_i, c_i$ of (7.21) by the same transformation which connects the eigenfunctions $\delta(\mathbf{r}-\mathbf{R})$ to the $\phi_k(\mathbf{r})$ of (7.1). Thus, we have

$$\delta(\mathbf{r}-\mathbf{x}_i) = \sum_k A_{ik}\phi_k(\mathbf{r})$$

or

$$|00...1_{x_i}00...\rangle = \sum_k A_{ik}|00...1_k00...\rangle$$

or

$$c^\dagger_{x_i}|000...\rangle = \sum_k A_{ik} c^\dagger_k|000...\rangle,$$

whence

$$\psi^\dagger(\mathbf{x}_i) = \sum_k A_{ik} c^\dagger_k$$

$$\psi(\mathbf{x}_i) = \sum_k A^*_{ik} c_k. \qquad (7.85)$$

The coefficient A_{ik} is

$$A_{ik} = \int d^3\mathbf{r}\, \phi^*_k(\mathbf{r})\, \delta(\mathbf{r}-\mathbf{x}_i) = \phi^*_k(\mathbf{x}_i). \qquad (7.86)$$

By using this transformation, it is easy to show that the Hamiltonian (7.51) (assume no external perturbing potential) may be written in terms of the field operators like this:

$$H = \int d^3\mathbf{x}\,\psi^\dagger(\mathbf{x})\left[-\frac{1}{2m}\nabla^2_x + U(\mathbf{x})\right]\psi(\mathbf{x}) +$$

$$+\tfrac{1}{2}\int\int d^3\mathbf{x}\, d^3\mathbf{x}'\, \psi^\dagger(\mathbf{x})\psi^\dagger(\mathbf{x}')V(\mathbf{x}-\mathbf{x}')\psi(\mathbf{x}')\psi(\mathbf{x}). \qquad (7.87)$$

For example, substituting (7.85) into the first term of H (call $\mathbf{x} \equiv \mathbf{x}_i$):

$$H_0 = \int d^3\mathbf{x}_i \sum_{k,l} A_{ik} c_k^\dagger \left[-\frac{1}{2m}\nabla^2_{x_i} + U(\mathbf{x}_i) \right] A_{il}^* c_l, \qquad (7.88)$$

and using (7.86), we find

$$H_0 = \sum_{k,l} c_k^\dagger c_l \int d^3\mathbf{x}_i \phi_k(\mathbf{x}_i)\, \epsilon_l\, \phi_l^*(\mathbf{x}_i)$$

$$= \sum_k \epsilon_k c_k^\dagger c_k \qquad (7.89)$$

same as in (7.47).

The transformation (7.85) allows us to break $\psi^\dagger$, ψ up into 'particle' and 'hole' parts, thus:

$$\psi^\dagger(\mathbf{x}_i) = \sum_{k>k_F} A_{ik} a_k^\dagger + \sum_{k<k_F} A_{ik} b_k$$

$$= \psi^\dagger_{part.}(\mathbf{x}_i) + \psi_{hole}(\mathbf{x}_i)$$

$$\psi(\mathbf{x}_i) = \psi^\dagger_{hole}(\mathbf{x}_i) + \psi_{part.}(\mathbf{x}_i) \qquad (7.90)$$

where $\psi^\dagger_{part.}(\mathbf{x}_i)$ creates a particle at point $\mathbf{x}_i$, $\psi_{hole}(\mathbf{x}_i)$ destroys a hole, etc.

7.7 Bosons

The occupation number story can easily be re-written with the boson as protagonist. The results we get are just like those for phonons presented in appendix $(\mathscr{A}.37)$–$(\mathscr{A}.42)$, since phonons are an example of bosons. We find:

(1) The $\Phi_{k_1, \ldots, k_N}(\mathbf{r}_1, \ldots, \mathbf{r}_N)$ of (7.6) becomes replaced by the symmetrized

$$\Phi_{k_1, k_2, \ldots, k_N}(\mathbf{r}_1, \ldots, \mathbf{r}_N) = \sqrt{\left(\frac{n_1! n_2! \ldots}{N!} \right)} \sum_P (+1)\, P\, [\phi_{k_1}(\mathbf{r}_1) \ldots \phi_{k_N}(\mathbf{r}_N)]$$

$$= \Phi_{n_1, \ldots, n_i, \ldots}(\mathbf{r}_1, \ldots, \mathbf{r}_N) \equiv \langle \mathbf{r}_1, \ldots, \mathbf{r}_N | n_1, \ldots, n_i, \ldots \rangle$$

where: $\qquad n_i = 0, 1, 2, 3, \ldots. \qquad (7.91)$

(2) The $c_i^\dagger$, c_i operators are re-defined by

$$c_i^\dagger | n_1 \ldots n_i \ldots \rangle = \sqrt{(n_i+1)} | n_1, \ldots, n_i+1, \ldots \rangle$$

$$c_i | n_1 \ldots n_i \ldots \rangle = \sqrt{n_i} | n_1, \ldots, n_i-1, \ldots \rangle. \qquad (7.92)$$

(3) The commutation relations, (7.32) are replaced by

(a) $\qquad [c_l, c_k^\dagger]_- = c_l c_k^\dagger - c_k^\dagger c_l = \delta_{lk}$

(b) $\qquad [c_l, c_k]_- = 0,$

(c) $\qquad [c_l^\dagger, c_k^\dagger]_- = 0. \qquad (7.93)$

(4) There are no holes in the boson case, hence no particle–hole formalism.

(5) The one- and two-body operators are the same as in the fermion case, hence also the expression for the Hamiltonian. (This is not true for phonons, where the interaction terms may involve the product of any number of creation and destruction operators (see Van Hove, 1961, p. 24, and also chapter 16).)

Further reading

Raimes (1972), p. 21 ff.
Landau and Lifschitz (1958), p. 215 ff.
Dirac (1947, 58), p. 225 ff., p. 248 ff.
Schweber (1961), chap. 6.
Schrieffer (1964a), appendix, p. 257.

Exercises

7.1 Find $c_1 c_3^\dagger c_2 |111000...\rangle$.

7.2 Find $\langle \Psi | \hat{N} | \Psi \rangle$, where $|\Psi\rangle = A|100...\rangle + B|111000...\rangle$, and $\hat{N}$ is in (7.30). (Remember that $\langle \Psi | = \overline{|\Psi\rangle}$ (see appendix (A.18)), and use (7.18).)

7.3 Verify the commutation rules (7.32) in the special case where $|\Psi\rangle$ is as in Ex. 7.2 above (i.e., show, for example, that $(c_1 c_2^\dagger + c_2^\dagger c_1)|\Psi\rangle = 0$, etc.).

7.4 Write out the Hamiltonian in second quantized form for the system described in Ch. 3, Ex. 3.1. Choose the states $\phi_n(x)$ as the single-particle eigenstates.

7.5 The system of Ch. 3, Ex. 3.1, is subjected to the perturbation of Ch. 3, Ex. 3.2. Find the form of the perturbation in occupation number formalism.

7.6 Calculate the interaction term in occupation number formalism for a fermion system (no external potential) in which the interaction between particles has the form $A\delta(\mathbf{r}_i - \mathbf{r}_j)$. Neglect spin.

7.7 Verify the particle–hole commutation rules in (7.73).

7.8 Verify (7.75).

7.9 Use occupation number formalism to prove the first line of (7.78b).

7.10 Verify that for a two-particle system, the matrix elements of the two-body operator $\mathcal{O}^{\text{occ}}$ in (7.43) between the two-particle states $\langle 0 \ldots 1_p \ldots 1_q \ldots |$ and $|0\ldots 1_r \ldots 1_s \ldots\rangle$ are the same as the matrix elements of $\mathcal{O}$ in (7.41) taken between the corresponding two-particle Slater determinants. (Use argument analogous to the one-particle case (7.36)–(7.38).)

7.11 Prove that the components of the total spin operator, **S**, in second quantized form are (use appendix K):

$$S_x = \frac{1}{2} \sum_k [c_{k\uparrow}^\dagger c_{k\downarrow} + c_{k\downarrow}^\dagger c_{k\uparrow}]$$

$$S_y = -\frac{i}{2} \sum_k [c_{k\uparrow}^\dagger c_{k\downarrow} - c_{k\downarrow}^\dagger c_{k\uparrow}]$$

$$S_z = \frac{1}{2} \sum_k [c_{k\uparrow}^\dagger c_{k\uparrow} - c_{k\downarrow}^\dagger c_{k\downarrow}].$$

Chapter 8

More about Quasi Particles

8.1 Introduction

Our first application of occupation number formalism will be to bring into focus some parts of the quasi particle picture which were left hazy in the semi-qualitative talk in the first half of the book. For one thing, we recall that in appendix $\mathscr{A}$ (which the reader should look at now if he hasn't already done so) it is stated that in most many-body systems, one can transform from the original Hamiltonian with strong interactions between particles,

$$H = \sum_i H_0(\mathbf{p}_i, \mathbf{r}_i) + \tfrac{1}{2} \sum_{i,j} V(\mathbf{r}_i, \mathbf{r}_j, \mathbf{p}_i, \mathbf{p}_j),$$

$$\left(\text{or} \quad H = \sum_k \epsilon_k c_k^\dagger c_k + \tfrac{1}{2} \sum_{k,l,m,n} V_{klmn} c_l^\dagger c_k^\dagger c_m c_n \right) \tag{8.1}$$

to a Hamiltonian of the form

$$H' = E_0 + \sum_q \epsilon_q' A_q^\dagger A_q + \underbrace{f(\ldots, A_q, \ldots, A_q^\dagger, \ldots)}_{small}. \tag{8.2}$$

This latter expression describes a set of approximately independent elementary excitations of energy ϵ_q' above a ground state of energy E_0, interacting weakly by means of the small term, f. This is illustrated for collective excitations by the phonon transformation. However, in general is it too difficult to go from (8.1) to (8.2) by means of a canonical transformation, and therefore we introduced the quantum field theoretical method of getting ϵ_q' directly from the poles of the propagator $G(q, \omega)$. In particular, all our quasi particle examples were solved by the field theoretic method.

In order to appreciate precisely the relation between the canonical transformation and quantum field theoretical methods, it would be valuable to have a simple example showing how the transformation (8.1) $\rightarrow$ (8.2) goes in the case of quasi particle excitations in Fermi systems. In §8.2 we introduce just such an example, a soluble model system with a 'pure Hartree' Hamiltonian which can be easily calculated exactly both by the field theoretic and the transformation method. Both methods give the same answer: The elementary excitations turn out to be a set of n quasi particles and an equal number of quasi holes, together with an interaction term.

142

In §8.3, we will do a crude calculation of the lifetime of quasi particles and show that this lifetime is inversely proportional to the square of the distance from the Fermi surface. Hence the quasi particle picture breaks down if we are too far from the Fermi surface.

Another limitation on the quasi particle picture is that the picture is invalid in the time interval just after the bare particle is introduced into the system, since it takes some time for it to get dressed. This requires adding a correction term to the propagator, which will be discussed in the last section.

An important point which can generate considerable confusion is that the word quasi particle, in the case of Fermi systems, is used in two different senses. The word quasi particle as we use it here means an elementary excitation above the ground state in the sense of (8.2), and there are two types of fermion quasi particles: quasi particles and quasi holes. However, there is an 'intuitive' definition of quasi particle introduced by Landau (Abrikosov (1965)), which is very close to the picture of the classical quasi-ions in Fig. 2.1. Landau visualizes the whole interacting system in its ground state as filled with quasi particles up to the Fermi surface. There are N Landau quasi particles, one for each bare particle. Excited states are formed by lifting Landau quasi particles out of the Fermi sea, thus creating quasi particles and holes, same as those we deal with.

This picture is fine provided we are acutely aware of its limitations: first, it makes sense only for Landau quasi particles near the Fermi surface, since the lifetime is otherwise too short. Thus even though we say 'N Landau quasi particles', this has only formal meaning, since we can really only talk about those with $|\mathbf{k}|$ near k_F. This also implies that if n_k is the number of Landau quasi particles in state $\mathbf{k}$, the statement: 'The function $n_k=0$ for $|\mathbf{k}|>k_F$, and $=1$ for $|\mathbf{k}|<k_F$' has meaning only near k_F. Finally, even though we talk of the ground state as 'filled with N Landau quasi particles up to k_F', we cannot get the properties of the ground state from such a model; in particular, the ground state energy is not equal to the sum of the energies of the Landau quasi particles.

8.2 A soluble fermion system: The pure Hartree model

Imagine that we have an N-fermion system with no external potential, and with a pure forward-scattering interaction between particles of the form

$$V_{klmn} = V_{klkl}\delta_{mk}\delta_{nl}, \tag{8.3}$$

(cf. (4.66) where we had an approximate forward-scattering interaction). Placing this in the general Hamiltonian (7.51) yields

$$H = \sum_k \epsilon_k c_k^\dagger c_k + \tfrac{1}{2}\sum_{kl} V_{klkl} c_l^\dagger c_k^\dagger c_k c_l. \tag{8.4}$$

For a system with the full H in (7.51), we can get only approximate solutions for the energies of quasi particles. But with the simple model H in (8.4), we can get exact solutions, as will be seen.

This model H will be called the 'pure Hartree' Hamiltonian, since, as we shall show, the only terms in it are those giving rise to the 'Hartree effective field' discussed after (4.73). Our object is to get a solution to the problem in the form of (8.2), i.e., the ground state energy plus a set of approximately independent elementary excitations (quasi particles in this case) above the ground state. We first do this by the straightforward diagrammatic method, then get the same result by the canonical transformation technique.

The interaction (8.3) has only the simple forms

where we have used (4.44). Hence the only graphs occurring in the series for the ground state energy (5.42) are

since a little experimenting shows that none of the other graphs in (5.42) can be drawn using only (8.5). (Note that just as in (4.54), (4.55), the propagator lines in these diagrams are all hole lines. Observe also that the diagrams in brackets violate the exclusion principle since there are simultaneously two hole lines in state $\mathbf{k}$.) This may be evaluated with the aid of (5.44) giving

$$E_0 = \sum_{k<k_F} \epsilon_k + \tfrac{1}{2} \sideset{}{'}\sum_{k,l<k_F} V_{klkl}. \tag{8.6}$$

(The $\mathbf{k=l}$ graphs cancel, as seen from (5.44). The prime means $\mathbf{k}\neq\mathbf{l}$.)

Now let us get the quasi particle energies, ϵ_k', from the poles of the Green's function. In this case, the propagator is given exactly by the sum over just bubble graphs:

since none of the other diagrams in (4.63) can be drawn using only (8.5). (Compare this with (4.67) where the propagator is only approximately given by the sum over bubble graphs.)

Series (8.7) was summed in (4.67) and gives the result (4.70), for quasi particle energy

$$\epsilon_k' = \epsilon_k + \sum_{l<k_F} V_{klkl}, \quad k > k_F$$

$$\tau_k = \infty. \tag{8.8}$$

In the case of quasi holes we just sum

$$(8.9)$$

The bracketed diagrams cancel and we get the result:

$$\epsilon_k' = \epsilon_k + \sum_{l<k_F}{}' V_{klkl}, \quad k < k_F$$

$$\tau_k = \infty. \tag{8.10}$$

Finally, we need the interaction between quasi particles (f-term in (8.2)). This can be obtained from the various two-particle propagators defined in §1.5. Consider the *particle–particle* propagator first. In the present case, this is given exactly by the sum:

$$(8.11)$$

The crossed 'exchange' diagrams in the brackets in (8.11) contribute only when $\mathbf{k}=\mathbf{l}$, because the labels on the incoming and outgoing lines in each diagram must match those of G_2 on the left. Since these diagrams are negative (see exercise 13.7), they cancel all the uncrossed diagrams when $\mathbf{k}=\mathbf{l}$, so $G_2 = 0$ for $\mathbf{k}=\mathbf{l}$.

We can greatly simplify (8.11) in the following way: Consider the diagram subset consisting of more and more bubbles inserted into the first propagator, i.e.:

$$\text{[diagram]} = \text{[diagram]} + \text{[diagram]} + \text{[diagram]} + \text{[diagram]} + \cdots. \qquad (8.12)$$

This can be easily summed:

$$\text{[diagram]} = \left[\text{[diagram]} + \text{[diagram]} + \text{[diagram]} + \cdots \right] \times \text{[diagram]} = \text{[diagram]}. \qquad (8.13)$$

Similarly, we can sum over all bubble insertions in all bare propagators, leading to a sum in which all propagators are clothed, i.e., in which all propagators are the quasi particle propagators (8.7):

$$\text{[diagram]} = \text{[diagram]} + \text{[diagram]} + \text{[diagram]} + \text{[diagram]} + \cdots. \qquad (8.14)$$

In this form, we can see that the quasi particle interaction is: $V = V_{klkl}(l \neq k)$, $V = 0(l = k)$. Similar arguments applied to the particle–hole and hole–hole propagators yield this same interaction.

We can now combine these results into a Hamiltonian of form (8.2). The expressions for E_0 and ϵ'_q are in (8.6), (8.8), and (8.10). The interaction term f in (8.2) will have the form (7.48), since the quasi particles here are fermions.

Letting $A_k^\dagger$, A_k, $B_k^\dagger$, B_k, be the quasi particle and quasi hole operators we find

$$H' = \left[\sum_{k<k_F} \epsilon_k + \tfrac{1}{2} \sum_{kl<k_F}' V_{klkl} \right] +$$

$$+ \sum_{k<k_F} \left(\epsilon_k + \sum_{l<k_F} V_{klkl} \right) A_k^\dagger A_k - \sum_{k<k_F} \left(\epsilon_k + \sum_{l<k_F}' V_{klkl} \right) B_k^\dagger B_k +$$

$$+ \tfrac{1}{2} \sum_{\substack{l,\,k>k_F}}' V_{klkl} A_l^\dagger A_k^\dagger A_k A_l - \sum_{\substack{k>k_F \\ l<k_F}} V_{klkl} B_l^\dagger A_k^\dagger A_k B_l +$$

$$+ \tfrac{1}{2} \sum_{\substack{k<k_F \\ l<k_F}}' V_{klkl} B_l^\dagger B_k^\dagger B_k B_l \tag{8.15}$$

Observe that in the particle–particle and hole–hole interaction terms, it is necessary to put in a factor $\tfrac{1}{2}$ to avoid counting interactions twice when we sum freely over **k** and **l**. Note that the $(-)$ sign in the $B_k^\dagger B_k$ term is put in because by (4.2) the hole energies (and therefore the quasi hole energies) are negative. The $(-)$ in the $B_l^\dagger A_k^\dagger A_k B_l$ term occurs for the following reason: The energy of a quasi particle in, say, state $k_1 > k_F$, includes interactions with *all* particles in the filled Fermi sea. But if there is a hole in, say, state $l_1 < k_F$, then the corresponding energy, $V_{k_1 l_1 k_1 l_1}$, does not exist, and should be subtracted from the quasi particle energy. The term $-V_{k_1 l_1 k_1 l_1} B_{l_1}^\dagger A_{k_1}^\dagger A_{k_1} B_{l_1}$ takes care of this subtraction (see exercise 8.5).

Now let us try to get this result by the canonical transformation method. What we want is a transformation which takes (8.4) into something of the form (8.15). This will be a special case of the transformation (8.1) → (8.2). The difficulty with the transformation method is immediately obvious. Where should we begin? Unlike the diagrammatic method, we have no cookbook rules to guide us. However, in the present case, it turns out that the required transformation is extremely simple—in fact it is just the transformation of (8.4) to ordinary particle–hole operators given in (4.19):

$$c_i = \theta_{k_i - k_F} a_i + \theta_{k_F - k_i} b_i^\dagger$$
$$c_i^\dagger = \theta_{k_i - k_F} a_i^\dagger + \theta_{k_F - k_i} b_i. \tag{8.16}$$

Before carrying out the transformation, let us do a little preliminary juggling of (8.4) to get it into a more transparent form. By the commutation rules (7.32),

$$c_i^\dagger c_k^\dagger c_k c_l = -c_k^\dagger c_l \delta_{kl} + c_k^\dagger c_k c_l^\dagger c_l. \tag{8.17}$$

Substituting this in (8.4) gives

$$H = \sum_k \epsilon_k c_k^\dagger c_k - \tfrac{1}{2} \sum_k V_{kkkk} c_k^\dagger c_k +$$

$$+ \tfrac{1}{2} {\sum_{kl}}' V_{klkl} c_k^\dagger c_k c_l^\dagger c_l + \tfrac{1}{2} \sum_k V_{kkkk} c_k^\dagger c_k c_k^\dagger c_k$$

$$= \sum_k \epsilon_k c_k^\dagger c_k + \tfrac{1}{2} {\sum_{kl}}' V_{klkl} c_k^\dagger c_k c_l^\dagger c_l. \tag{8.18}$$

(The second and fourth terms cancel in a Fermi system because $c_k^\dagger c_k c_k^\dagger c_k$ has exactly the same effect as $c_k^\dagger c_k$ when operating on an arbitrary wave function like (7.20), since $n_k = 0$, or 1 only.) Substituting the c_k's from (8.17) into (8.18), using the fact that $b_k b_k^\dagger = 1 - b_k^\dagger b_k$ and collecting terms produces

$$H' = \left[\sum_{k<k_F} \epsilon_k + \tfrac{1}{2} {\sum_{k,l<k_F}}' V_{klkl} \right] +$$

$$+ \sum_{k>k_F} \left(\epsilon_k + \sum_{l<k_F} V_{klkl} \right) a_k^\dagger a_k - {\sum_{k<k_F}}' \left(\epsilon_k + \sum_{l<k_F} V_{klkl} \right) b_k^\dagger b_k +$$

$$+ \left[\tfrac{1}{2} {\sum_{kl>k_F}}' V_{klkl} a_k^\dagger a_k a_l^\dagger a_l - \sum_{\substack{k>k_F \\ l<k_F}} V_{klkl} a_k^\dagger a_k b_l^\dagger b_l + \right.$$

$$\left. + \tfrac{1}{2} {\sum_{k,l<k_F}}' V_{klkl} b_k^\dagger b_k b_l^\dagger b_l \right]. \tag{8.19}$$

Comparing this with (8.15), shows that the quasi particle operators $A_k^\dagger$, A_k in this simple case are just the ordinary particle operators $a_k^\dagger$, a_k, and the quasi hole operators are just $b_k^\dagger$, b_k. Thus, we see that the result of the canonical transformation (8.19) is the same as that obtained graphically. This shows the equivalence of the diagrammatic and transformation methods.

It is a good idea here to show explicitly why one must be careful when using the 'Landau quasi particle' model described in §8.1. Suppose we in a naïve way regard the interacting system in its ground state as 'filled up' to the Fermi surface with quasi particles of energy ϵ_k' (as given in (8.10)), in the same way that the non-interacting system was filled up with bare particles. Then we get for the ground state energy

$$E_0 = \sum_{k<k_F} \epsilon_k + {\sum_{k,l<k_F}}' V_{klkl} \tag{8.20}$$

which is wrong as seen by comparison with (8.6). The contribution from the interaction part is twice as big as it should be. This is just another manifestation of the 'double counting' talked about after (0.2). It is due to the fact that the interaction V_{klkl} has been counted twice, once as the effective field at k due to l, and once as the field at l due to k.

8.3 Crude calculation of quasi particle lifetime

At the beginning, we mentioned that the quasi particle picture breaks down if the energy is too far away from the Fermi energy. In order to understand this, look first at the criterion (3.17). In a Fermi system, since one deals with particle-like excitations above ϵ'_F (the Fermi energy of the interacting system), and hole-like ones below, the criterion is taken relative to the Fermi energy, i.e.:

$$\frac{1}{\tau_k} \ll \epsilon'_k - \epsilon'_F. \tag{8.21}$$

In the pure Hartree model, by (8.10), $\tau_k = \infty$, so this is satisfied for any k. But this is not true in general. In fact, we are now going to show that in most Fermi systems the quasi particle lifetime obeys

$$\frac{1}{\tau_k} \propto (\epsilon'_k - \epsilon'_F)^2. \tag{8.22}$$

Hence, for ϵ'_k too far from ϵ'_F, τ_k becomes too short to satisfy (8.21) and the quasi particle is no longer a valid concept. Equation (8.22) will be discussed more rigorously in chapter 11. Here we will just give a crude quasi-proof of it.

The lifetime of a quasi particle in momentum state $\mathbf{k}$ will be the inverse of the transition probability per second that the quasi particle will be scattered out of state $\mathbf{k}$ by collisions with other quasi particles. Let us pretend that quasi particle collisions are like those between real particles (they are not, actually, since quasi particles can have a 'retarded', i.e., time-dependent, interaction even when the bare particles interact instantaneously—see §10.4, also §10.6) and calculate the transition probability out of state $\mathbf{k}_1$ for a particle in state $\mathbf{k}_1$, where $|\mathbf{k}_1| \geqslant k_F$. In a typical interaction, the particle will collide with a particle in state $|\mathbf{k}_2| \leqslant k_F$ and the final state will be a particle in $\mathbf{k}_3$ and one in $\mathbf{k}_4$, where by conservation of momentum

$$\mathbf{k}_4 = \mathbf{k}_1 + \mathbf{k}_2 - \mathbf{k}_3. \tag{8.23}$$

The transition probability is

$$W_{k_1} \propto \int d^3\mathbf{k}_2 \int d^3\mathbf{k}_3 \, |V_{k_3, \, k_1+k_2-k_3, \, k_1, \, k_2}|^2 \tag{8.24}$$

where $V_{k, \, l, \, m, \, n}$ is given in (4.42).

To evaluate (8.24) we note that by the Pauli principle all states under k_F are occupied so that

$$|\mathbf{k}_3| \geqslant k_F, \quad |\mathbf{k}_4| \geqslant k_F \tag{8.25}$$

and by conservation of energy,

$$k_1^2 + k_2^2 = k_3^2 + k_4^2. \tag{8.26}$$

Equations (8.25), (8.26) imply together that

$$k_1^2 + k_2^2 \geqslant 2k_F^2. \tag{8.27}$$

Consider first the limiting case when $|\mathbf{k}_1| = k_F$. Then by (8.27), $k_2^2 \geqslant k_F^2$. But since $|\mathbf{k}_2| \leqslant k_F$, this implies $|\mathbf{k}_2| = k_F$. Similarly, $|\mathbf{k}_3| = |\mathbf{k}_4| = k_F$. That is, all momenta lie on the Fermi sphere. Suppose now that $|\mathbf{k}_1| = k_F + \delta$ where $k_F \gg \delta > 0$. Then by (8.27), $|\mathbf{k}_2| \geqslant k_F - \delta$. Similarly, since $|\mathbf{k}_2| \leqslant k_F$, we have that in order to satisfy (8.26), $|\mathbf{k}_3|$, $|\mathbf{k}_4|$ must be less than $k_F + \delta$. Hence all momenta lie in a shell of thickness $\delta = |\mathbf{k}_1| - k_F$ around the Fermi sphere. Assuming there's nothing peculiar about the behaviour of V, the integral over the $\mathbf{k}_2$-shell gives a factor $\propto 4\pi k_F^2(|\mathbf{k}_1| - k_F)$, and the same for the $\mathbf{k}_3$-shell, so

$$W_{k_1} \propto (|\mathbf{k}_1| - k_F)^2. \tag{8.28}$$

But

$$\epsilon_{k_1} - \epsilon_F \propto (k_1^2 - k_F^2) = (|\mathbf{k}_1| - k_F)(|\mathbf{k}_1| + k_F)$$
$$\approx 2k_F(|\mathbf{k}_1| - k_F). \tag{8.29}$$

Whence

$$\frac{1}{\tau_k} = W_k \propto (\epsilon_k - \epsilon_F)^2 \tag{8.30}$$

which 'proves' (8.22), assuming that quasi particles interact roughly like real particles.

In this calculation, we obtained the lifetime by applying the 'Golden Rule' to find the transition probability corresponding to the matrix element given by the diagram

$$(8.30')$$

This same result may also be obtained by evaluating the imaginary part of the proper self-energy diagram:

$$(8.30'')$$

as will be shown in §9.8.

Note that in the special case of the model interaction (8.3), we have that $\mathbf{k}_1 = \mathbf{k}_3$, and it is easily shown that the argument above leads to a region of integration of zero volume, so that $W = 0$ and the lifetime is infinite, in agreement with (8.10).

8.4 General form of quasi particle propagator

The quasi particle formula in (3.15) was good enough for describing the 'quantum pinball' and Hartree–Fock cases, but needs to be generalized slightly to cope with the more complicated situations to come. Before presenting the more general expression, let us recapitulate.

According to (3.2) and (4.29) the single-particle propagator or 'Green's function' is defined as

$$G(k_2, k_1, t_2 - t_1) = G^+(k_2, k_1, t_2 - t_1)_{t_2 > t_1} + G^-(k_2, k_1, t_2 - t_1)_{t_2 \leqslant t_1} \quad (8.31)$$

where

$$iG^+(k_2, k_1, t_2 - t_1) = \text{probability amplitude that if at time } t_1 \text{ we}$$

add a particle in ϕ_{k_1} to the interacting system in its ground state, then at time t_2 the system will be in its ground state with an added particle in ϕ_{k_2}, (8.32a)

$$-iG^-(k_2, k_1, t_2 - t_1) = \text{probability amplitude that if at time } t_2 \text{ we}$$

add a hole in ϕ_{k_2} to the interacting system in its ground state, then at time t_1 the system will be in its ground state with an added hole in ϕ_{k_1}. (8.32b)

(The $-$ sign on G^- is for fermions; bosons have a $+iG^-$ instead.)

The form taken by $G^\pm$ in the free particle case was given in (3.9, 13) and (4.31, 32). These results may be written in a compact way by introducing the functions

$$\theta_x \begin{cases} = 1, & \text{for } x > 1 \\ = 0, & \text{for } x < 1 \end{cases}; \qquad \delta_k \begin{cases} = +\delta, & \epsilon_k > \epsilon_F \\ = -\delta, & \epsilon_k < \epsilon_F. \end{cases} \quad (8.33)$$

This gives, using (8.31), and letting $t = t_2 - t_1$,

$$G_0(k, t) = -i[\theta_t \theta_{\epsilon_k - \epsilon_F} e^{-i\epsilon_k t} - \theta_{-t} \theta_{\epsilon_F - \epsilon_k} e^{-i\epsilon_k t}], \quad t \neq 0$$
$$= +i\theta_{\epsilon_F - \epsilon_k}, \quad t = 0 \quad (8.34)$$

and

$$G_0(k, \omega) = \frac{1}{\omega - \epsilon_k + i\delta_k} \quad (8.35)$$

(see appendix I regarding the $i\delta$-factor).

Now in chapter 3 we argued physically that in systems describable by quasi particles, G will look like G_0 except for replacing ϵ_k by ϵ'_k, introducing a lifetime τ_k, and an amplitude factor, Z_k as in (3.15). However, this is not quite right, because it neglects the fact that when the bare particle is first put into the system, it will take some finite time, say t_c, for it to become 'clothed' so it will not act like a quasi particle until $t_2 - t_1 > t_c$. Further, it can be shown that the quasi particle expression for the propagator is no longer valid when $t_2 - t_1 \gg \tau_k$, where τ_k is the lifetime. For this reason it is necessary to write the propagator as the sum of a pure quasi particle part plus a correction term which will be important for $t < t_c$ and $t > \tau_k$.

It is also a good idea to include a 'quasi hole' part in the definition, similar to the way in which the hole part was included in the free propagator (8.34, 35). These considerations yield (setting $t = t_2 - t_1$), for $t \neq 0$:

$$G_{\substack{\text{quasi}\\\text{particle}}}(k, t) = -iZ_k[\theta_t\,\theta_{\epsilon_{k'}-\epsilon_{F'}}\,e^{-i(\epsilon_{k'}-i\tau_k^{-1})t} -$$
$$-\,\theta_{-t}\,\theta_{\epsilon_{F'}-\epsilon_{k'}}\,e^{-i(\epsilon_{k'}+i\tau_k^{-1})t}] + F(k, t). \qquad (8.36)$$

where $0 < Z_k \leqslant 1$ (Z_k independent of t!), ϵ_F' is the Fermi energy of the interacting system and $F(k, t)$ is the correction term. Taking the Fourier transform:

$$G_{\substack{\text{quasi}\\\text{particle}}}(k, \omega) = \frac{Z_k}{\omega - \epsilon_k' + i[2\theta_{\epsilon_{k'}-\epsilon_{F'}} - 1]\tau_k^{-1}} + F(k, \omega). \qquad (8.37)$$

It must be remembered that these are bona fide quasi particles only if the condition (8.21) is satisfied, i.e.,

$$\frac{1}{\tau_k} \ll \epsilon_k' - \epsilon_F'. \qquad (8.38)$$

Finally, a word about the suspicious-looking correction term, F. Sceptics may feel that so many sins are packed into F that the quasi particle expression is useless. This is not true, because we of course require that F does not contain a piece cancelling the Z_k term (!!). There are also certain 'sum rules' that it has got to satisfy (see (9.25)), but otherwise there are no particular restrictions on it. It is presumably well-behaved, and, hopefully, small. It can be shown to describe collective excitations in the system, but this will not be discussed here.

Further reading

Nozières (1964), chap. 1 (Landau theory of quasi particles).
Abrikosov (1965), chap. 1.
Falicov (1961).

Exercises

8.1 Using the rules for diagrams, Table 4.3, prove that the bracketed diagrams in (8.9) cancel.

8.2 Verify (8.19).

8.3 The energy and lifetime of quasi particles in the electron gas are given (in RPA) in (10.46). Taking $r_s = 1$ (high density case), calculate the order of magnitude of how far out from the Fermi surface the quasi-particle picture is still valid.

8.4 Verify (8.37) from (8.36).

8.5 Consider a three-particle Fermi system with interaction as in (8.4), in the excited state $|\Psi\rangle = |0_1 0_2 1_3 0_4 1_5 1_6 0_7 0_8, \ldots,\rangle$. Show using (8.4) that the energy of this state is $E = \epsilon_3 + \epsilon_5 + \epsilon_6 + V_{35} + V_{56} + V_{36}$. Rewrite $|\Psi\rangle$ in particle–hole notation and show that H' in (8.19) yields the same energy.

Chapter 9

The Single-Particle Propagator Re-Visited

9.1 Second quantization and the propagator

In the early chapters we used an intuitive approach to the quantum propagator, defining it physically and obtaining its diagrammatic perturbation expansion by analogy to the pinball case. While such an approach is a good way to get a foot in the front door of the field, it has obvious limitations. Therefore, in this chapter we are going to take a more rigorous look at the propagator with the aid of the occupation number formalism.

By introducing creation and destruction operators $c_k^\dagger(t)$, $c_k(t)$ which create and destroy a particle at time t, it is possible to write a simple, compact mathematical expression for the propagator. Once we get used to it, this expression is a lot easier to work with than the verbal definition (8.31, 32), and greatly facilitates understanding the properties of the propagator. The expression also reveals the origin of the diagram expansion, since it may be expanded in a perturbation series, each term of which corresponds to one of the graphs we have been drawing.

In §9.5 we use the idea of 'diagram topology' to simplify the rules for drawing and evaluating diagrams. These rules are then applied to the simplest example of a true many-body calculation, i.e., a calculation in which we go beyond Hartree–Fock and include correlation effects. The example consists of finding the contribution to the quasi particle energy and lifetime coming from the single pair-bubble self-energy diagram:

9.2 Mathematical expression for the single-particle Green's function propagator

The closed mathematical expression for the propagator, G, appears usually in one of two forms:

$$G(k_2, k_1, t_2 - t_1) = -i \langle \Psi_0 | T\{c_{k_2}(t_2) c_{k_1}^\dagger(t_1)\} | \Psi_0 \rangle \qquad (9.1)$$

or

$$G(\mathbf{r}_2, \mathbf{r}_1, t_2 - t_1) = -i \langle \Psi_0 | T\{\psi(\mathbf{r}_2, t_2) \psi^\dagger(\mathbf{r}_1, t_1)\} | \Psi_0 \rangle. \qquad (9.2)$$

We shall only consider the first form—the second may be analysed in a similar way. We'll explore the meaning of each term in (9.1) and show that it is precisely the same as (8.31, 32).

154

First, Ψ_0 is the exact normalized wave function of the ground state of the interacting N-particle system. The operators $c_k(t)$, $c_k^\dagger(t)$ respectively, destroy and create a particle in state k at time t. More precisely, they are the ordinary c_k, $c_k^\dagger$ transformed to 'Heisenberg picture', defined by (see Schrieffer (1964a), p. 104):

$$c_{k_1}^\dagger(t_1) = e^{+iHt_1} c_{k_1}^\dagger e^{-iHt_1}$$
$$c_{k_2}^\dagger(t_2) = e^{+iHt_2} c_{k_2} e^{-iHt_2} \tag{9.3}$$

where H is the Hamiltonian of the interacting system as in (7.51). The exponential operator in (9.3) is defined in appendix (B.3).

Finally the Wick time-ordering operator, T, is defined by

$$
\begin{aligned}
T\{A(t_1)B(t_2)\ldots\} = {} & (-1)^P \times \text{operators rearranged so that time} \\
& \text{decreases from left to right, assuming} \\
& \text{no two times are equal,} \\
= {} & (-1)^P \times \text{operators rearranged so all } c^\dagger\text{'s (or} \\
& a^\dagger\text{'s or } b\text{'s) stand to the left of } c\text{'s (or} \\
& a\text{'s, or } b^\dagger\text{'s) for the case of equal} \\
& \text{times (see end of appendix F),} \tag{9.4}
\end{aligned}
$$

where P is the number of interchanges of operators required to get the operators in the proper time order, starting with the order given in the brackets. Thus,

$$
\begin{aligned}
T\{c_{k_2}(t_2)\,c_{k_1}^\dagger(t_1)\} &= c_{k_2}(t_2)\,c_{k_1}^\dagger(t_1) \quad \text{for } t_2 > t_1 \\
&= -c_{k_1}^\dagger(t_1)\,c_{k_2}(t_2) \quad \text{for } t_2 \leqslant t_1. \tag{9.5}
\end{aligned}
$$

(Note: The factor of $(-1)^P$ is not present in the boson case.) Hence G may be re-written

$$
\begin{aligned}
G &= G^+(k_2, k_1, t_2 - t_1) = -i\langle \Psi_0|\, c_{k_2}(t_2)\, c_{k_1}^\dagger(t_1)\,|\Psi_0\rangle, \quad t_2 > t_1 \\
&= G^-(k_2, k_1, t_2 - t_1) = +i\langle \Psi_0|\, c_{k_1}^\dagger(t_1)\, c_{k_2}(t_2)\,|\Psi_0\rangle, \quad t_2 \leqslant t_1. \tag{9.6}
\end{aligned}
$$

Consider the $t_2 > t_1$ case first. Substituting (9.3) gives

$$G^+ = -i\langle \Psi_0|\, \underbrace{e^{iHt_2} c_{k_2}}_{B^\dagger}\, \underbrace{e^{-iH(t_2-t_1)} c_{k_1}^\dagger e^{-iHt_1}\,|\Psi_0\rangle}_{A}. \tag{9.7}$$

Now $\exp(-iHt)$ is the time development operator (see appendix B), so that $\exp(-iHt_1)|\Psi_0\rangle$ is the ground state at time t_1, and $c_{k_1}^\dagger \exp(-iHt_1)|\Psi_0\rangle$ is the state with one particle in ϕ_{k_1} added to the ground state at time t_1. Hence

$$A = e^{-iH(t_2-t_1)} c_{k_1}^\dagger e^{-iH t_1}\,|\Psi_0\rangle \tag{9.8}$$

is the state of the system at time t_2 when a particle in ϕ_{k_1} was added at t_1. (Note that this state is not normalized. See Nozières (1964), p. 60.)

The meaning of $B^\dagger$ is obtained from

$$B^\dagger = \overline{c_{k_2}^\dagger e^{-iHt_2}\,|\Psi_0\rangle} \tag{9.9}$$

(see appendix (A.18)). This is evidently the complex conjugate of the state
with one particle in ϕ_{k_2} added to the ground state at time t_2. Hence we obtain

$$G^+ = B^\dagger A = \text{component of } B \text{ along } A$$

$$= \text{probability amplitude that the state of the system}$$
$$\text{at } t_2, \text{ when a particle in } \phi_{k_1} \text{ was added to the}$$
$$\text{ground state at } t_1, \text{ is the state with one particle in}$$
$$\phi_{k_2} \text{ added to the ground state at time } t_2 \qquad (9.10)$$

which is evidently just a mouth-full-of-marbles way of saying (8.32a). Applying the same method to G^- in (9.6) yields (8.32b). This proves the complete
equivalence of (9.1) and (8.32). Equation (9.2) involves identical arguments
applied to the operators $\psi(\mathbf{r})$, $\psi^\dagger(\mathbf{r})$ defined in (7.83).

The formalism here reveals that the process of 'adding a particle in state
ϕ_{k_1}' is not as simple as it sounds. For example, consider the special case of a
one-particle system with exact ground state

$$|\Psi_0\rangle = B|1000...\rangle + D|0100...\rangle \qquad (9.11)$$
where
$$|B|^2 + |D|^2 = 1, \quad |B|^2 > 0, \qquad (9.12)$$

and imagine that we try to add a particle in the lowest single particle state ϕ_1
at time $t_1 = 0$. We have

$$c_1^\dagger e^{-iH\cdot 0}|\Psi_0\rangle = c_1^\dagger|\Psi_0\rangle = 0 + D|1100...\rangle. \qquad (9.13)$$

This shows that 'adding a particle' requires sweeping out the old piece of
particle sitting in ϕ_1 in the B-component of $|\Psi_0\rangle$ and adding one only to the
empty place in the D-component.

Now let $t_1 = 0$, $t_2 = t$, and calculate $G(1,1,t)$ just after the particle in ϕ_1 is
added to the system, i.e., as $t \to 0^+$:

$$\lim_{t\to 0^+} G(1,1,t) = \lim_{t\to 0^+} -i\langle\Psi_0|c_1(t)c_1^\dagger(0)|\Psi_0\rangle$$
$$= -i\langle\Psi_0|c_1 c_1^\dagger|\Psi_0\rangle$$
$$= -iD^2\langle 0100...|0100...\rangle = -iD^2 \qquad (9.14)$$

where (7.18) has been used.

Since by (9.12), D^2 is less than 1, it is clear that even immediately after the
particle is added to the system in ϕ_1, the probability of observing an added
particle in ϕ_1 is less than 1 because ϕ_1 was already occupied with prob-
ability $= B^2$ in $|\Psi_0\rangle$. In fact, if $D = 0$, (9.14) shows $G = 0$ (so the particle does
not propagate) simply because it is completely impossible to add a particle in
state ϕ_1 to $|\Psi_0\rangle$, i.e.: filled bus, no ride.

Of course, the one-particle system here is a special case, but it is easy to
show (by using (7.20) for Ψ_0 instead of (9.11), and breaking it up into one

part with $n_{k_1}=1$ and another part with $n_{k_1}=0$) that exactly the same considerations apply in the many-body case. This is the reason for inclusion of the amplitude factor, Z_k, in the quasi particle expression (8.36).

It is a good brain-building exercise to show how (9.1) boils down to the expression for the free propagator (8.34), in the non-interacting case. The non-interacting Hamiltonian and ground state are given by

$$H_0 = \sum_p \epsilon_p c_p^\dagger c_p, \quad H_0|\Phi_0\rangle = \sum_{p<k_F} \epsilon_p |\Phi_0\rangle, \quad |\Phi_0\rangle = |1\,1\,1\ldots1_F000\ldots\rangle.$$
$$(9.15)$$

Let us calculate just G_0^+ setting $t_1=0$, $t_2=t$:

$$G_0^+(k,t) = -i\langle\Phi_0|\,e^{+iH_0 t}\,c_k e^{-iH_0 t}\,c_k^\dagger|\Phi_0\rangle\,\theta_t. \qquad (9.16)$$

In an obvious notation,

$$c_k^\dagger|\Phi_0\rangle = (-1)^N|\Phi_0, 1_k\rangle\,\theta_{\epsilon_k - \epsilon_F}. \qquad (9.17)$$

Thus k must be greater than k_F. Now

$$H_0|\Phi_0, 1_k\rangle = \sum_p \epsilon_p c_p^\dagger c_p|\Phi_0, 1_k\rangle = \left[\sum_{p<k_F} \epsilon_p + \epsilon_k\right]|\Phi_0, 1_k\rangle \qquad (9.18)$$

so that by appendix (B.7)

$$c_k e^{-iH_0 t}|\Phi_0, 1_k\rangle = (-1)^N|\Phi_0\rangle \exp\left\{-i\left[\sum_{p<k_F} \epsilon_p + \epsilon_k\right]t\right\}. \qquad (9.19)$$

Finally operating with $\exp(iH_0 t)$ produces

$$G_0^+(k,t) = -i\theta_{\epsilon_k - \epsilon_F}\,\theta_t\,e^{-i\epsilon_k t} \qquad (9.20)$$

confirming (8.34).

Using (9.1), it is easy to obtain the ground state expectation value of any single-particle operator (7.37) in terms of the propagator, thus:

$$\langle\Psi_0|\mathcal{O}^{occ}|\Psi_0\rangle = -i\sum_{kl} \mathcal{O}_{kl}\lim_{t\to 0^-} G(l,k;t). \qquad (9.20')$$

The propagator may also be used to find the ground state energy (see, Fetter and Walecka (1971), p. 70, or Schultz, 1964, p. 77).

In this section, we have actually defined three types of single-particle propagators: the 'time-ordered' propagator G in (9.1), the 'retarded' propagator G^+ in (9.6) and the 'advanced' propagator G^- in (9.6). There are two other types of retarded and advanced propagators which are particularly useful because of their simple mathematical properties. These are defined by:

Retarded: $G^R(k_2, k_1, t_2 - t_1) = -i\theta_{t_2 - t_1}\langle\Psi_0|[c_{k_2}(t_2), c_{k_1}^\dagger(t_1)]_+|\Psi_0\rangle$ (9.20'')

Advanced: $G^A(k_2, k_1, t_2 - t_1) = +i\theta_{t_1 - t_2}\langle\Psi_0|[c_{k_2}(t_2), c_{k_1}^\dagger(t_1)]_+|\Psi_0\rangle$ (9.20''')

where $[\;]_+$ means anticommutator, as in (7.32). These are discussed in detail in appendix L.

9.3 Spectral density function

One of the most useful instruments in the toolbox of the many-body physicist is the 'spectral density function' (also called the 'weight function' or 'strength function'). First of all, it is indispensable for analysing the mathematical properties of propagators, especially their analytic properties. Secondly, it is extremely convenient to use in many-body calculations which involve diagrams containing 'dressed' or 'renormalized' propagators (see chapter 11). We present a brief introduction to the subject here. There are more details in appendices H and L. (See also Fetter and Walecka (1971), pp. 72–82, and Pines (1961), pp. 29–34.)

The idea is similar to the spectral decomposition of a time-dependent function $f(t)$ into the sum of its components at various frequencies:

$$f(t) = \int_{-\infty}^{+\infty} F(\omega) e^{i\omega t} d\omega \qquad (9.21)$$

where $F(\omega)$ gives the 'spectrum' of $f(t)$. The corresponding expression for the propagator is (see appendix H), for a system with no external potential:

$$G(\mathbf{k}, t) = -i \int_0^\infty d\omega\, A^+(\mathbf{k}, \omega) e^{-i(\omega+\mu) t}, \quad t > 0$$

$$= +i \int_0^\infty d\omega\, A^-(\mathbf{k}, \omega) e^{+i(\omega-\mu) t}, \quad t \leqslant 0 \qquad (9.22)$$

where μ is the chemical potential:

$$\mu = \begin{bmatrix} \text{ground state energy} \\ \text{of interacting } N \\ \text{particle system} \end{bmatrix} - \begin{bmatrix} \text{ground state energy} \\ \text{of interacting } N-1 \\ \text{particle system} \end{bmatrix} = E_0^N - E_0^{N-1}. \quad (9.23)$$

The $A^\pm(\mathbf{k}, \omega)$ is the 'spectral density function', analogous to $F(\omega)$ in (9.21). The Fourier transform of (9.22) yields

$$G(\mathbf{k}, \omega) = \int_0^\infty d\omega' \left\{ \frac{A^+(\mathbf{k}, \omega')}{\omega - \omega' - \mu + i\delta} + \frac{A^-(\mathbf{k}, \omega')}{\omega' + \omega - \mu - i\delta} \right\} \qquad (9.24)$$

which is the so-called 'Lehmann representation' of the propagator, especially useful for discussing analytic properties (see appendix L). The spectral density has the important properties that (see appendix H, Ex. H.2)

(a) $A^{\pm}(\mathbf{k}, \omega) \geqslant 0,$ real

(b) $\int_0^{\infty} [A^+(\mathbf{k}, \omega) + A^-(\mathbf{k}, \omega)]\, d\omega = 1$ ('sum rule').   (9.25)

For free particles the spectral density is a δ-function:

$$A_0^{\pm}(\mathbf{k}, \omega) = \delta(\pm \omega - \epsilon_k + \mu) \qquad (9.26)$$

which gives $G_0(\mathbf{k}, t)$, $G_0(\mathbf{k}, \omega)$ when substituted in (9.22, 24). For quasi particles, the δ-function gets broadened out and we find the Lorentz form

$$A_{\substack{\text{quasi} \\ \text{particle}}}^{\pm}(\mathbf{k}, \omega) = \frac{1}{\pi} \frac{(1/\tau_k) Z_k}{[\omega \mp (\epsilon_k' - \mu)]^2 + (1/\tau_k)^2} + D(\mathbf{k}, \omega), \qquad (9.27)$$

where $D(\mathbf{k}, \omega)$ is a correction required so that the sum rule is satisfied. When this is substituted into (9.22, 24) it yields just the quasi particle propagator (8.36). (See Schultz (1964), p. 29 ff., Fetter and Walecka (1971), p. 80 ff.)

Finally, we note that if (3.76) is applied to (9.24) we obtain the following expressions for calculating $A^{\pm}$ from the propagator:

$$A^+(\mathbf{k}, \omega - \mu) = -\frac{1}{\pi} \operatorname{Im} G(\mathbf{k}, \omega), \quad \omega > \mu$$

$$A^-(\mathbf{k}, \mu - \omega) = +\frac{1}{\pi} \operatorname{Im} G(\mathbf{k}, \omega), \quad \omega < \mu. \qquad (9.28)$$

9.4 Derivation of the propagator expansion in the many-body case

In chapters 3 and 4, the perturbation expansion of the propagator was obtained mainly by analogy with the pinball case. In order to make sure that all the arguments by monkey did not give us the idea that the diagram subject is either (a) all kindergarten stuff, or (b) based on black magic, we gave a rough idea of the rigorous derivation of the expansion for the single-particle case in §3.4. The argument in the many-body case is much more complicated. However, it is not necessary for understanding the rest of the book, so it has been relegated to appendixes B through G. (We will from time to time refer to certain parts of it, which may be read separately.) An outline of the argument appears in appendix B which shows how the whole diagram expansion comes from the time-dependent Schrödinger equation.

Probably most readers who glance at the appendix now will find it sufficiently repulsive-looking to keep them content with the monkey argument.

9.5 Topology of diagrams

In order to develop general methods for working with diagrams, we need a systematic way of drawing all graphs in nth order. This will be discussed now.

Let us begin with the simplest situation first, e.g., N fermions in an external potential. The series for G in this case appears in (4.34). To get all nth-order diagrams here, draw n fat 'vertex' dots labelled $t'_1, ..., t'_n$, in a vertical row, with two skinny fixed points', labelled t_1, t_2 as shown in (9.28'a)

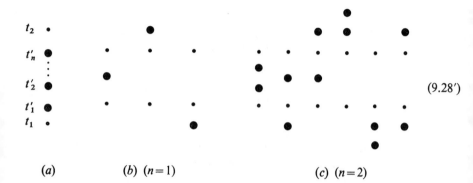

t_2 t'_n t'_2 t'_1 t_1 (9.28')

(a)   (b) (n = 1)   (c) (n = 2)

The fat dots may have any position along the vertical relative to the fixed points. The various possibilities are shown for $n = 1$ in (9.28'b) and for $n = 2$ in (9.28'c). Join the dots with directed lines in all possible 'topologically different' (in the Goldstone sense—see below) connected ways such that one line enters and one leaves each dot. (Note: 'connected' or 'linked' means diagram consists of only one piece. See appendix (G.1) ff.) In (9.29) we see all possible diagrams through first order and five of the twelve possible diagrams in second order:

$$k_2 \begin{vmatrix} t_2 \\ \\ k_1 \end{vmatrix} t_1 = \quad + \quad \bullet \quad + \quad) \quad + \quad) \quad + \quad + \quad + \quad + \quad + \quad + \cdots$$

(9.29)

(a)   (b)   (c)   (d)   (e)   (f)   (g)   (h)   (i)

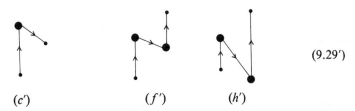

$$(c') \qquad\qquad (f') \qquad (h') \qquad\qquad (9.29')$$

By 'topologically different' (in the Goldstone sense) we mean this: visualize the dots and lines as wads of hardened chewing gum connected by rubber bands with attached arrowheads. Then two diagrams are topologically identical (in the Goldstone sense) if one can be distorted into the other without changing the vertical ordering of the dots and points. Two diagrams are different if they are not topologically identical. Thus (c), (f), and (h) in (9.29) are topologically identical to respectively (c'), (f'), and (h') in (9.29'). Note that 'all topologically different ways of joining the n dots' corresponds to 'all physically different ways the particle can propagate through the system, scattering n times'.

The above 'Goldstone method' of drawing the diagrams may be greatly simplified if we associate the full propagator, $G = G^+ + G^-$ (see (8.31) and (9.6)) with directed lines instead of just G^+ or G^-. This is the 'Feynman method'. Then in the integrals over intermediate times we automatically get $G_0(t'-t) = G_0^+$ when $t' > t$ and $G_0(t'-t) = G_0^-$ when $t' \leqslant t$. Thus it is no longer necessary to draw any hole lines since a directed line is a particle line for $t' > t$ and a hole line for $t' < t$. Thus the time order of the dots is no longer important. Hence (9.29) becomes:

$$\begin{array}{c} k_2 \\ \\ k_1 \end{array} = \begin{array}{c} k_2 \\ \\ k_1 \end{array} + \quad + \quad + \quad + \quad + \cdots . \qquad (9.30)$$

In other words, Goldstone diagrams are time-ordered, whereas Feynman diagrams are not. (In the Goldstone method, time integrations are over the regions allowed by the time order in the diagram, while in the Feynman method they are from $-\infty(1 - i\eta)$ to $+\infty(i - i\eta)$. (See paragraph in [ ] following (4.37) regarding these limits.))

This saves an enormous amount of art work—for example, the single two dot diagram in (9.30) is topologically equivalent (in the Feynman sense) to the 12 diagrams, (e), (f), ... in (9.29)! The rule for drawing the nth-order diagram is now trivial—just draw n dots and two external points with a vertical line through them all. Equation (9.30) may be translated into (omit ω's for brevity):

$$G(k_2, k_1) = \delta_{k_1 k_2} G_0(k_1) + G_0(k_1) V_{k_2 k_1} G_0(k_2) +$$
$$+ \sum_{\text{all } q} G_0(k_1) V_{q k_1} G_0(q) V_{k_2 q} G_0(k_2) + \cdots. \qquad (9.31)$$

Drawing diagrams in the case of mutually interacting fermions with no external potential (see (4.63)) is considerably more difficult. To get all nth-order Goldstone diagrams, draw n wiggly lines each with vertex dots at both ends, and two fixed points, thus (note that the wiggles may have any vertical position relative to the fixed points):

$$\left. \begin{array}{c} \cdot \\ \sim\sim\sim \\ \sim\sim\sim \\ \vdots \\ \sim\sim\sim \\ \cdot \end{array} \right\} n. \qquad (9.32)$$

Join all points to each other with directed lines in all linked topologically distinct (in the Goldstone sense) ways such that one line enters and one leaves each vertex point, and a line enters one external point and leaves the other. Note that by (4.60) twisted diagrams like (4.61) are not topologically distinct in the Goldstone sense, and only one of them should be drawn.

To illustrate, consider the second-order case. The number of distinct linked diagrams coming out of (9.32) even in this low order is dazzling, or perhaps depressing, depending on whether your viewpoint is aesthetic or practical. A few typical ones are (9.33).

Sitting in seclusion with these surrealisms for a while shows that many of them can be eliminated. First because of conservation of momentum, graphs (b), (d), (h), (i), (j), (l) have a particle and a hole both in the same momentum state. But this is impossible by (7.72). Therefore these graphs do not occur. (Such graphs are called 'anomalous' or 'momentum non-conserving'. They do give a contribution when the system is non-isotropic, e.g. external field present, or at finite temperatures.) Second, if we agree to use the Feynman convention (see (9.30)) in which the full G is associated with each line and time order has no significance, then $(f) \equiv (e)$, $(g) \equiv (k)$. Hence the only survivors of (9.33) are (9.34) or (9.35), where (9.35) is another conventional way of drawing the diagrams (obtained by making the diagrams out of rubber bands and pulling at top and bottom until the main line is straight). Equations (9.34) and (9.35) are evidently topologically equivalent in the Feynman sense. Note that with the Feynman method, the arrows no longer designate particles or holes but just show the direction of momentum flow. Observe the importance of the arrows! It is because of them that diagram (2) in (9.34) is topologically different from diagram 3!

$$G_{2\text{nd order}} =$$

(a)   (b)   (c)   (d) $+ \cdots$

(e)   (f)   (g) $+ \cdots +$ (h)

(i) $+ \cdots$ (j)   (k)   (l) $+ \cdots .$

$$(9.33)$$

$$G_{2\text{nd order}} =$$

(1)   (2)   (3)   (4)   (5)   (6)

$$(9.34)$$

or

$$G_{2\text{nd order}} =$$

$$(9.35)$$

Let us evaluate a typical second-order diagram using the Feynman method. Employing Table 4.3, we have in $(\mathbf{k}, t)$-space (regarding the limits on the time integration, see paragraph in [[ ]] following (4.37)):

$$= (-1) \sum_{q,p} \int_{-\infty(1-i\eta)}^{+\infty(1-i\eta)} dt\, dt' [iG_0(\mathbf{k}, t-t_1)] [-iV_q] \times$$
$$\times [iG_0(\mathbf{k}-\mathbf{q}, t'-t)] [iG_0(\mathbf{p}, t-t')] \times$$
$$\times [iG_0(\mathbf{p}+\mathbf{q}, t'-t)] [-iV_q] [iG_0(\mathbf{k}, t_2-t')]. \qquad (9.36)$$

Note that the order of the times in G is always: time at end of directed line minus time at beginning. Remember that (9.36) is really (e) and (f) in (9.33) lumped into one.

The (-1) is for the fermion loop: . Fermion loops were defined just after (4.56). For example, in (9.33), a, b, h have two loops, c, d, e, f, i, j have one, the others have zero. (See end of appendix G.)

The Fourier transform of (9.36) is

$$= (-1) [iG_0(\mathbf{k}, \omega)]^2 \sum_{p,q} \int \frac{d\epsilon\, d\beta}{2\pi\, 2\pi} [iG_0(\mathbf{k}-\mathbf{q}, \omega-\epsilon)] \times$$
$$\times [iG_0(\mathbf{p}, \beta)] [iG_0(\mathbf{p}+\mathbf{q}, \beta+\epsilon)] [-iV_q]^2. \qquad (9.37)$$

It is seen that the frequencies, ω, ϵ, β, etc., are conserved at the vertices, just like the momentum. This comes about because of the appearance of δ-functions similar to the $2\pi\delta(\omega'-\omega)$ in (2.23) when the transform is carried out.

Note that a frequency is associated with interaction wiggles as well as propagators. In some cases V_q will actually depend on ϵ (this is a 'frequency dependent' or, in $(\mathbf{q}, t)$-space 'retarded', interaction). Finally, observe that one integrates or sums over all intermediate momenta and frequencies where 'intermediate' means excluding the $\mathbf{k}$, ω of the incoming and outgoing lines.

One more thing. We can avoid treating the 'non-propagating' lines as a special case by including a convergence factor $\exp(i\omega 0^+)$ when translating these lines into functions, where 0^+ is a positive infinitesimal such that $0^+ \times \infty = \infty$ (see end of appendix I), thus:

$$\mathbf{l}, \epsilon \; \bigcirc \quad \text{or} \quad \underset{\mathbf{l},\epsilon}{\smile} \equiv iG_0(\mathbf{l}, \epsilon) \exp(i\epsilon 0^+) \qquad (9.38)$$

(Note that this factor may also be included in $G_0(\mathbf{l}, \epsilon)$ itself—see Schrieffer (1964), pp. 108–9.) Hence the integral over the intermediate frequency ϵ gives:

$$\int_{-\infty}^{+\infty} \frac{d\epsilon}{2\pi} \frac{i\,e^{i\epsilon 0^+}}{\omega - \epsilon_l + i\delta_l} = \begin{array}{ll} -1 & \text{for } l < k_F \\ 0 & \text{for } l > k_F \end{array} \tag{9.38'}$$

(The integral is done by residues as in appendix I. The contour is closed in the upper half-plane where the convergence factor makes the integral vanish.) Thus, for example:

$$= [iG_0(\mathbf{k}, \omega)]^2 \sum_l \int \frac{d\epsilon}{2\pi} [-iV_{lkkl}] [iG_0(\mathbf{l}, \epsilon)] e^{i\epsilon 0^+}$$

$$= [iG_0(\mathbf{k}, \omega)]^2 \sum_{l < k_F} (-iV_{lkkl})(-1)$$

same as in (4.62).

The same factor must be used when the non-propagating line is an exact propagator:

$\equiv iG(\mathbf{l}, \epsilon) \exp(i\epsilon 0^+).$ \tag{9.39}

9.6 Diagram rules for single-particle propagator

We have now reached the point where we can present in summary form the rules for drawing and evaluating the type of graphs which will be used in the next two chapters. These are the diagrams describing a system of mutually interacting fermions with no external field and they will always be drawn in $(\mathbf{k}, \omega)$-space, using the Feynman method. The rules are:

(1) In nth order, draw n wiggly lines with vertex dots and two external points as in (9.32).

(2) Join all vertex dots and external points to each other with directed lines in all linked topologically distinct (in the Feynman sense) ways, with one line entering and one leaving each vertex dot and a line entering one external point and leaving the other. Two diagrams are topologically distinct if they are visualized as made of rubber bands, and one cannot be deformed into the other.

(3) Label each line and wiggle with a momentum, $\mathbf{k}$ (short for $\mathbf{k}, \sigma$, where $\sigma =$ spin), and frequency ω, such that the sum of momenta (and frequencies) entering each vertex = sum of those leaving. Eliminate all 'anomalous'

or 'momentum-non-conserving' diagrams, i.e., which have a hole and a particle in the same state. Rules (1)–(3) yield the series:

$$(9.40)$$

(4) Evaluate graphs by means of the dictionary in Table 9.1.

These rules will be seen in action in §9.8. Before going on to this, however, it is a good idea to point out that not everybody draws graphs the same way we do here. Some graphologists use abbreviated diagrams in which the interaction wiggles are compressed to points or little squares. Thus, for example

$$(9.41)$$

Note that each abbreviated diagram stands for several Feynman diagrams. Those drawn with points are called 'Hugenholtz diagrams' (see Van Hove (1961), p. 171, or Hugenholtz (1965)), while those with squares are 'Abrikosov diagrams' (Abrikosov (1963), p. 71 ff.).

Table 9.1 *Diagram dictionary for interacting many-fermion system with no external potential (Feynman method)*

Diagram element	Factor
$\mathbf{k}, \omega$ ⬆ or $\mathbf{k}, \omega$ ⬇	$iG(\mathbf{k}, \omega)$
$\mathbf{k}, \omega$ ↑ or $\mathbf{k}, \omega$ ↓	$iG_0(\mathbf{k}, \omega) = \dfrac{i}{\omega - \epsilon_k + i\delta_k}, \quad \begin{array}{l} \delta_{k>k_F} = +\delta \\ \delta_{k<k_F} = -\delta \end{array}$
⟲ or ⟲ $\mathbf{k}, \omega$	$iG(\mathbf{k}, \omega) \exp(i\omega 0^+)$ $(0^+ \times \infty = \infty)$
○ or ⌣ $\mathbf{k}, \omega$	$iG_0(\mathbf{k}, \omega) \exp(i\omega 0^+)$ $\left(\text{so that: } \displaystyle\int_{-\infty}^{+\infty} \frac{d\omega}{2\pi} iG_0(\mathbf{k}, \omega) \times \right.$ $\left. \times \exp(i\omega 0^+) = -\theta_{k_F - k} \right)$
$\mathbf{k} \quad\quad \mathbf{l}$ $\sim\sim\sim$ $\mathbf{q}, \epsilon$ $\mathbf{m} \quad\quad \mathbf{n}$	$-iV_{klmn}$ or $-iV_q$ (use $V_{klmn}(\epsilon)$ or $V_q(\epsilon)$ for time-dependent interaction)
Each fermion loop Example:	(-1)
Each intermediate frequency ω	$\displaystyle\int \frac{d\omega}{2\pi}$
Each intermediate momentum, $\mathbf{k}$	$\displaystyle\sum_{\mathbf{k}}$ or $\displaystyle\int \frac{d^3\mathbf{k}}{(2\pi)^3}$ (for $\Omega = 1$) (include sum over spins)

9.7 Modified propagator formalism using chemical potential, μ

The formalism just described can be somewhat inconvenient in actual calculations because, unless special precautions are taken, it may produce approximations for G which yield the wrong total number of particles for the system.

In order to understand this, let us first derive the relation between the propagator G and the total particle number N. The quantity N is the expectation value of the total number operator $2 \sum_k c_k^\dagger c_k$ (factor of 2 for spin) in the interacting ground state:

$$N = \langle \Psi_0 | 2 \sum_k c_k^\dagger c_k | \Psi_0 \rangle = 2 \sum_k \langle \Psi_0 | c_k^\dagger c_k | \Psi_0 \rangle. \qquad (9.42)$$

The summand is easily expressed in terms of $G(k_2, k_1, t_2 - t_1)$ by setting $t_2 = t$, $t_1 = 0$, $k_1 = k_2 = \mathbf{k}$ in (9.6), then letting t approach zero from the left. This yields

$$N = 2 \sum_k (-i) \times \lim_{t \to 0^-} G(\mathbf{k}, t) = -2i \lim_{t \to 0^-} \int \frac{d^3 \mathbf{k}}{(2\pi)^3} \int \frac{d\omega}{2\pi} G(\mathbf{k}, \omega) e^{-i\omega t}, \qquad (9.43)$$

which is the desired relation.

Suppose now that there are N_0 particles in the particular system we are dealing with. Imagine that we calculate an approximate G for the system by a partial summation over some types of diagrams in (9.40), and then place this G in (9.43) to check and see if it yields $N = N_0$. Evidently, G(approx.) will be a function of N_0, because each G_0 entering the calculation of G(approx.) depends on k_F, and k_F is related to N_0 by

$$N_0 = 2 \sum_{k < k_F} 1 = \frac{2}{(2\pi)^3} \int_{k < k_F} d^3 \mathbf{k} = k_F^3 / 3\pi^2, \quad \text{or} \quad k_F = (3\pi^2 N_0)^{\frac{1}{3}}. \qquad (9.44)$$

(Note that the Fermi energy is

$$\epsilon_F = k_F^2 / 2m = \frac{(3\pi^2 N_0)^{\frac{2}{3}}}{2m}. \Big) \qquad (9.45)$$

Hence N will be a function of N_0. But, since G is only approximate, there is no guarantee that N will equal N_0.

It is possible to remove this difficulty without changing the formalism developed up to now, either by using the spectral density method discussed in §9.3 or the self-consistent renormalization of §11.1. However, it turns out to be simpler just to modify the formalism slightly. The method of doing this is to use the 'grand canonical ensemble' at zero temperature (see chapter 14 for the finite temperature case). In this method we no longer regard the system as isolated, with definite particle number N_0, but instead put it in contact with a particle reservoir, so that it can gain or lose particles. Thus, particle number, N, is variable throughout the calculation. The chemical potential of the system, μ (see (9.23)), is fixed, but unknown; its value is determined at the end of the calculation by setting the total particle number equal to N_0. The modified Hamiltonian for this case is (Nozières (1964), p. 307):

$$H' = H - \mu N = H'_0 + H_1$$

where

$$H'_0 = \sum_k (\epsilon_k - \mu) c_k^\dagger c_k$$

$$H_1 = \tfrac{1}{2} \sum_{klmn} V_{klmn} c_l^\dagger c_k^\dagger c_m c_n, \tag{9.46}$$

where N is the total particle number operator.

The ground state of the modified unperturbed Hamiltonian, H'_0, is obtained by selecting that number of particles, and that way of filling the energy levels which minimizes the energy. It is easily seen that this means all levels filled up to $\epsilon_k = \mu$, i.e., up to $k_F^\mu = \sqrt{(2m\mu)}$. The corresponding particle number is $N = (3\pi^2)^{-1} \times (2m\mu)^{\frac{3}{2}}$. The free propagator for H'_0 is

$$G_0^\mu(\mathbf{k}, \omega) = \frac{1}{\omega - (\epsilon_k - \mu) + i\delta_k^\mu}, \quad \delta_k^\mu = \begin{cases} +\delta & \text{for } \epsilon_k > \mu. \\ -\delta & \text{for } \epsilon_k < \mu. \end{cases} \tag{9.47}$$

The rules for diagrams are the same as those in Table 9.1, except that G_0 is replaced by G_0^μ.

It will be convenient in chapter 11, where we use this formalism, to define a new ω such that $\omega_{\text{new}} \equiv \omega + \mu$. In addition, in order to get the correct result when we do self-consistent Hartree–Fock (see exercise 11.3) it is necessary to re-write the infinitesimal in the form i sign $(\omega - \mu)\delta$ or $i(\omega - \mu)\delta$ for short, where $\omega \equiv \omega_{\text{new}}$. These two changes yield ($\omega \equiv \omega_{\text{new}}$):

$$G_0^\mu(\mathbf{k}, \omega) = \frac{1}{\omega - \epsilon_k + i(\omega - \mu)\delta} \tag{9.47'}$$

Note that the poles of (9.47') are at $\omega = \epsilon_k - i(\epsilon_k - \mu)\delta$ which, (if we set $\omega = \omega + \mu$) is the same as the location of the poles of (9.47).

With this modified formalism, it is found that $G(\text{approx.})$ is a function of μ so that when it is placed in (9.43), N becomes a function of μ. If μ is now determined by setting

$$N(\mu) = N_0 \tag{9.48}$$

we guarantee that $G(\text{approx.})$ yields the correct number of particles. (Note that the value of μ obtained from (9.48) depends on the interaction, H_1. For a non-interacting system ($H_1 = 0$), the exact propagator G is just equal to G_0^μ, and we obtain, using (9.47), (9.43), (9.48):

$$N(\mu) = \frac{(k_F^\mu)^3}{(3\pi)^2} = \frac{(2m\mu)^{\frac{3}{2}}}{3\pi^2} = N_0 \tag{9.49}$$

so that $\mu = \epsilon_F$, the Fermi energy of the non-interacting system as given in (9.45).)

[9.8 Beyond Hartree–Fock: the single pair-bubble approximation]

In Hartree–Fock approximation, while the incoming particle in state **k** propagates through the system, the other particles are considered to be 'static', i.e., remain in their original unperturbed stationary states, that is, the eigenstates ϕ_{k_i} in (7.4). But in reality, we know that these particles are perturbed by the incoming particle in such a way that their motion 'follows' or 'is correlated with' the motion of the incoming particle. That is, they act in a *dynamic* rather than a static way, i.e., in a time-dependent fashion. The simplest process showing this effect is the single pair-bubble self-energy part:

$$\tag{9.50}$$

If we include in our approximation for G all bubbles, open oysters, and pair-bubbles thus:

$$\tag{9.51}$$

we find easily (see (10.5) for details):

$$\text{---} = \frac{1}{\left[^{-1} - \left[\text{---}\bigcirc + \text{---} + \text{---}\right]\right]} .\qquad (9.52)$$

We will now evaluate the single pair-bubble self-energy part to see its effect on the quasi particle energy and lifetime.

We first note in (9.50) that because of the conservation of momentum and frequency at each vertex, the outgoing momentum and frequency equal the incoming ones. Using the dictionary Table 9.1 we find (note: factor of 2 is for sum over spin directions in pair-bubble):

$$\mathbf{l},\beta = \int \frac{d^3 q}{(2\pi)^3} \int \frac{d\alpha}{2\pi}\; iG_0(\mathbf{k}-\mathbf{q}, \omega-\alpha) \times (-iV_q)^2 \times$$

$$\times (-1) \times 2 \int \frac{d^3 l}{(2\pi)^3} \int \frac{d\beta}{2\pi}\; iG_0(\mathbf{l},\beta) \times iG_0(\mathbf{l}+\mathbf{q},\beta+\alpha)$$

$$\underbrace{\phantom{\int \frac{d^3 l}{(2\pi)^3} \int \frac{d\beta}{2\pi}\; iG_0(\mathbf{l},\beta) \times iG_0(\mathbf{l}+\mathbf{q},\beta+\alpha)}}_{-i\pi_0(\mathbf{q},\alpha)}$$

$$(9.53)$$

where $-i\pi_0$ is the pair bubble

$$-i\pi_0(\mathbf{q},\alpha) = \quad \mathbf{l},\beta \qquad (9.54)$$

Let us start off by evaluating the frequency integral over β in the pair-bubble. There are four cases. In the first case we have $|\mathbf{l}| > k_F$, $|\mathbf{l}+\mathbf{q}| < k_F$. Then

$$\bigcirc = (-1) \times 2 \times \int \frac{d^3 l}{(2\pi)^3} \int \frac{d\beta}{2\pi} \frac{i}{\beta - \epsilon_l + i\delta} \times \frac{i}{\beta + \alpha - \epsilon_{l+q} - i\delta} \qquad (9.55)$$

The integrand has poles at

$$\beta = \epsilon_l - i\delta$$

$$\beta = -\alpha + \epsilon_{l+q} + i\delta \qquad (9.56)$$

The integral may be evaluated by contours, completing the contour in either half of the complex β-plane. We choose the upper:

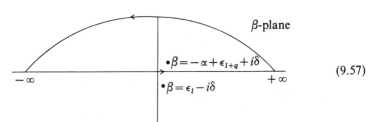

$$\hspace{10cm} (9.57)$$

We have

$$\oint = \int + \int_{-\infty}^{+\infty} = 2\pi i \sum \text{residues} \hspace{3cm} (9.58)$$

and

$$\int \propto \lim_{R\to\infty} \int \frac{d(Re^{i\theta})}{2\pi} \times \frac{i}{Re^{i\theta}} \times \frac{i}{Re^{i\theta}} \propto \lim_{R\to\infty} \frac{1}{R} = 0 \hspace{2cm} (9.59)$$

Hence

$$\int_{-\infty}^{+\infty} \frac{d\beta}{2\pi} \cdots = 2\pi i \,(\text{Residue at the point: } \beta = -\alpha + \epsilon_{l+q} + i\delta)$$

$$= \frac{2\pi i}{2\pi} \lim_{(\beta \to -\alpha + \epsilon_{l+q} + i\delta)} \left\{ \frac{\beta - [-\alpha + \epsilon_{l+q} + i\delta]}{\beta + \alpha - \epsilon_{l+q} - i\delta} \times \frac{1}{\beta - \epsilon_l + i\delta} \right\}$$

$$= \frac{i}{\alpha - \epsilon_{l+q} + \epsilon_l - i\delta} . \hspace{3cm} (9.60)$$

The next two cases are $|\mathbf{l}|$, $|\mathbf{l}+\mathbf{q}| > k_F$, and $|\mathbf{l}|$, $|\mathbf{l}+\mathbf{q}| < k_F$. Here both poles are in the same half-plane, so closing the contour in the other half-plane we immediately see that the result is zero.

The final case is for $|\mathbf{l}| < k_F$, $|\mathbf{l}+\mathbf{q}| > k_F$. A procedure similar to that used in the first case yields

$$\int \frac{d\beta}{2\pi} = \frac{-i}{\alpha - \epsilon_{l+q} + \epsilon_l + i\delta} \hspace{3cm} (9.61)$$

Thus we get for the pair-bubble

$$-i\pi_0(\mathbf{q}, \alpha) = \underbrace{\int_{\substack{l<k_F \\ |l+q|>k_F}} \frac{d^3l}{(2\pi)^3} \frac{2i}{\alpha - \epsilon_{l+q} + \epsilon_l + i\delta}}_{-i\pi_0^A} - \underbrace{\int_{\substack{l>k_F \\ |l+q|<k_F}} \frac{d^3l}{(2\pi)^3} \frac{2i}{\alpha - \epsilon_{l+q} + \epsilon_l - i\delta}}_{-i\pi_0^B}.$$

(9.62)

Putting this result into (9.53) we have

$$\mathbf{1}, \beta = \int \frac{d^3q}{(2\pi)^3} \int \frac{d\alpha}{2\pi} (-iV_q)^2 \frac{i}{\omega - \alpha - \epsilon_{k-q} + i\delta_{k-q}} \times$$

$$\times [-i\pi_0^A(\mathbf{q},\alpha) + i\pi_0^B(\mathbf{q},\alpha)].$$

(9.63)

Let us first examine the π_0^A-term:

$$= (-2) \int \frac{d^3q}{(2\pi)^3} V_q^2 \int_{\substack{l<k_F \\ |l+q|>k_F}} \frac{d^3l}{(2\pi)^3} \int \frac{d\alpha}{2\pi} \times$$

$$\times \frac{1}{\omega - \alpha - \epsilon_{k-q} + i\delta_{k-q}} \times \frac{1}{\epsilon_l - \epsilon_{l+q} + \alpha + i\delta}.$$

(9.64)

The poles here, in the α-integration are at

$$\alpha = -\epsilon_l + \epsilon_{l+q} - i\delta$$

$$\alpha = \omega - \epsilon_{k-q} + i\delta_{k-q}.$$

(9.65)

The integration is done by contours just as before. We see that $|\mathbf{k}-\mathbf{q}|$ must be $> k_F$, otherwise both poles are in the lower half-plane and we get zero. The result is

$$\equiv -i\Sigma^A(\mathbf{k}, \omega) = (-2i) \int \frac{d^3q}{(2\pi)^3} \int_{\substack{l<k_F \\ |l+q|>k_F \\ |k-q|>k_F}} \frac{d^3l}{(2\pi)^3} |V_q|^2 \frac{1}{\omega + \epsilon_l - \epsilon_{l+q} - \epsilon_{k-q} + i\delta}.$$

(9.66)

In a similar fashion, we find for the other term

$$\text{(diagram)} \equiv -i\Sigma^B(\mathbf{k}, \omega) = (-2i) \int \frac{d^3q}{(2\pi)^3} \int \frac{d^3l}{(2\pi)^3} |V_q|^2 \frac{1}{\omega + \epsilon_l - \epsilon_{l+q} - \epsilon_{k-q} - i\delta}$$

$$l > k_F, \ |l+q| < k_F$$
$$|k-q| < k_F$$

$$(9.67)$$

Analogous to the Hartree-Fock result, we can interpret $\Sigma^{A, B}(\mathbf{k}, \omega)$ here as coming from an effective external potential. However, in this case, the potential is dependent on the frequency, ω, hence would be time-dependent if we Fourier transformed it. This shows the *dynamical* effect we mentioned before, i.e., the other particles 'follow' the extra propagating particle, so the effective potential coming from them must be time-dependent.

As indicated in (3.71), we need the real and imaginary parts of the total self energy, Σ, in order to find the new energy ϵ_k' and lifetime τ_k for quasi particles. These can be found from applying (3.76″) to Σ. For Σ^A we find

$$\text{Re}\,\Sigma^A = +2P \int \frac{d^3q}{(2\pi)^3} \int \frac{d^3l}{(2\pi)^3} |V_q|^2 \frac{1}{\omega + \epsilon_l - \epsilon_{l+q} - \epsilon_{k+q}} \qquad (9.68)$$

$$l < k_F, \ |l+q| > k_F$$
$$|k-q| > k_F$$

$$\text{Im}\,\Sigma^A = -2\pi \int \frac{d^3q}{(2\pi)^3} \int \frac{d^3l}{(2\pi)^3} |V_q|^2 \delta(\omega + \epsilon_l - \epsilon_{l+q} - \epsilon_{k-q}). \qquad (9.69)$$

$$l < k_F, \ |l+q| > k_F$$
$$|k-q| > k_F$$

Assuming V is small, we can get a simple approximation for the contribution of $\text{Re}\,\Sigma^A$ to the quasi particle energy by using (3.71), i.e. we place $\omega \approx \epsilon_k$ in (9.68). The result agrees with what one obtains from ordinary second-order perturbation theory.

We now consider the imaginary part. Making the transformation of variables $\mathbf{q} = \mathbf{n} - \mathbf{l}, \mathbf{l} = \mathbf{l}, \mathbf{k} - \mathbf{q} = \mathbf{k} - \mathbf{n} + \mathbf{l}$:

$$\text{Im}\,\Sigma^A = -2\pi \int \frac{d^3n}{(2\pi)^3} \int \frac{d^3l}{(2\pi)^3} |V_{n-l}|^2 \delta(\omega + \epsilon_l - \epsilon_n - \epsilon_{k-n+l}). \qquad (9.70)$$

$$n > k_F, \ l < k_F, \ |k-n+l| > k_F$$

Let

$$\epsilon_l = \epsilon_F - t_l, \quad \epsilon_n = \epsilon_F + t_n, \quad \epsilon_{k-n+l} = \epsilon_F + t_{k-n+l} \qquad (9.71)$$

where all t's are positive. We now find the maximum value the t's may have. To get a contribution from the δ-function, we must have

$$\omega - \epsilon_F = t_l + t_n + t_{k-n+l}. \qquad (9.72)$$

To satisfy (9.72) it is necessary that $\omega - \epsilon_F > 0$. For a given $\omega - \epsilon_F$, the maximum value t_1 may have is evidently $\omega - \epsilon_F$ (in which case t_n, t_{p-n+1} have their minimum value, i.e., zero). Hence the l-integration is in a shell of energy thickness = $\omega - \epsilon_F$ about the Fermi surface. The corresponding thickness in momentum space for $\omega - \epsilon_F \ll \epsilon_F$ is $dl = (m/k_F)(\omega - \epsilon_F)$. The same holds true for the n-integration. Hence we have (let Ω stand for angular variables)

$$\text{Im} \, \Sigma^A = -2\pi \int\limits_{k_F}^{[k_F + (m/k_F)(\omega - \epsilon_F)]} \frac{dn}{(2\pi)^3} \int\limits_{[k_F - (m/k_F)(\omega - \epsilon_F)]}^{k_F} \frac{dl}{(2\pi)^3} \int d\Omega_n \times$$

$$\times \int d\Omega_l \, n^2 \, l^2 \, V^2(n, l, \Omega_n, \Omega_l) \tag{9.73}$$

Since $\mathbf{n}$, $\mathbf{l}$, are in a thin shell about the Fermi surface, we can set $n = l = k_F$, and obtain immediately

$$\text{Im} \, \Sigma^A \propto (\omega - \epsilon_F)^2. \tag{9.74}$$

The same result is easily shown to hold true for $\text{Im} \, \Sigma^B$.

In the case where the interaction V is small, so that Σ is small, we may use (3.71), which yields for the reciprocal lifetime

$$\tau_k^{-1} = -\text{Im} \, \Sigma(k, \epsilon_k) \propto (\epsilon_k - \epsilon_F)^2 \tag{9.75}$$

This extremely important result shows that in single pair-bubble approximation, the quasi particle lifetime becomes infinite as we approach the Fermi surface. From (9.75) we see that

$$\tau_k^{-1} \ll \epsilon_k - \epsilon_F \tag{9.76}$$

in agreement with the criterion (8.21), since for V small, $\epsilon_k' \approx \epsilon_k$, $\epsilon_F' \approx \epsilon_F$. (If this criterion were not satisfied, it would make no sense to talk of quasi particles here.) Note that our derivation here is equivalent to the 'Golden Rule' argument in §8.4, as was mentioned around (8.30'), (8.30'').

Further reading

Fetter and Walecka (1971), chap. 3.
Pines (1961), chap. 2, pp. 26–34, 48–51.
Schultz (1964), chap. 3, pp. 13–62.
Schrieffer (1964a), chap. 5, pp. 103–36 (includes phonons).
Thouless (1972).
Thouless (1964).
Nozières (1964), chap. 3.

Anderson (1963), chap. 3, pp. 104–12.
Abrikosov (1965).
Klein (1962).
Mills (1969), chap 3, 4.

Exercises

9.1 Find $T\{c_k(t_1)c_l^\dagger(t_2)c_m^\dagger(t_3)\}$ where $t_2 > t_3$ and $t_3 = t_1$.

9.2 Calculate the hole part of the propagator (9.6) in the non-interacting case, and show that it is equal to the hole part of (8.34) (use argument like that in (9.15)–(9.20)).

9.3 Verify that when (9.26) is substituted into (9.22, 24) we obtain the correct expressions for the free propagators.

9.4 Show that diagrams (j) and (l) in (9.33) contain a particle and hole in the same state.

9.5 Which of the following diagrams are topologically equivalent in the Feynman sense?

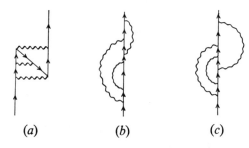

(*a*)   (*b*)   (*c*)

9.6 Using labels showing momentum conservation explicitly, write out the expression for diagram (g) in (9.33), in $(\mathbf{k}, \omega)$-space, employing Feynman convention.

9.7 Verify (9.20′).

9.8 Verify (9.61), closing the contour in the upper half-plane. Show that you get the same result if the contour is closed in the lower half-plane.

9.9 Draw the pair-bubbles for the four cases described in §9.8. Why are the cases $|\mathbf{l}| > \mathbf{k}_F$, $|\mathbf{l}+\mathbf{q}| > \mathbf{k}_F$ and $|\mathbf{l}| < \mathbf{k}_F$, $|\mathbf{l}+\mathbf{q}| < k_F$ automatically zero?

9.10 Verify (9.28).

9.11 Verify (9.38′).

Chapter 10

Dyson's Equation, Renormalization, RPA and Ladder Approximations

10.1 General types of partial sums

We have pointed out that ordinary perturbation theory is helpless when confronted with the big, badly-behaved interactions in many-body systems. In typical cases, like the electron gas and nuclear matter, nearly all terms in the propagator and vacuum amplitude perturbation expansions are divergent. To get any sensible results, one is therefore forced to use a method which goes beyond ordinary perturbation theory, e.g., the method of partial summation. We have seen simple examples of the method in the first half of the book.

In this chapter and the next one we are going to reveal certain general partial summation tricks which are a routine part of most propagator calculations. These tricks can have a rather disconcerting effect on the uninitiated, since infinite numbers of infinite series of diagrams appear to successively vaporize into thin air until only two diagrams are left! However, the method behind the magic is extremely simple—just summing one geometric series after the other—it is only the accompanying mumbo-jumbo that makes things look mystical.

The general types of partial sums for calculating the Green's function for a system of interacting fermions with no external field are listed in Table 10.1.

The process of carrying out summations (2), (3), and (4) is often called 'renormalization' (see (0.2)). The result of renormalizing interaction lines, for example, is a simplified series in which no interaction lines have any inserted 'polarization parts' and all interaction lines are dressed. Similar statements hold true for renormalizing particle lines and vertices.

Superficially, the most striking thing about these summations is the enormous simplification which they produce in the appearance of the series. However, it must not be supposed on this account that it is a matter of art for art's sake. In fact, the really important thing is that infinite summations performed on a series of divergent terms produce a new series in which the terms are each finite. We will see this remarkable effect in the case of the high-density electron gas.

Another significant feature of the general partial sums is that they have an

177

Table 10.1 *General types of partial sums*

General type of diagrams summed over	Result
(1) All diagrams containing repeated proper (or 'irreducible') self-energy parts. (Summation is complete)	Dyson's equation
(2) All diagrams with 'polarization parts' inserted in interaction lines	'dressed', 'effective' or 're-normalized' interactions
(3) All diagrams with 'self-energy parts' inserted in free particle and hole lines	'dressed' or 'renormalized' particle and hole lines (self-consistent renormalization)
(4) All diagrams with 'irreducible vertex parts' inserted in place of a vertex	dressed vertices

immediate physical interpretation—for example, the dressed interaction is just a screened interaction while the dressed particle line is just a quasi particle.

Finally, it should be mentioned that Table 10.1 looks a bit more impressive than it actually is because the word 'all diagrams' in each case turns out in practice to be replaced by 'one or two types of diagrams'. In certain limits, this turns out to be an excellent approximation—for instance, in the high-density electron gas, nearly the whole contribution comes from the 'ring' diagrams ('random phase approximation' or 'RPA') while in a low-density fermion gas with short-range interactions, like nuclear matter, the major contribution comes from just 'ladder' diagrams. And even in non-limiting cases, the sum over just one or two types of diagrams is valuable, since it is sufficient to remove the divergences.

10.2 Dyson's equation

The partial sum technique used in the Hartree–Fock (4.76) and single pair-bubble (9.52) cases can be generalized to yield an extremely convenient exact expression for the propagator which is known as Dyson's equation. The partial sum was possible in those cases because we were dealing with repeated simple parts of diagrams, like ⌒⌒○ and ⌣⌣, hanging on the main directed $(\mathbf{k}, \omega)$-line like pearls on a string. Examination of the full propagator (9.40) shows that it too consists of strings of repeated simple parts and can be summed in a similar fashion.

Let us state more carefully what is meant by a 'simple part' or, as it is called in the fancy language of the literature, 'proper self-energy part' or 'irreducible self-energy part'. First we define:

Self-energy part: Any diagram without external (i.e., incoming and outgoing) lines, which can be inserted into a particle (or hole) line.

Examples:

$$(10.1)$$

(1)   (2)   (3)   (4)   (5)

Note that two little extra stumps of line are drawn on each part to show where it is to be inserted in a particle line. Equation (9.40) shows these diagrams inserted in various way into particle lines. Then we have

Proper (or 'irreducible') self-energy part: A self-energy part which cannot be broken into two unconnected self-energy parts by removing one particle or hole line.

Examples:

$$(10.2)$$

Parts which can be so broken—like diagrams 3 and 5 in (10.1)—are called 'improper' or 'reducible'.

Now in (4.67) we summed over all diagrams containing the repeated proper self-energy part and got

$$(10.3)$$

In (4.76) the sum over all repetitions of the two irreducible parts and gave

$$(10.4)$$

In (9.51), (9.52) we summed over all repetitions of three irreducible parts,

$$\text{(diagram)}, \quad \text{(diagram)} \quad \text{and} \quad \text{(diagram)} :$$

$$\text{(diagram series)} \approx \text{(diagram series)} + \cdots$$

$$= \text{(diagram)} \times \left[1 + \text{(diagram)} \times \left(\text{(diagrams)} \right) + \text{(diagram)}^2 \times \left(\text{(diagrams)} \right)^2 + \cdots \right.$$

$$= \frac{1}{\text{(diagram)}^{-1} - \left(\text{(diagrams)} \right)}$$

(10.5)

And in general it is possible to sum over all repetitions of *all* irreducible self-energy parts:

$$\text{(diagram series)} = \text{(diagram series)} + \cdots$$

$$= \text{(diagram)} \times \left[1 + \text{(diagram)} \times \left(\text{(diagrams)} + \cdots \right) \right.$$

$$+ \text{(diagram)}^2 \times \left(\text{(diagrams)} + \cdots \right)^2 + \cdots \bigg]$$

$$= \frac{1}{\text{(diagram)}^{-1} - \left(\text{(diagrams)} + \cdots \right)},$$

(10.6)

or

$$\text{(diagram)} = \frac{1}{\text{(diagram)}^{-1} - \text{(}\Sigma\text{)}},$$

(10.7)

where $\overset{|}{\textcircled{Σ}}$ is the sum of all proper (irreducible) self-energy parts or '*irreducible self-energy*':

$$\overset{|}{\underset{|}{\textcircled{Σ}}} = \text{[diagrams]} + \cdots . \tag{10.8}$$

Translated into functions with the aid of Table 9.1 this becomes:

$$G(\mathbf{k}, \omega) = \frac{1}{\omega - \epsilon_k - \Sigma(\mathbf{k}, \omega) + i\delta_k} \tag{10.9}$$

where

$$-i\Sigma(\mathbf{k}, \omega) \equiv \overset{|}{\underset{|}{\textcircled{Σ}}}. \tag{10.10}$$

For non-interacting systems, $\Sigma(\mathbf{k}, \omega) = 0$. Note that in all of these summations, $(10.3) \rightarrow (10.6)$, it was necessary to restrict the sum to just repeated *proper* parts. If we had summed over repeated *improper* parts as well, diagrams would have been counted twice, since as seen in diagrams 3, 5 of (10.1), the improper parts themselves contain repetitions.

Equation (10.7) or (10.9) is called Dyson's equation and is the basic equation from which most propagator calculations start. It is exact since all the diagrams in (9.40) are composed of either proper parts or repetitions of proper parts, and we have summed over them all. That is, the summation here is complete rather than just partial. But don't be fooled into thinking that because it is exact (10.7) is the answer to our problem! All that has been done is to sum over *repeated* proper parts; the sum (10.8) over the proper parts themselves is still left to do, and has the unfortunate quality of being in general impossible. It can, however, be evaluated to various degrees of approximation. For example, the Hartree–Fock is the lowest-order approximation for $\overset{|}{\underset{|}{\textcircled{Σ}}}$:

$$\overset{|}{\underset{|}{\textcircled{Σ}}} \approx \text{[diagrams]} . \tag{10.11}$$

It is easy to see the physical interpretation of $\Sigma(\mathbf{k}, \omega)$ by comparing the exact (10.7) with the Hartree approximation in (10.3), or the Hartree–Fock in (10.4). By analogy with the argument around (4.73), $\Sigma(\mathbf{k}, \omega)$ is a generalized 'effective field' or 'effective potential' which the particle in state **k** sees because of its interaction with all the other particles of the system. This field is of course considerably more complicated than the Hartree–Fock field because of its ω-dependence, which describes the motion of the quasi-particle cloud (see second paragraph after (4.95), also after (9.67)).

It is important to note that the form of the Dyson equation in (10.7) is only valid in the special case (with which we shall be mainly concerned) of a system with no external potential and with diagrams calculated in $(\mathbf{k}, \omega)$-space. There is, however, a more general form of the Dyson equation which holds whenever expansion (9.40) holds; the general form is (cf. (3.36'))

$$\text{(diagram)} \qquad (10.12)$$

This may be proved by iteration:

$$(10.13)$$

Equation (10.12) boils down to (10.7) in the above special case because the value of each diagram is then the algebraic product of the values of its parts; thus we have

$$\text{(diagram)}$$

or

$$G(\mathbf{k}, \omega) = G_0(\mathbf{k}, \omega) + G(\mathbf{k}, \omega) \, \Sigma(\mathbf{k}, \omega) \, G_0(\mathbf{k}, \omega), \qquad (10.14)$$

which are easily solved to yield (10.7) or (10.9). But (10.12) is also valid when the diagrams do not factor. For example, in $(\mathbf{k}, t)$-space we find

$$\tag{10.15}$$

or

$$
iG(\mathbf{k}, t_2 - t_1) = iG_0(\mathbf{k}, t_2 - t_1) +
$$

$$
+ \int \int dt' \, dt'' \, iG_0(\mathbf{k}, t_2 - t'')(-i) \, \Sigma(\mathbf{k}, t'' - t') \, iG(\mathbf{k}, t' - t_1)
\tag{10.16}
$$

which is an integral equation for G, unlike the algebraic (10.14).

[Another example of (10.12) is the case of a system with an external potential. Then it is found that

$$
iG(k_2, k_1; \omega) = iG_0(k_2, \omega) \delta_{k_2 k_1} +
$$

$$
+ \sum_{k, k'} iG_0(k_2, \omega) \delta_{k_2 k'}(-i) \, \Sigma(k', k; \omega) \, iG(k, k_1; \omega).
\tag{10.17}
$$

Note that now anomalous graphs must be included in (9.40) as mentioned on p. 141. If $G(k_2, k_1; \omega)$ is regarded as the (k_2, k_1)th element of a matrix then (10.17) may be written as a matrix equation, for which form (10.7) holds (Luttinger (1960b)).]

The self-energy diagrams may be evaluated in a straightforward way by using dictionary Table 9.1 (see for example (10.29)). However, we then have to perform all the integrations over intermediate energy parameter. A short-cut method which avoids this involves a new dictionary similar to that used for the ground state energy diagrams, Table 12.1. The rules are described in Thouless (1964) (p. 65), Klein (1962) and Luttinger (1961).

10.3 Quasi particles in low-density Fermi system (ladder approximation)

As an example of the calculation of the proper self-energy, we will describe briefly the theory of Galitski (Fetter and Walecka (1971), p. 128 ff.) for a system of particles interacting by means of short-range repulsive forces having range a, and with average distance between particles r_0. By 'low density' is meant that $a/r_0 \ll 1$. This can also be stated in terms of k_F since n, the number of particles/cm^3 is equal to $\frac{1}{3}\pi^{-2}k_F^3$ (see (9.44)) and $n = 1/r_0^3$ so $1/r_0 \sim k_F$. Hence the low-density criterion is that $k_F a \ll 1$. Such a theory can be applied in a qualitative way to the case of nuclear matter (see §12.4), where $a/r_0 \sim \frac{1}{3}$, provided we neglect the attractive part of the nuclear potential. It does not hold for He^3, where $a/r_0 > 1$.

Let us first analyse the self-energy diagrams to see which are most important. Regarding the diagrams as time-ordered, we observe that whenever there is a hole line labelled, say, **p**, in a diagram, there is an associated $\int d^3\mathbf{p}$ over all $|\mathbf{p}| < k_F$. Particle lines, on the other hand, have $\int d^3\mathbf{p}$ over $|\mathbf{p}| > k_F$. Now as mentioned above, $n \sim k_F^3$, so low density means small k_F. Thus in the low-density case, the contribution from the hole line integrals will be very small compared with those from the particle lines. (A more careful analysis shows that the criterion for hole contribution $\ll$ particle contribution is that $k_F \ll 1/a$, i.e., just the low density criterion stated above. This is illustrated in detail by exercise 10.7.) Therefore, in series (10.8), the dominant diagrams will be those with the *least* number of hole lines, i.e., one. For example, the last three diagrams in (10.8) may be neglected in the low-density limit since they have two hole lines. We find for the sum of graphs containing just one hole line the following (note that these graphs are time-ordered!):

$$(10.18)$$

These are the 'ladder graphs'.

The sum of ladder graphs may be carried out with the aid of the so-called 't' or 'K'-matrix. This is defined by the time-ordered graphs:

$$(10.19)$$

The K-matrix obeys the Dyson-like integral equation

$$(10.20)$$

as is seen immediately by iterating (just as was done in (10.13)). Translation of (10.20) gives (note that the K-diagram is translated as $-iK$):

$$K(\mathbf{p}'\,\epsilon', \mathbf{p}\epsilon; \mathbf{q}, \omega) = V_{p-p'} + i \int \frac{d^3\mathbf{p}''\,d\epsilon''}{(2\pi)^4}\, V_{p''-p'}\, G_0^+(\mathbf{p}'', \epsilon'')\, G_0^+(\mathbf{q}-\mathbf{p}'', \omega-\epsilon'')$$

$$\times\, K(\mathbf{p}'', \epsilon'', \mathbf{p}, \epsilon; \mathbf{q}, \omega). \tag{10.21}$$

This equation is solved for two special potentials in exercises 10.4, 10.7.

The K-matrix may be generalized to include hole–hole scattering by replacing G_0^+ by G_0. In the low-density case, the contribution from hole–hole scattering is negligible. (see exercise 10.7.) For further generalization, see §10.6.

If (10.19) is substituted in (10.18) we obtain (note $(-)$ for fermion loops):

$$\tag{10.22}$$

We first solve approximately for K, using (10.21). The result is substituted into (10.22), and the value for Σ is put into the Dyson equation. This yields a propagator having the quasi particle form (8.37), for k near k_F. The results for the quasi particle effective mass and lifetime near the Fermi surface are

$$m^* = m\left[1 + \frac{8}{15\pi^2}(7\ln 2 - 1)\, k_F^2 a^2\right]; \quad \frac{1}{\tau_k} = \frac{1}{\pi} k_F^2 a^2 (k - k_F)^2. \tag{10.23}$$

Note that the result for the lifetime agrees with (8.30) and (8.29).

10.4 Quasi particles in high-density electron gas (random phase approximation)

The electron gas was introduced in §4.9 as a theoreticians 'dream metal' consisting of N electrons moving against a smeared-out positive charge background. At zero temperature, the gas is characterized by a single parameter, r_s, which is roughly the average distance between electrons. More precisely, r_s is given by

$$\frac{1}{n}\left(\frac{\mathrm{cm}^3}{\mathrm{electron}}\right) = \tfrac{4}{3}\pi(r_s a_0)^3 \tag{10.24}$$

where n = electron density, and a_0 = Bohr radius = $\hbar^2/me^2$. Several different regions may be distinguished, as shown in Fig. 10.1.

So far, two regions have been tackled with success. The first is the high density one where the kinetic energy (KE) of the electrons is much higher than their mutual potential energy (PE), so the latter acts as a relatively small perturbation. The other is the low-density region, where the PE is so much greater than KE that the electron density becomes non-uniform, and the points of maximum density form a body-centred cubic lattice called the Wigner lattice.

It is important to observe here that the low density region of a system with long-range interaction, like the electron gas, is physically quite different from the low-density region of a system with short-range interaction, like nuclear matter. Hence ladder approximation is not applicable to a low-density electron gas.

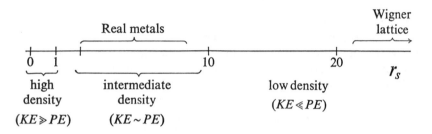

Fig. 10.1 *Classification of Density Regions in the Electron Gas*

The interesting intermediate density region where real metals lie, i.e., $r_s =$ 1·8–5·6, has turned out to be quite stubborn. To paraphrase E. Montroll, the physicist here is like a mouse nibbling away at the ends of a stale cheese, unable to get at the good part in the middle. In this section, we will have a look at the left-hand end of the cheese (high-density limit) using the many-body technique. Those who are interested in seeing what progress has been made in the mid-cheese region (metallic densities) should have a look at Hedin and Lundqvist (1969), Vashishta and Singwi (1973), Fishlock and Pendry (1973), and the references contained in these articles.

The Hamiltonian for the electron gas in the smeared out positive background is given by

$$H = \sum_k \epsilon_k c_k^\dagger c_k + \tfrac{1}{2} \sum_{klmn} V_{klmn} c_l^\dagger c_k^\dagger c_m c_n$$

$$+ H_{\substack{\text{positive}\\ \text{background}}} + H_{\substack{\text{electron-positive}\\ \text{background}}} \qquad (10.25)$$

where $V_{klmn} = V_q$ is in (7.71). This may be simplified since the V_0 part of the second sum cancels the last two terms, as follows: The V_0 may be evaluated most easily by (7.69), dropping spin factors for simplicity:

$$V_0 = V_{klkl} = \frac{1}{\Omega^2} \int d^3 \mathbf{r} \, d^3 \mathbf{r}' \frac{e^2}{|\mathbf{r} - \mathbf{r}'|}. \qquad (10.26)$$

Thus the V_0 term becomes:

$$\frac{V_0}{2}\sum_{kl} c_l^\dagger c_k^\dagger c_k c_l = \frac{V_0}{2}\sum_{kl} c_l^\dagger c_l c_k^\dagger c_k - \frac{V_0}{2}\sum_{k} c_k^\dagger c_k$$

$$= \frac{V_0}{2}(N^2 - N) = \frac{1}{2}\frac{N^2 e^2}{\Omega^2}\int \frac{d^3 \mathbf{r}\, d^3 \mathbf{r}'}{|\mathbf{r} - \mathbf{r}'|} \qquad (10.27)$$

(Note: we have used (8.13) in deriving this result.) Also, we have employed (7.30) and used the fact that $N^2 \gg N$ in a large system. Equation (10.27) is just the (infinite) self-energy of a static uniform negative charge distribution. The static positive background is exactly equal to (10.27) while the electron-background term gives a contribution which is twice (10.27) and of opposite sign. That is, we have a uniform positive charge with an equal uniform negative charge on top of it, and the two cancel each other. Hence (10.25) may be re-written

$$H = \sum_k \epsilon_k c_k^\dagger c_k + \tfrac{1}{2}\sum_{\substack{q,m,n \\ (q \neq 0)}} V_q c_{n-q}^\dagger c_{m+q}^\dagger c_m c_n. \qquad (10.28)$$

We shall now look for quasi particles in the electron gas, using Dyson's equation (10.7). A typical irreducible self-energy part in (10.8) is:

$$= (-1)\sum_{q,p}\int \frac{d\epsilon\, d\beta}{(2\pi)^2} iG_0(\mathbf{k}-\mathbf{q}, \omega-\epsilon) \times iG_0(\mathbf{q}+\mathbf{p}, \beta+\epsilon) \times$$

$$\times iG_0(\mathbf{p}, \beta) \times \frac{(4\pi e^2)^2}{\Omega^2 q^4}. \qquad (10.29)$$

Changing from a sum over $\mathbf{q}$ to an integral reveals that the above expression diverges at $\mathbf{q}=0$ because of the q^4 in the denominator (see argument leading to (12.19)) so that

$= \infty . \qquad (10.30)$

A look at the higher orders shows the same behaviour: most of the diagrams diverge. This appears to be a slightly discouraging start. However, the situation is saved by the partial summation method as follows: First of all, we observe that each term in the self-energy is proportional to some power of r_s ($\sim 1/k_F$—see §10.3). To see this, consider the second-order self-energy term in (9.66). To make the integral dimensionless, express quantities in units of k_F, i.e., set $q = q'k_F$, $\epsilon_l = \epsilon'_l k_F^2$, etc. Then we find that (9.66) $\sim k_F^3 \times k_F^3 \times k_F^{-4} \times k_F^{-2} = 1$, so that (9.66) has no explicit dependence on k_F. Now look at a third-order term—e.g., the fifth or sixth diagram on the right of (10.8). After doing the frequency integrals, we find, compared with the second-order term, an extra V_q ($\sim k_F^{-2}$), an extra energy denominator ($\sim k_F^{-2}$) and an extra integral over k ($\sim k_F^3$), so that we have an extra factor $k_F^{-1} \sim r_s$. Thus all third-order terms are $\sim r_s$. In general, any nth-order term $\sim r_s^{n-2}$.

Next, we arrange the diagrams according to degree of divergence (= number of factors q^2 in the denominator of the integrand) and dependence on r_s:

$$(10.31)$$

The first-order bubble has $\mathbf{q}=0$, so it vanishes by (10.28); that is, the bubble is cancelled by the positive background. We see that the digarams can be arranged so that in each order of divergence they form a power series in r_s (see along diagonal lines). Thus, for small r_s, i.e., high-density limit, the dominant infinite terms are just those of lowest order in r_s. Hence in the high density limit, the self-energy series is just the sum

$$= \text{finite} \ + \ \infty \ + \ \infty \ + \ \infty \ + \cdots \tag{10.32}$$

i.e., the sum over all diagrams of the repeated pair-bubble or 'ring' type. The remarkable thing is that this sum over an infinite number of infinite terms can actually be carried out, and it gives a finite result! Approximation (10.32) for the self-energy is called the 'random phase approximation' or 'RPA', for historical reasons.

The sum over rings is easy. Factoring out a free propagator from each diagram in (10.32) (this is not quite straightforward for the oyster, since it is a special case, by the rules of Table 9.1; however, it is legitimate) gives

$$\tag{10.33}$$

The double wiggle is the 'effective interaction' in RPA, which was mentioned in §4.9, and interpreted as a 'screened' interaction between two particles:

$$\tag{10.34}$$

Diagrams of form (1), (2), (3), ..., having one interaction line entering and one leaving are called '*polarization diagrams*'. The reason for this is that they show how the interaction causes the medium to become 'virtually polarized' in all possible ways. For example, if we regard diagram (2) in (10.34) as drawn in $(\mathbf{r}, t)$-space, with the lower vertex of the 'pair bubble' at $\mathbf{r}_1, t_1$ and the upper at $\mathbf{r}_2, t_2$, then for $t_1 < t < t_2$ there are a negative electron and a positive hole separated in space forming a 'virtual dipole'. (Of course, the position co-ordinates of the particle and hole are no longer sharp for $t > t_1$ but this makes no difference—the virtual dipole just becomes 'fuzzy'.)

Equation (10.34) may be written in functional form as

$$V_{\text{eff(RPA)}}(\mathbf{q}, \omega) = \frac{V_q}{1 + V_q \pi_0(\mathbf{q}, \omega)} \equiv \frac{V_q}{\epsilon_{\text{RPA}}(\mathbf{q}, \omega)} \qquad (10.35)$$

where

$$-i\pi_0(\mathbf{q}, \omega) \equiv \begin{matrix} \mathbf{k} + \mathbf{q} \\ \epsilon + \omega \end{matrix} \bigcirc \mathbf{k}, \epsilon \;. \qquad (10.36)$$

This has the form of an interaction taking place between two charges in a dielectric, with

$$\epsilon_{\text{RPA}}(\mathbf{q}, \omega) = 1 + V_q \pi_0(\mathbf{q}, \omega) \qquad (10.37)$$

being the frequency-dependent or so-called 'generalized' dielectric constant. Of course, this is no coincidence. The dielectric properties of a medium arise just because of the polarization of the medium by a field, and (10.34) is just the sum of diagrams representing the polarization of the electron gas by the field of one of the electrons in the gas itself. Note that $V_{\text{eff}}(\mathbf{q}, \omega)$ depends on ω, unlike the bare V_q. If V_{eff} is Fourier transformed to $(\mathbf{q}, t)$-space, it will thus be a time-dependent interaction; this is due to the inertia of the polarization charge.

We may evaluate $\pi_0(\mathbf{q}, \omega)$ by using the rules for graphs, Table 9.1, yielding

$$-i\pi_0(\mathbf{q}, \omega) = 2 \times (-1) \int \frac{d^3\mathbf{k}\, d\epsilon}{(2\pi)^4} \frac{i}{\omega + \epsilon - \epsilon_{k+q} + i\delta_{k+q}} \times \frac{i}{\epsilon - \epsilon_k + i\delta_k}, \qquad (10.38)$$

where the factor of 2 comes from the sum over spins and the (-1) from the fermion loop. The integral over ϵ is in (9.55–62) and $\int d^3\mathbf{k}$ is done in §10.7. In the limit when $\omega = 0$ and $\mathbf{q}$ is small, it is found that

$$\pi_0(q \ll k_F, \omega = 0) = \frac{\lambda^2}{4\pi e^2}, \quad \lambda^2 = \frac{6\pi n e^2}{\epsilon_F} = \frac{4me^2 k_F}{\pi} = \left(\frac{2}{\pi}\right)\left(\frac{4}{9\pi}\right)^{1/3} r_s k_F^2 \qquad (10.39)$$

where $n = $ electron density $= \frac{1}{3}\pi^{-2}k_F^3$, $\epsilon_F = k_F^2/2m$.

It is now possible to calculate $V_{\text{eff}}(\mathbf{q}\text{ small},0)$. Setting $\Omega = 1$ and omitting spins for simplicity in (7.71), we have

$$V_q = \frac{4\pi e^2}{q^2}. \qquad (10.40)$$

Substituting this together with (10.39) into (10.35) yields

$$V_{\text{eff(RPA)}}(\text{small } \mathbf{q}, 0) = \frac{4\pi e^2}{q^2 + \lambda^2} \qquad (10.41)$$

(valid in the limit $(\lambda/k_F)^2 \ll 1$, i.e., $r_s \ll 1$. See 10.82). Hence, assuming (10.41) holds for all q,

$$V_{\text{eff}}(r) \propto \frac{e^2}{r} e^{-\lambda r} \qquad (10.42)$$

which is a shielded Coulomb interaction. (See (10.83), for a more correct expression!) This reveals the physical significance of the effective interaction: the bare interaction (10.40) virtually polarizes the medium, and the polarization cloud in turn shields the bare interaction converting it into the much weaker effective interaction. (This effective interaction turns out to be just the effective interaction between quasi particles (see Fig. 0.12), as discussed in Falicov (1961).)

We can now go on to the evaluation of $\Sigma_{\text{RPA}}(\mathbf{k}, \omega)$ as it appears in (10.33). Translating this into functions with the aid of (10.35) gives

$$-i \sum_{\text{RPA}} (\mathbf{k}, \omega) = \sum_{\mathbf{q}} \int \frac{d\gamma}{2\pi} [(-i) V_{\text{eff(RPA)}}(\mathbf{q}, \omega)] [i G_0(\mathbf{k}-\mathbf{q}, \omega-\gamma)]$$

$$= \int \frac{d^3 q}{(2\pi)^3} \int \frac{d\gamma}{2\pi} \frac{4\pi e^2}{q^2 \epsilon_{\text{RPA}}(\mathbf{q}, \gamma)} \times \frac{1}{\omega - \gamma - \epsilon_{k-q} + i\delta_{k-q}}. \qquad (10.43)$$

Despite the fact that, excepting for the oyster, all diagrams in (10.32) which we added to get this result were infinite, (10.43) is finite! This is due to the fact that unlike the bare interaction which goes to ∞ as $\mathbf{q} \to 0$, the effective interaction remains finite as $\mathbf{q} \to 0$, as shown in (10.41).

Let us first examine (10.43) in the simple limit where we use the static approximation (10.41) for V_{eff}. With the aid of Table 9.1 we find that

$$-i \sum_{\text{RPA}} (\mathbf{k}, \omega) = -i \int \frac{d^3 q}{(2\pi)^3} \left(\frac{4\pi e^2}{q^2 + \lambda^2}\right) \int \frac{d\gamma}{2\pi} i G_0(\mathbf{k}-\mathbf{q}, \omega-\gamma)$$

$$= -i \int\limits_{|\mathbf{k}-\mathbf{q}|<k_F} \frac{d^3 q}{(2\pi)^3} \left(\frac{4\pi e^2}{q^2 + \lambda^2}\right)(-1)$$

$$= -i \int\limits_{|\mathbf{l}|<k_F} \frac{d^3 l}{(2\pi)^3} \frac{4\pi e^2}{[(\mathbf{k}-\mathbf{l})^2 + \lambda^2]}. \qquad (10.44)$$

Placing this in the Dyson equation (10.9) reveals that we have a quasi particle of energy

$$\epsilon_k' = \frac{k^2}{2m} - \int\limits_{|l|<k_F} \frac{d^3 l}{(2\pi)^3} \frac{4\pi e^2}{[(\mathbf{k}-\mathbf{l})^2 + \lambda^2]}. \tag{10.45}$$

The effective mass for this case, in the (rather unrealistic) large λ limit, was given in (4.105).

Evaluating (10.43) using the full frequency-dependent $V_{\text{eff}}(\mathbf{q}, \omega)$ is an enormously complicated business which will not be discussed here (see Pines (1961), p. 63, for reference). When the final result for Σ is substituted back into the Dyson equation, it is found that $G(\mathbf{k}, \omega)$ has the quasi particle form (8.37) for k near k_F, with

$$\epsilon_k' = \frac{k^2}{2m} - 0.166 r_s (\ln r_s + 0.203) \frac{kk_F}{2m} + \text{constant}$$

$$\frac{1}{\tau_k} = 0.252 r_s^{\frac{1}{2}} \frac{(k-k_F)^2}{2m}. \tag{10.46}$$

Using (4.94) yields for the effective mass near the Fermi surface

$$m^* = m[1 - 0.083 r_s (\ln r_s + 0.203)]^{-1}. \tag{10.47}$$

Further, the criterion (8.38) for the existence of quasi particles is satisfied since

$$\lim_{k \to k_F} \frac{\tau_k^{-1}}{\epsilon_k' - \epsilon_F'} \propto \frac{(k-k_F)^2}{(k-k_F)} = (k-k_F) \to 0. \tag{10.48}$$

The result (10.46) is the *quasi electron* mentioned in §0.2, consisting of the bare electron plus positive screening cloud.

10.5 The general 'dressed' or 'effective' interaction

The obvious way to generalize the concept of the 'effective interaction in RPA' is to include *all* possible polarization diagrams (diagrams with one interaction wiggle entering and one leaving) in the partial sum for V_{eff}, instead of just the 'pair loops' of (10.34). Thus, we get the general effective (or 'dressed' or 'renormalized') interaction:

$$- i V_{\text{eff}}(\mathbf{q}, \omega) \equiv \text{\small\ \ } = \text{\small\ \ } + \text{\small\ \ } + \text{\small\ \ }$$

$$+ \text{\small\ \ } + \text{\small\ \ } + \cdots \tag{10.49}$$

It is possible to sum (10.49) in the same manner that the propagator series (9.40) was summed to yield the Dyson equation (10.7). We first define the 'polarization part' by

Polarization part: Any diagram without external interaction lines which may be inserted into an interaction line.

For example:

$$\text{(1)} \quad \text{(2)} \quad \text{(3)} \quad \text{(4)} \quad \text{(5)} \quad \text{(6)} \tag{10.50}$$

and the irreducible polarization part by

Proper (or 'irreducible') polarization part: A polarization part which cannot be broken into two unconnected polarization parts by removing one interaction line, such as:

$$\cdots. \tag{10.51}$$

(Note that ⟨⟩ is not a polarization part!) According to this, diagrams (2) and (5) in (10.50) are reducible polarization parts; the others are irreducible. Representing the sum over all irreducible polarization parts by

$$- i \pi(\mathbf{q}, \epsilon) \equiv \left(\pi \right) = \left(\ \right) + \left(\ \right) + \left(\ \right) + \left(\ \right) + \cdots \tag{10.52}$$

it is easy to derive the Dyson-like equation

$$(10.53)$$

or

$$(10.54)$$

analogous to (10.7) or (10.12). The functional form of this is (for a system with no external potential, and with diagrams calculated in $(\mathbf{q}, \omega)$-space):

$$V_{\text{eff}}(\mathbf{q}, \omega) = \frac{V_q}{1 + V_q \pi(\mathbf{q}, \omega)} \equiv \frac{V_q}{\epsilon(\mathbf{q}, \omega)}. \qquad (10.55)$$

Equations (10.53), (10.55) are the generalizations of the RPA results (10.34), (10.35).

Note that the RPA effective interaction, which uses ⟨⟩ as the approximation for ⟨π⟩, is valid only if the exchange interaction is much smaller than the direct one. If exchange interaction is important, we must include in π diagrams like (3) in (10.50), since it is the exchange diagram corresponding to (2) in (10.50). Thus we must have the following approximation for π:

$$(10.55')$$

Using the effective interaction, the appearance of the expansion for the proper self-energy may be enormously simplified. Consider some arbitrary irreducible self-energy part, say the second-order exchange diagram

$$(10.56)$$

(These are all the same diagram—see (9.35).) In the expansion for Σ, besides this diagram, there will occur a class of diagrams just the same as (10.56) except that one of the interaction wiggles—say the upper one—has been replaced by something more complicated; this subset may be collected in brackets thus:

(10.57)

Since the value of a diagram for specified momenta and frequencies is the product of the values of its parts, we can factor out the unchanged part in each diagram in this subset, obtaining:

(10.58)

where we have used (10.49). We can also sum over all inserts in the bottom wiggle of the diagram in (10.58). In the same way, we may carry out all possible sums like that in (10.57) by just replacing the true interaction by the effective interaction in the first diagram of each such series. Thus,

(10.59)

and so on, leading to

(10.60)

That is, we now have a much simpler series in which no interaction lines contain any polarization parts, and bare interaction lines have become dressed. It must be emphasized that when drawing such diagrams, no polarization parts may be included, since these have already been included in the sum which replaced $\sim\!\!\sim\!\!\sim$ by $\approx\!\!\approx\!\!\approx$. Thus, a diagram like

$$(10.61)$$

is *strictly illegal* since these diagrams have already been included in

$$(10.62)$$

10.6 The scattering amplitude

It is possible to go one step further than we did in §10.3 and §10.5, and generalize the concept of the K-matrix or the effective interaction to include *all* processes in which two particles interact with each other in the presence of the many-body system. This generalized K-matrix, or generalized effective interaction is usually called the 'scattering amplitude' or 'quasi-particle scattering amplitude' (some authors, e.g., Landau, call it the 'vertex part', but we reserve this term for an entirely different entity—see §11.4). It includes the ordinary effective interaction and the K-matrix as special cases, and it is closely related to the two-particle propagator (see §13.6).

The scattering amplitude is given by the series:

$$(10.63)$$

Note that although there may be any number of self-energy parts in the internal lines of each diagram, there are no self-energy parts in the incoming or outgoing lines. Thus the scattering amplitude is really the amplitude for scattering of two quasi particles, since if we attach incoming and outgoing quasi-particle lines (i.e., exact propagators) to Γ, these lines will themselves include all possible self-energy parts.

Observe that viewed in **r**-space, Γ is a *non-local* interaction, since it is not an interaction between a particle at $\mathbf{r}_1$ and another at $\mathbf{r}_2$, but rather involves two particles starting their interaction at $\mathbf{r}_1$ and $\mathbf{r}_2$ respectively, but ending it at two other points, $\mathbf{r}_3$ and $\mathbf{r}_4$.

It is useful to note that the proper self-energy may be expressed in terms of the scattering amplitude thus:

$$(10.64)$$

[10.7 Evaluation of the pair bubble; Friedel oscillations]

We will first find the pair bubble, $\pi_0(\mathbf{q}, \omega)$, in the limit $q \ll k_F$, arbitrary ω. (The evaluation for all $\mathbf{q}$, ω, is in Fetter and Walecka (1971), pp. 158–163.) We've already done part of the work involved, in (9.62). If in (9.62) we change the label α to ω, and change the variables in π_0^B to make the integration region in π_0^B the same as in π_0^A thus:

$$\text{first set } \mathbf{l} + \mathbf{q} = \mathbf{m}, \text{ then set } \mathbf{m} = -\mathbf{l} \qquad (10.65)$$

we obtain:

$$\pi_0(\mathbf{q}, \omega) = -2 \int\limits_{\substack{|\mathbf{l}| < k_F \\ |\mathbf{l}+\mathbf{q}| > k_F}} \frac{d^3 l}{(2\pi)^3} \left\{ \frac{1}{\omega - \epsilon_{l+q} + \epsilon_l + i\delta} - \frac{1}{\omega - \epsilon_l + \epsilon_{l+q} - i\delta} \right\} \qquad (10.66)$$

or (note that the vector **q** defines the z-direction):

$$\pi_0(\mathbf{q}, \omega) = \frac{+2}{(2\pi)^3} \int\limits_0^{2\pi} d\phi \int\limits_{-1}^{+1} \underbrace{d(\cos\theta)}_{x} \int\limits_{\substack{|\mathbf{l}| < k_F \\ |\mathbf{l}+\mathbf{q}| > k_F}} dl\, l^2 \times$$

$$\times \underbrace{\left\{ \frac{1}{\omega - \dfrac{1}{2m}[2lqx + q^2] + i\delta} - \frac{1}{\omega + \dfrac{1}{2m}[2lqx + q^2] - i\delta} \right\}}_{F(l,q,x)}. \qquad (10.67)$$

The region of integration is shown in Fig. 10.2. (This figure is cylindrically symmetric about the z-axis.) Thus, $\int dl$ goes from $k_F - qx$ to k_F, where $x = \cos\theta$. For $q \ll k_F$, the region of integration is a thin shell about the *top* half

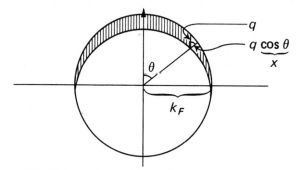

Fig. 10.2. *Region of Integration (Shaded) for* (10.67)

of the Fermi sphere. We may thus set $l = k_F$ in the integrand and drop the q^2 terms in the denominator. Hence the $\int dl$ becomes:

$$\int dl = \int_{k_F - qx}^{k_F} dl\, l^2\, F(lqx) = qx k_F^2\, F(k_F qx) \qquad (10.68)$$

$$= qx k_F^2 \left\{ \frac{1}{\omega - \dfrac{k_F qx}{m} + i\delta} - \frac{1}{\omega + \dfrac{k_F qx}{m} - i\delta} \right\}. \qquad (10.69)$$

Now let

$$\zeta = \frac{m\omega}{k_F q}. \qquad (10.70)$$

This yields

$$\int dl = m x k_F \left\{ \frac{1}{\zeta - x + i\delta} - \frac{1}{\zeta + x - i\delta} \right\}. \qquad (10.71)$$

Substituting this in (10.67) and remembering that by Fig. 10.2 $\theta = 0 \to \pi/2$ so $x = +1 \to 0$, and integrating over ϕ, yields:

$$\pi_0(q, \omega) = \frac{4\pi m k_F}{(2\pi)^3} \int_0^1 dx \cdot x \left\{ \frac{1}{\zeta - x + i\delta} - \frac{1}{\zeta + x - i\delta} \right\}. \qquad (10.72)$$

Applying (3.76) to the first integral gives

$$\int_0^1 dx \frac{x}{\zeta - x + i\delta} = P \int_0^1 dx \frac{x}{\zeta - x} - i\pi \int_0^1 dx\, x\, \delta(\zeta - x) \qquad (10.73)$$

$$P \int_0^1 dx \frac{x}{\zeta - x} = -P \int_0^1 dx \frac{x - \zeta}{x - \zeta} - \zeta P \int_0^1 \frac{d(x - \zeta)}{x - \zeta} = -1 - \zeta \ln \left| \frac{1 - \zeta}{-\zeta} \right| \qquad (10.74)$$

$$-i\pi \int_0^1 dx\, x\, \delta(\zeta - x) = -i\pi\zeta, \quad \text{for } 0 < \zeta < 1$$
$$= 0, \; \zeta \text{ outside } (0, 1). \qquad (10.75)$$

Similarly, for the second integral we get:

$$-\int_0^1 dx \frac{x}{\zeta + x - i\delta} = -1 + \zeta \ln \left| \frac{1 + \zeta}{\zeta} \right| \underbrace{- i\pi|\zeta|}_{\text{for } -1 < \zeta < 0,} \qquad (10.76)$$
$$= 0 \text{ otherwise.}$$

Adding the two results yields for π_0:

$$\pi_0(q, \omega) = \frac{mk_F}{\pi^2} \left[1 - \frac{\zeta}{2} \ln \left| \frac{1 + \zeta}{1 - \zeta} \right| + i \frac{\pi}{2} |\zeta|\, \theta(1 - |\zeta|) \right] \qquad (10.77)$$

where $\theta(x > 0) = 1$, and $\theta(x < 0) = 0$.

We now examine the static limit, i.e., $\omega = 0$ so $\zeta = 0$. Then

$$\pi_0(q, \omega) = \frac{mk_F}{\pi^2} = \frac{3n}{2\epsilon_F}, \qquad (10.78)$$

(n = electron density), which confirms (10.39). The frequency-dependent $\pi_0(q, \omega)$ is discussed in chapter 13, in connection with plasmons.

It is also straightforward to evaluate $\pi_0(\mathbf{q}, \omega)$ for $\pi = 0$ and arbitrary $\mathbf{q}$. Starting with (10.66) and writing the l-integration limits in terms of the step function θ, and applying (3.76), we find for the real and imaginary parts:

$$\text{Re}\,\pi_0(\mathbf{q}, 0) = -4P \int \frac{d^3 l}{(2\pi)^3} \theta(k_F - |\mathbf{l}|)[1 - \theta(k_F - |\mathbf{l} + \mathbf{q}|)] \frac{1}{\epsilon_l - \epsilon_{l+q}} \qquad (10.79)$$

$$\text{Im}\,\pi_0(\mathbf{q}, 0) = 4\pi \int \frac{d^3 l}{(2\pi)^3} \theta(k_F - |\mathbf{l}|)[1 - \theta(k_F - |\mathbf{l} + \mathbf{q}|)] \delta(\epsilon_l - \epsilon_{l+q}). \qquad (10.80)$$

The imaginary part is obviously zero since ϵ_l cannot be equal to ϵ_{l+q}. The second term of $\mathrm{Re}\,\pi_0$—call it A—can be seen to vanish by making the transformation $\mathbf{l}' = -(\mathbf{l}+\mathbf{q})$ and noting that this yields $A = -A$, so $A = 0$. We have left

$$\mathrm{Re}\,\pi_0(\mathbf{q},0) = +4mk_F P \int \frac{d^3\mathbf{l}}{(2\pi)^3} \frac{\theta(1-|\mathbf{l}|)}{\mathbf{q}\cdot\mathbf{l}+\frac{1}{2}\mathbf{q}^2} \tag{10.81}$$

where all vectors are measured in terms of k_F. The integration is easy and yields

$$\mathrm{Re}\,\pi_0(\mathbf{q},0) = \frac{mk_F}{2\pi^2}\left[1 - \frac{1}{q}(1-\tfrac{1}{4}q^2)\ln\left|\frac{1-\frac{1}{2}q}{1+\frac{1}{2}q}\right|\right]. \tag{10.82}$$

The singularity (infinite slope) of (10.82) at $q=2$ (i.e., at $2k_F$) has extremely important consequences in the electron gas. When $\mathrm{Re}\,\Pi_0(\mathbf{q},0)$ is placed in the expression for the effective interaction (10.35), it is found that in **r**-space for large **r**:

$$V_{\mathrm{eff}}(r) \propto \frac{\cos(2k_F r)}{r^3} \tag{10.83}$$

(see Fetter and Walecka, (1971) p. 179). This is a much longer range interaction than our previous result (10.42). The oscillations of wavevector $2k_F$ are known as *Friedel oscillations* and have been observed experimentally.

Further reading

Fetter and Walecka (1971), chap. 3, pp. 105–111 chap. 4.
Schultz (1964), chap. 3, pp. 62–8, 91–105.
Pines (1961), chap. 2, pp. 52–4; chap. 3, pp. 55–69.
Pines (1963), chap. 3.
Thouless (1972).
Thouless (1964).
Abrikosov (1965), chap. 2.
Schrieffer (1964a), chap. 6, pp. 137–48.
Klein (1962).
Kittel (1963), chap. 5.
Bjorken (1965), p. 284 ff.   } Renormalization in quantum electrodynamics
Schweber (1961), pp. 607–15 $\int$ —basic definitions

Exercises

10.1 Which of the following diagrams are self-energy parts? Of these which are proper and which are improper?

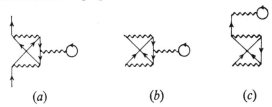

(a)   (b)   (c)

10.2 Write Dyson's equation diagrammatically (in $\mathbf{k}, \omega$-space), together with an expression for the proper self-energy, for a system of non-interacting fermions in an external potential. Translate the result into functions. (Hint: use (9.30), together with considerations similar to those in (10.12, 13, 17).)

10.3 Draw the first few terms in the perturbation expansion of the propagator in RPA.

10.4 Suppose we have a system of fermions interacting by means of a repulsive potential of the form $A\delta(\mathbf{r}_i - \mathbf{r}_j)$. Using the result of Ch. 7, Ex. 7.6, find an explicit expression for the K-matrix. (Hint: Show first that in this case, K is a function only of $\mathbf{q}, \omega$.)

10.5 Which of the following diagrams are polarization parts? Of these which are proper and which are improper?

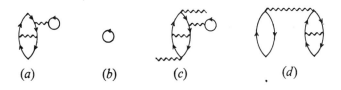

(a)   (b)   (c)   (d)

10.6 Write the approximation to the proper polarization part sum (10.55′) in terms of the K-matrix, and translate the result into functions. (Note: You will need the 'particle–hole' K-matrix, which is just (10.19) but with one side of the ladder a hole line.)

10.7 (a) Solve the generalized K-matrix equation [i.e., (10.21) with G_0^+ replaced by G_0] for K when $V_{p-p'}$ is approximated by a factorizable potential, i.e., $V_{p-p'} \approx A u_p u_{p'}$. (Hint: Try guessing the form of K. If you get stuck, go back to (10.19) and carry out the sum directly.)

(b) In the integral which occurs in the solution for K in (a)—call this integral $I(\mathbf{q}, \omega)$—carry out the integration over frequency ϵ''. Show that I is the sum of two contributions: I^+ coming from G_0^+ (i.e., particles) and I^- coming from G_0^- (i.e., holes).

(c) Evaluate I^+ and I^- in (b) in the case $\omega = 0$, $\mathbf{q} = 0$. Take $u_p = 1$ for $0 < |\mathbf{p}| < w$ ($w > k_F$) and $u_p = 0$ for $|\mathbf{p}| > w$. Use theorem (3.76).

(d) Show that in the case of low density, $I^+ \gg I^-$, i.e., contribution from particle lines much greater than that from hole lines.

10.8 Consider two electrons in a cubic box with sides of length L. Obtain *order of magnitude* expressions for the kinetic and potential energy of the system in terms of r_s (see (10.24)) and use these to find a criterion for the high and low density cases in terms of r_s.

10.9 Verify (10.79)–(10.82) in detail. Write out $V_{eff}(\mathbf{q},0)$ for this case in terms of λ (see (10.39)) and find the limiting behaviour for $q \ll 2$, $q \gg 2$ ($\mathbf{q}$ in units of k_F). What is dV_{eff}/dq at $q = 2$?

Chapter 11

Self-Consistent Renormalization and the Existence of the Fermi Surface

11.1 Dressed particle and hole lines, or 'clothed skeletons'

We saw in the last chapter how the series for the propagator could be beautified by expressing it in terms of the proper self-energy, Σ, and then writing Σ in terms of the dressed or 'effective' interaction. Now we are going to do still another face-lifting operation on Σ by partially summing over all diagrams in which there are propagator lines containing self-energy parts. This produces a series in which no propagator lines have self-energy parts and all free propagator lines have been replaced by clothed propagators. The result is called 'self-consistent renormalization'. It will be used to derive the conditions for the existence of quasi particles, and to demonstrate how it is possible to have a sharp Fermi surface in a strongly interacting system. At the end of the chapter, we will discuss a further simplification of the propagator expansion by means of a partial sum over so-called 'vertex parts'.

In the series for Σ, there will be subsets of diagrams like

$$(11.1)$$

in which each of the propagator lines has more and more self-energy parts inserted into it. Analogous to (10.59), it is easily seen that the sum in the brackets in (11.1) is carried out by just replacing the free propagators by the exact propagator in the first diagram, thus:

$$(11.2)$$

The first diagram of such a series is often called a 'skeleton', so that the renormalization process here consists of putting self-energy flesh on the bones of the skeleton—a sort of butchering in reverse.

This trick can be performed on the series as a whole, leading to the simplified series

$$(11.3)$$

(1)   (2)   (3)   (4)

Again, it must be remembered that no diagram may be counted twice—thus

is illegal since all these diagrams are already included in diagram (3) of (11.3).

Of course, the exact propagators used in (11.3) are not known—they must be obtained from the Dyson equation (10.7). This means that (11.3) and (10.7) must be regarded as simultaneous equations to be solved 'self-consistently' for the exact propagator. That is, we first calculate (11.3) using bare propagators, then substitute the result into (10.7) to get the first approximation for the clothed propagator, then re-calculate (11.3) using this approximate clothed propagator, etc., etc., until things stop changing and the calculation is 'self-consistent'. This procedure is called '*self-consistent renormalization*'. In first order, we have

$$(11.4)$$

which is the field-theoretic form of the Hartree–Fock equation in the general case. It is thus applicable to a system with an external potential, as discussed just after (4.74). [Regarding the proof that (11.4) is fully equivalent to the usual Hartree–Fock equation, see Hedin (1965), p. A797, for reference.] In the special case of no external potential, the propagator (4.77) turns out to be the self-consistent solution of (11.4), so there is no real self-consistency problem.

It is also possible to renormalize simultaneously both the interactions, as in (10.60), and the propagators, as in (11.3), leading to the series

$$
\Sigma = \text{[diagram]} + \text{[diagram]} + \text{[diagram]} + \cdots . \tag{11.5}
$$

In the literature, the double lines are often omitted, and it is simply stated that all interactions and propagators are dressed. A self-consistent calculation can be carried out for example to first order in (11.5); this yields

$$
\Sigma \approx \text{[diagram]} + \text{[diagram]} \tag{11.6}
$$

which may be substituted in (10.7), producing

$$
\text{[diagram]} = \frac{1}{\text{[diagram]}^{-1} - \left[\text{[diagram]} + \text{[diagram]} \right]} . \tag{11.7}
$$

This is essentially the Hartree–Fock approximation using a frequency-dependent interaction instead of a static one. When $\sim\!\sim\!\sim\!\sim$ is in RPA, this is called the 'time-dependent Hartree–Fock approximation' (see Thouless (1972)).

11.2 Existence of quasi particles when the perturbation expansion is valid

It was shown in chapter 10 that quasi particles exist in high- and low-density Fermi gases. Do they exist in other systems, like an electron gas at ordinary metallic densities, or in liquid He^3? With the aid of the clothed skeleton series (11.3), it is possible to prove that there are quasi particles in any Fermi system with an interaction such that the perturbation expansion holds (i.e., such that the propagator series (9.40) is valid). These are the so-called 'normal' systems—the 'abnormal' ones, like superconductors and ferromagnets are discussed in §15.4, and in chapter 17. We will first find the general form of G near the Fermi energy, in a normal system, then prove the existence of quasi particles.

The self-consistent equation for the propagator is, by (10.7),

$$
\text{[diagram]} = \frac{1}{\text{[diagram]}^{-1} - \Sigma} , \quad \text{or} \quad G(\mathbf{k}, \omega) = \frac{1}{\omega - \epsilon_k - \Sigma_R(\mathbf{k}, \omega) - i\Sigma_I(\mathbf{k}, \omega)} \tag{11.8}
$$

with $\left(\overset{|}{\underset{|}{\Sigma}}\right)$ as in (11.3). The quantities Σ_R, Σ_I are respectively the real and imaginary parts of the proper self-energy. We will now obtain the form of Σ_I near the Fermi energy by assuming it has a certain form, then showing this assumption yields a self-consistent solution for G in (11.8), (11.3).

In order to get a clue regarding what form to assume, let us evaluate the contribution of the first few diagrams in the self-energy expansion (10.8). We will use the modified propagator formalism involving the chemical potential, μ, described in §9.7, since this turns out to be the easiest route to the self-consistent result. (It will turn out that μ is just ϵ'_F, the Fermi energy of the interacting system.) The bubble and open oyster diagrams have zero imaginary part. Using the same argument as that leading to (9.74), with $G_0(\mathbf{k}, \omega)$ replaced by $G_0^\mu(\mathbf{k}, \omega)$ (see (9.47')), the first of the second-order diagrams yields:

$$\lim_{\omega \to \mu} \text{Im} \;\; \text{} \;\; \propto \;\; (\omega - \mu)^2.$$

Using this as a starting point, let us assume (Luttinger (1961)) that Σ_I has the limiting form

$$\lim_{\omega \to \mu} \Sigma_I(\mathbf{k}, \omega) = -\text{sgn}(\omega - \mu)\, C_k (\omega - \mu)^2 \quad \text{where } C_k \geq 0, \text{ real}$$

$$\text{sgn}(\omega - \mu) = +1, \quad \omega > \mu$$

$$= -1, \quad \omega < \mu. \tag{11.9}$$

(Regarding the sign change at $\mu = \omega$, see Nozières (1964), p. 195. Note that this agrees with the sign of the infinitesimal in (9.47').) Then (11.8) may be written

$$\underset{\omega \to \mu}{G}(\mathbf{k}, \omega) = \frac{1}{\omega - \epsilon_k - \Sigma_R(\mathbf{k}, \omega) + i\,\text{sgn}(\omega - \mu)\, C_k(\omega - \mu)^2}. \tag{11.10}$$

We can show that (11.10) is a self-consistent solution for G by using it as the dressed propagator in the clothed skeleton diagrams of (11.3). When this is done we find [Luttinger (1961)]

$$\lim_{\omega \to \mu} \text{Im} \;\; \text{} \;\; \propto \;\; (\omega - \mu)^2, \tag{11.11}$$

and, in general

$$\lim_{\omega \to \mu} \left(\begin{array}{c} \text{any} \\ \text{clothed} \\ \text{skeleton} \end{array} \right) \propto (\omega - \mu)^{2m}, \quad m = 1, 2, 3, \ldots . \quad (11.12)$$

Hence, the imaginary part of the proper self-energy has the form

$$\lim_{\omega \to \mu} \Sigma_I \propto (\omega - \mu)^2$$

just as was assumed in (11.9). This shows that (11.10) is indeed the desired self-consistent solution for the propagator. The form (11.10) is valid for any system in which the perturbation expansion (9.40) is valid.

When (11.10) holds, it is easy to show that quasi particles exist in the system (Mattuck (1964)). We will first get the single-particle excitation energies, E_k, then the lifetime of the excitations, τ_k, and show that τ_k obeys (8.38).

By (3.14) the excited energies of the system may be obtained by solving the equation for the poles of (11.10):

$$\omega - \epsilon_k - \Sigma_R(\mathbf{k}, \omega) + i \, \text{sgn} \, (\omega - \mu) \, C_k (\omega - \mu)^2 = 0. \quad (11.13)$$

First let us get a zeroth-order approximation for the poles. For ω very near μ (the case we are interested in) we can neglect the imaginary part and get

$$\omega_{\text{pole}} \equiv E_k = \epsilon_k + \Sigma_R(\mathbf{k}, E_k) \quad (11.14)$$

To obtain a first-order solution, expand Σ_R about the zeroth-order solution, E_k:

$$\Sigma_R(\mathbf{k}, \omega) = \Sigma_R(\mathbf{k}, E_k) + \Sigma_R'(\mathbf{k}, E_k) \times (\omega - E_k) + \cdots \quad (11.15)$$

and put this in the pole equation:

$$\omega - [\epsilon_k + \Sigma_R(\mathbf{k}, E_k)] - \Sigma_R'(\mathbf{k}, E_k) \, (\omega - E_k) + i\lambda \, \text{sgn} \, (\omega - \mu) \, C_k (\omega - \mu)^2 = 0 \quad (11.16)$$

where we have put in λ to keep track of the order. Let

$$\omega = E_k + \lambda \omega_1$$

to first order. Then (11.16) becomes

$$\lambda \omega_1 - \Sigma_R' \lambda \omega_1 + i\lambda \, \text{sgn} \, (E_k + \lambda \omega_1 - \mu) \, C_k (E_k - \mu + \lambda \omega_1)^2 = 0.$$

Dropping higher-order terms in λ:

$$\omega_1 = -i \, \text{sgn} \, (E_k - \mu) \frac{C_k(E_k - \mu)^2}{(1 - \Sigma_R')}. \tag{11.17}$$

Therefore (set $\lambda = 1$)

$$\omega = \omega_0 + \omega_1 = E_k - i \, \text{sgn} \, (E_k - \mu) \frac{C_k(E_k - \mu)^2}{(1 - \Sigma_R')}. \tag{11.18}$$

Thus the energy and reciprocal lifetime are given by

$$\epsilon_k' = E_k = \epsilon_k + \Sigma_R(\mathbf{k}, \epsilon_k')$$

$$\tau_k^{-1} = \text{sgn} \, (\epsilon_k' - \mu) \frac{C_k(\epsilon_k' - \mu)^2}{(1 - \Sigma_R')}. \tag{11.19}$$

It is seen that we have true quasi particles here because the energy level width τ_k^{-1} is much less than the energy, in this case measured relative to μ. (Note that in (8.22), energy is taken relative to ϵ_F', the Fermi energy of the interacting system. It will be seen in the next section that μ and ϵ_F' are the same.)

We can get an expression for G which is valid near the poles by substituting (11.15) into (11.10) for the real part of Σ and evaluating the small imaginary part of Σ at E_k. This yields

$$G(\mathbf{k}, \omega) =$$

$$= \frac{1}{\omega - \underbrace{(\epsilon_k + \Sigma_R(\mathbf{k}, E_k))}_{E_k} - \Sigma_R'(\mathbf{k}, E_k)(\omega - E_k) + i \, \text{sgn} \, (E_k - \mu) \, C_k(E_k - \mu)^2} + F(\mathbf{k}, \omega). \tag{11.20}$$

The $F(\mathbf{k}, \omega)$ term is a correction containing everything left out of the first term. Equation (11.20) may be written

$$G(\mathbf{k}, \omega) = \frac{Z_k}{\omega - \epsilon_k' + i\tau_k^{-1}} + F(\mathbf{k}, \omega) \tag{11.21}$$

where the amplitude factor, Z_k is:

$$Z_k = [1 - \Sigma_R'(\mathbf{k}, E_k)]^{-1}. \tag{11.22}$$

This has just the quasi particle form (8.37). Thus we have proved that if the interaction is such that perturbation theory holds, then quasi particles exist in the system. The only other assumption involved is that $\Sigma(\mathbf{k}, \omega)$ can be expanded in a power series.

11.3 Existence of the Fermi surface in an interacting system

In order to appreciate the mystery concerning the existence of the Fermi surface, look at the momentum distribution function n_k for a system of non-interacting fermions ($T = 0$, no external potential) in Fig. 11.1. This function is the probability that the state $\mathbf{k}$ is occupied by a (bare) particle. It is seen that there is a sharp discontinuity at the Fermi surface, $|\mathbf{k}| = k_F$. It is because of this discontinuity that it is meaningful to say there is a surface here (in $\mathbf{k}$-space). Now, in a typical interacting system like, say, the electron gas, the average interaction energy between any pair of particles is comparable to the Fermi energy ($V \sim 7$ eV, $\epsilon_F \sim 5$ eV in an electron gas). Naïve physical intuition would say that turning on such a strong interaction would cause collisions, knocking particles out of the occupied states below k_F to the unoccupied ones above, resulting in a complete 'smearing out' of the discontinuity at the Fermi surface, as shown in Fig. 11.2. This is roughly analogous to the smearing out of the water surface directly under the Niagara falls. However, experiments on electrons in metals indicate the presence of a discontinuity at $|\mathbf{k}| = k_F$ which is sharp to within 10^{-4} eV!

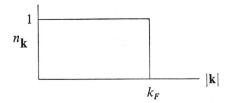

Fig. 11.1 *Momentum Distribution Function for Non-interacting Fermions*

The mystery was cleared up by Migdal (Pines (1961), p. 34) and Luttinger (1960b). They showed that there is a discontinuity in n_k in the interacting system but that the magnitude of the discontinuity is < 1. Their argument makes use of the fact that n_k may be obtained directly from the single-particle Green's function. This is easily seen as follows: n_k is the expectation value of

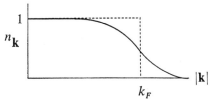

Fig. 11.2 *Naïve Guess at Form of Momentum Distribution Function in Interacting Fermi System*

the number operator, $c_k^\dagger c_k$, in the interacting ground state $|\Psi_0\rangle$:

$$n_k = \langle \Psi_0 | c_k^\dagger c_k | \Psi_0 \rangle. \qquad (11.23)$$

Using the closed expression for the propagator (9.6), with $t_1 = 0$, $t_2 = t$, $k_1 = k_2$, this may be expressed in terms of the propagator by

$$n_k = -i \lim_{t \to 0^-} G(\mathbf{k}, t) \qquad (11.24)$$

where 0^- means an infinitesimal time interval before $t = 0$.

Let us evaluate (11.24) in the case of a system where perturbation theory is valid. Then $G(\mathbf{k}, \omega)$ has the form (11.19) and $G(\mathbf{k}, t)$ may be obtained from it by Fourier transform. This yields

$$G(\mathbf{k}, t) = -iZ_k\{\theta_t \theta_{E_k - \mu} e^{-iE_k t} e^{-t/\tau_k} - \theta_{-t} \theta_{\mu - E_k} e^{-iE_k t} e^{+it/\tau_k}\} + F(\mathbf{k}, t) \qquad (11.25)$$

where $F(\mathbf{k}, t)$ is the transform of the correction $f(\mathbf{k}, \omega)$.

Substituting (11.25) into (11.24) yields

$$n_k = Z_k \theta_{\mu - E_k} - i F(\mathbf{k}, 0^-). \qquad (11.26)$$

Assuming $F(\mathbf{k}, 0^-)$ is continuous across $\mathbf{k} = \mathbf{k}_\mu$, where $\mathbf{k}_\mu$ is the momentum corresponding to energy μ, this reveals a discontinuity of magnitude Z_{k_μ} on the surface $E_k = \mu$. In Fig. 11.3 we have plotted n_k, assuming isotropic interaction (so that by symmetry, $\mathbf{k}_\mu$ is spherical). Comparing this with Fig. 11.1 we see that it is reasonable to call $\mathbf{k}_\mu, \mu$ respectively the Fermi surface and Fermi energy of the interacting system. Hence, assuming the interaction is such that a perturbation expansion may be used, the Fermi surface exists in the interacting system.

It would be a convincing demonstration of the correctness of the above arguments if the curve in Fig. 11.3 could be measured experimentally. Calculations by Daniel and Vosko (1960) indicate that Z_{k_μ} differs greatly from 1. For example, an estimate for Na metal gives $Z_{k_\mu} \approx 0 \cdot 5$. However, one cannot carry out such a measurement by any low-energy methods—like, for example, soft X-ray spectra—since these give only the quasi particle distribution function (which is nearly that for free electrons, because the quasi particles are nearly independent—see last paragraph in §8.1).

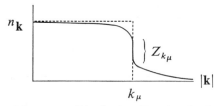

Fig. 11.3 *True Momentum Distribution Function in Interacting Fermi System*

To actually measure n_k requires energies which are large in comparison with the inter-electron interaction, i.e., large enough to 'undress' the electrons. This can be achieved using Compton scattering with high energy electrons (Platzman (1972)) and the experimental results for Na show good agreement with theory (Lundqvist (1971)).

11.4 ⟦Dressed vertices⟧

So far, we have seen how to simplify the series for the proper self-energy by 'renormalizing' the interactions (§10.5) and the propagators (§11.1). This section reveals a final simplification which boils the series down to two terms; it is carried out by renormalizing the vertices. The useful definitions here are:

> *Vertex part:* any diagram without external lines which may be inserted in place of a vertex (i.e., can be connected to two particle lines and one interaction line), such as, for example:

$$(11.27)$$

> *Proper (irreducible) vertex part:* a vertex part which cannot be broken into two disconnected pieces by removing *either* one particle (or hole) line *or* one interaction line; for example:

$$(11.28)$$

An example of the replacement of a vertex by an irreducible vertex part is

$$(11.29)$$

If we define the sum of all irreducible vertex parts by

$$(11.30)$$

then it is not hard to show that the irreducible self-energy is just given by

$$\text{(11.31)}$$

This is easily seen by expanding:

$$\text{(11.32)}$$

Note that only one vertex is dressed because of the fact that dressing both would count diagrams twice. For example

$$\text{and} \qquad \text{(11.33)}$$

look like a vertex insertion in the upper vertex and lower vertex respectively, but this is pure optical illusion. In reality they are topologically equivalent and are therefore the same diagram, as mentioned in (10.56). This underlines a necessary condition for playing the graph game: one must be completely sober to avoid double-counting diagrams!

Further reading

Pines (1961), chap. 2, pp. 31–4.
Falicov and Heine (1961).
Luttinger (1960c).
Thouless (1964).
Schweber (1961), pp. 607–15.
Bjorken (1965), p. 284 ff.

Exercises

11.1 Which of the following diagrams should not be included in the expansion (11.3), and why not? Which should not be included in (11.5), and why not?

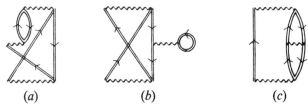

(a)   (b)   (c)

11.2 Suppose we have a hypothetical system in which the proper self-energy has the form $\Sigma(\mathbf{k}, \omega) \approx A\omega + i\,\mathrm{sgn}(\mu - \omega)\,B(\omega - \mu)^2$, with $B > 0$, $A < 0$, near the Fermi surface, where μ is assumed known. Find the form of the propagator. What is the discontinuity at the Fermi surface?

11.3 Assume that in a system with fixed chemical potential, μ, the solution to the self-consistent Hartree–Fock equation (11.4) (using the formalism in §9.7) has the form:

$$G(\mathbf{k}, \omega) = \frac{1}{\omega - \epsilon_k' + i(\omega - \mu)\,\delta}.$$

Use this assumed G to calculate the self-energy Σ in (11.6), place Σ in (11.7) and show that you get the assumed G back again provided that:

$$\epsilon_k' = \epsilon_k + \sum_{l\,[\epsilon_l' < \mu]} (V_{klkl} - V_{lkkl})$$

(Use the $\exp(i\omega 0^+)$ convergence factor in Table 9.1 when calculating the frequency integral.)

Chapter 12

Ground State Energy of Electron Gas and Nuclear Matter

12.1 Review

In chapter 5 we saw how to get a ground state energy perturbation series for the case of a single particle in an external potential. The method made use of the diagram expansion of the vacuum amplitude, from which the graphical series for the energy could be obtained by taking the limit (5.4). (The rigorous argument lying behind the pinball approach in chapter 5 is in appendices B→G.) It was shown how a partial sum over energy diagrams could be carried out in the simple case where only two levels were involved, and a brief glimpse of the many-body case was given.

Now we go into the details of finding the ground state energy in a many-body system with no external potential. The rules for the many-body diagrams are pretty much the same as for the single particle case, so they will simply be stated without fanfare, and we'll concentrate on the applications to the electron and nuclear cases. In the electron gas case, Gell-Mann and Brueckner (GB) showed that the 'ring diagrams'—all individually infinite— gave the dominant contribution in the high-density limit. They used a trick to sum over all diagrams of a given order: this produced a logarithmic series which could then be summed to infinity. The result was finite and later shown to be exact. The nuclear (low density) case will be treated only very briefly since this is a vastly more complicated problem. We'll simply show how a sum may be carried out over the individually infinite 'ladder diagrams' with the aid of the K-matrix and illustrate by means of a simple example how this can lead to a finite result.

12.2 Diagrams for the ground state energy

The ground state energy as presented in (5.42) appears in (12.1). In this expression we have omitted two types of diagrams: (1) All diagrams obtained by twisting one or more interactions through 180 degrees. This cancels the factor of $\frac{1}{2}$ in the interaction potential (see (4.60)). However, in the case where the diagram is completely symmetric, such as the double bubble, the oyster, and the two second-order diagrams, twisting all the interactions does not produce a new diagram, so one factor of $\frac{1}{2}$ is not cancelled. (2) All momentum-

$$E_0 = W_0 + \text{} + \text{} + \text{} + \text{}$$

$$+ \text{} + \text{} + \text{} + \text{}$$

$$+ \text{} + \cdots + \text{} + \cdots . \qquad (12.1)$$

non-conserving ('anomalous') graphs (see p. 162). Examples of omitted graphs:

| Twisted | Anomalous | (12.2) |

Note that as in §5.3, the diagrams here are Goldstone diagrams, so the time order is important. That is, two diagrams are topologically equivalent only if one can be distorted into the other without changing the time order of the interactions. Thus:

$$(12.3)$$

Analogous to the cookbook rules for evaluating the propagator, we give the recipe for solving the ground state energy problem. The actual proof of these rules is extremely complicated and tedious, but the rules themselves are so similar to what we have seen before that they shouldn't produce any traumatic effect on the reader.

Diagram rules for ground state energy (no external field)

(1) Draw N horizontal wiggles as in the propagator case but with no external points.

(2) Join all vertices to each other with directed lines, one line into and one out of each vertex. Do this in all connected topologically distinct ways. (Only distortions which preserve time order allowed.)

(3) Label each directed line with a momentum $\mathbf{k}$ ($\equiv \mathbf{k}, \sigma$). Conserve momentum at each interaction. Eliminate all anomalous graphs.

(4) Draw a light dotted horizontal line between each successive pair of interaction wiggles, thus

$$(12.4)$$

(5) Evaluate diagrams by dictionary in Table 12.1.

Table 12.1 *Diagram dictionary for ground state energy of interacting fermion system (no external potential)*

Diagram element	Factor
Dotted line:	$\left[\sum_H - \sum_P\right]^{-1}$
	$\sum_H$ = sum of ϵ_k's for all hole lines crossing dotted line
	$\sum_P$ = sum of ϵ_k's for all particle lines crossing dotted line
interaction:	V_{klmn} or V_q
each hole line:	-1
each fermion loop example:	-1
completely symmetric diagram. Example:	$\frac{1}{2}$
each particle (hole) momentum parameter $\mathbf{k}$	$\sum_k$ or $\int \frac{d^3\mathbf{k}}{(2\pi)^3}$ particle: $k > k_F$ $k > k_F$ hole: $k < k_F$ $k < k_F$

For example, we have that

(1) Double bubble = (two fermion loops, two hole lines, symmetric)

$$= (-1)^4 \times \tfrac{1}{2} \times \sum_{k,\,l<k_F} V_{klkl} \tag{12.5}$$

(2) Oyster = (one fermion loop, two hole lines, symmetric)

$$= (-1)^3 \times \tfrac{1}{2} \times \sum_{k,\,l<k_F} V_{lkkl} \tag{12.6}$$

(3) $$= (-1)^4 \times \frac{1}{2} \sum_{q,k,l} \frac{V_q^2}{(\epsilon_k + \epsilon_l - \epsilon_{k+q} - \epsilon_{l-q})} \tag{12.7}$$

$$(|\mathbf{k}|, |\mathbf{l}| < k_F; \; |\mathbf{k+q}|, |\mathbf{l-q}| > k_F)$$

(4) $$= (-1)^3 \times \frac{1}{2} \sum_{q,k,l} \frac{V_q V_{k-l+q}}{(\epsilon_k + \epsilon_l - \epsilon_{k+q} - \epsilon_{l-q})} \tag{12.8}$$

$$(|\mathbf{k}|, |\mathbf{l}| < k_F; \; |\mathbf{k+q}|, |\mathbf{l-q}| > k_F)$$

The diagram method of getting the ground state energy will now be applied to a high-density electron gas and to nuclear matter.

12.3　Ground state energy of high-density electron gas: theory of Gell-Mann and Brueckner

The electron gas was defined in §4.9 as the theoretician's 'dream metal', consisting of N electrons moving against a uniform positive charge background and interacting by pure Coulomb forces. It was characterized by the average inter-electron distance, r_s. The Hamiltonian was (10.25). We showed how to calculate the quasi particle effective mass and lifetime in the high density ($r_s < 1$) limit by a sum over ring diagrams. Here the ground state energy will be investigated in the same limit.

The simplest approximation to the ground state energy is the Hartree-Fock, which in this case, with no external potential, is simply the energy to first order of perturbation theory. In terms of diagrams it may be written (see (12.1))

$$E_{\mathrm{HF}} = W_0 + \text{} + \text{} \tag{12.9}$$

This crude approximation takes no account of the fact that because of Coulomb repulsion, electrons tend to keep away from each other, i.e., their motions are correlated. The *correlation energy*, E_c, is defined by:

$$E_c \equiv E_{\text{correlation}} = E_{(\text{exact})} - E_{\text{HF}}$$

$$= \text{(diagram)} + \text{(diagram)} + \text{(diagram)} + \cdots . \quad (12.10)$$

Let us first evaluate the Hartree–Fock energy. Using (7.75), (12.5) and (12.6), we find that (12.9) is

$$E_{\text{HF}} = \sum_{k<k_F} \epsilon_k + \tfrac{1}{2} \sum_{k,l<k_F} [V_{klkl} - V_{lkkl}]. \quad (12.11)$$

The first term is (putting spin sum in explicitly):

$$W_0 = \sum_{\substack{k<k_F \\ \sigma = \pm\frac{1}{2}}} \hbar^2 k^2 / 2m = 2 \times \frac{\hbar^2}{2m} \times \frac{\Omega}{(2\pi)^3} \int\limits_{k<k_F} d^3 k\, k^2 = \frac{\Omega \hbar^2}{2\pi^2 m} \times \tfrac{1}{5} k_F^5, \quad (12.12)$$

where the 2 is from the spin sum, $\Omega =$ crystal volume, and $\Omega/(2\pi)^3 =$ density of points in k-space. The Fermi momentum, k_F, may be found from

$$N = 2 \sum_{k<k_F} 1 = \frac{2\Omega}{(2\pi)^3} \int\limits_{k<k_F} d^3 \mathbf{k} = \frac{\Omega}{3\pi^2} k_F^3, \quad (12.13)$$

so that, using (10.24)

$$\frac{W_0}{N} = \frac{\Omega}{N} \frac{\hbar^2}{2m} \times \frac{1}{5} \left(3\pi^2 \frac{N}{\Omega} \right)^{\frac{2}{3}} = \frac{2 \cdot 21}{r_s^2} \frac{\text{rydberg}}{\text{electron}}. \quad (12.14)$$

The term in $V_{klkl} \equiv V_{q=0}$ is equal to zero by (10.28) (it is cancelled by the interaction with the positive charge background). The last term is (spin sums produce factor of 2):

$$\frac{1}{N} \times \text{(diagram)} = -\frac{2}{N} \times \frac{1}{2} \times \left[\frac{\Omega}{(2\pi)^3} \right]^2 \times \frac{4\pi e^2}{\Omega} \iint\limits_{k,l<k_F} \frac{d^3 \mathbf{k}\, d^3 \mathbf{l}}{|\mathbf{k}-\mathbf{l}|^2}$$

$$= -\frac{0 \cdot 916}{r_s} \frac{\text{rydberg}}{\text{electron}} \quad (12.15)$$

(see Raimes (1961), p. 171 ff.). Hence

$$\frac{E_{\text{HF}}}{N} = \frac{2 \cdot 21}{r_s^2} - \frac{0 \cdot 916}{r_s} \frac{\text{rydberg}}{\text{electron}}. \quad (12.16)$$

Now consider the correlation energy. The first graph in (12.10) may be evaluated from (12.7). The spin sum gives a factor of 4. Changing from sum to integral, expressing all momenta in terms of the Fermi momentum and using (12.13) yields (in rydbergs per electron):

$$\frac{1}{N} \times \text{[diagram]} = \frac{-3}{8\pi^5} \int \frac{d^3 q}{q^4} \int\limits_{\substack{|\mathbf{k}|<1 \\ |\mathbf{k}+\mathbf{q}|>1}} d^3 k \int\limits_{\substack{|\mathbf{l}|<1 \\ |\mathbf{q}-\mathbf{l}|>1}} d^3 l \frac{1}{q^2+\mathbf{q}\cdot(\mathbf{k}-\mathbf{l})}. \quad (12.17)$$

Similarly, (12.8) gives

$$\frac{1}{N} \times \text{[diagram]} = +\frac{3}{16\pi^5} \int \frac{d^3 q}{q^2} \int\limits_{\substack{|\mathbf{k}|<1 \\ |\mathbf{k}+\mathbf{q}|>1}} d^3 k \int\limits_{\substack{|\mathbf{l}|<1 \\ |\mathbf{q}-\mathbf{l}|>1}} d^3 l \frac{1}{(\mathbf{k}-\mathbf{l}+\mathbf{q})^2} \times$$

$$\times \frac{1}{q^2+\mathbf{q}\cdot(\mathbf{k}-\mathbf{l})}. \quad (12.18)$$

Look first at (12.17). The major contribution of the integrand is evidently from small $\mathbf{q}$. For small $\mathbf{q}$, because of the limits on the integration region, $\mathbf{k}$ and $\mathbf{l}$ lie in a shell about the Fermi surface, of thickness $\propto q$. The integrand $\propto q^{-1}$ for small q, and $\int d^3 k \int d^3 l \propto q^2$, so it follows that

$$\frac{1}{N} \times \text{[diagram]} \propto \int \frac{d^3 q}{q^3} \propto \int_0 \frac{dq}{q} = \infty. \quad (12.19)$$

The reason for this divergence is the long-range character of the Coulomb interaction. A shielded Coulomb interaction of the form (10.41) would have been finite at $\mathbf{q}=0$ and would have given a finite result.

The same argument yields for (12.18).

$$\frac{1}{N} \times \text{[diagram]} \propto \int \begin{bmatrix} \propto q & \text{for small } q \\ \propto q^{-4} & \text{for large } q \end{bmatrix} dq = \text{finite}$$

$$\left(= 0.046 \frac{\text{rydberg}}{\text{electron}} \right). \quad (12.20)$$

The reason why [diagram] diverges while [diagram] does not is that in the former there is the same momentum $\mathbf{q}$ transferred at each interaction, contributing a factor of $V_q^2 \propto 1/q^4$, while in the latter there is $\mathbf{q}$ transferred at only one wiggle, giving only $1/q^2$.

In third order we have the diagrams

$$E^{(3)} = \quad + \quad + \cdots + \quad + \cdots . \tag{12.21}$$

Note how they fall into classes: same momentum $\mathbf{q}$ at all three wiggles, at two wiggles, or at only one wiggle. Using the same analysis as in the second-order case reveals that for small q,

$$E^{(3)} = r_s\left\{A^{(3)} \int \frac{d^3\mathbf{q}}{q^5} + B^{(3)} \int \frac{d^3\mathbf{q}}{q^3} + C^{(3)} \int \frac{d^3\mathbf{q}}{q}\right\}, \tag{12.22}$$

where A, B, C are constants. Observe the dependence on r_s. The three terms come from the three types of diagrams, and we see again that the most divergent term comes from the diagram with the same $\mathbf{q}$ at each wiggle.

The same behaviour occurs in all orders, and it is found that the perturbation series for the correlation energy may be written

$$E_{\text{correl.}} = A^{(2)} \int \frac{d^3\mathbf{q}}{q^3} + B^{(2)} \int \frac{d^3\mathbf{q}}{q} \quad = 0\cdot046$$

$$+ \ r_s A^{(3)} \int \frac{d^3\mathbf{q}}{q^5} + r_s B^{(3)} \int \frac{d^3\mathbf{q}}{q^3} + r_s C^{(3)} \int \frac{d^3\mathbf{q}}{q}$$

$$+ \ r_s^2 A^{(4)} \int \frac{d^3\mathbf{q}}{q^7} + r_s^2 B^{(4)} \int \frac{d^3\mathbf{q}}{q^5} + r_s^2 C^{(4)} \int \frac{d^3\mathbf{q}}{q^3} + r_s^2 D^{(4)} \int \frac{d^3\mathbf{q}}{q}$$

$$+ \quad \vdots \quad \cdots \quad \vdots \quad \cdots . \tag{12.23}$$

The terms can be arranged (see diagonal lines) in such manner that in each order of divergence they form a power series in r_s (see after (10.30)). This means that for small r_s, i.e., the high-density limit, we may take just the terms of lowest order in r_s. Thus:

$$E_{\substack{\text{correl.}\\ \text{(high density)}}} = 0.046 + A^{(2)} \int \frac{d^3\mathbf{q}}{q^3} + r_s A^{(3)} \int \frac{d^3\mathbf{q}}{q^5} + r_s^2 A^{(4)} \int \frac{d^3\mathbf{q}}{q^7} + \cdots$$

$$= \quad + \quad + \quad + \quad +$$

$$+ \cdots + \quad + \cdots . \tag{12.24}$$

This is evidently a sum over diagrams of the 'ring' form, like those met in the corresponding propagator case in (10.31).

Because of the importance of the time ordering of the interactions in (12.24) the sum over rings here is not as straightforward as in the propagator case. GB accomplished it by a trick which enabled them to first sum over all diagrams within each order of perturbation theory. It was then straightforward to carry out the sum over all orders. The trick involves the function

$$F_q(t) = \int d^3\mathbf{p}\, e^{-i|t|(\frac{1}{2}q^2 + \mathbf{q}\cdot\mathbf{p})} \tag{12.25}$$

and its transform:

$$Q_q(u) = \int_{-\infty}^{+\infty} e^{ituq} F_q(t)\, dt. \tag{12.26}$$

(This turns out to be related to the lowest-order polarization part, $\pi_0(\mathbf{q}, \omega)$ (see (10.36)) by:

$$Q_q(u) = \frac{(2\pi)^3}{2} \pi_0(\mathbf{q}, iqu).) \tag{12.27}$$

GB showed that the total contribution from all ring diagrams in nth order was given in terms of $Q_q(u)$ by

$$\frac{E^{(n)}}{N} = \frac{B}{r_s^2} \int q\, d^3\mathbf{q} \int_{-\infty}^{+\infty} du\, \frac{(-1)^n}{n} \left[\frac{C r_s\, Q_q(u)}{q^2} \right]^n \tag{12.28}$$

where B, C are numerical factors.

The correlation energy is then just the sum over all orders:

$$\frac{E_{\text{correl.}}}{N} = 0.046 + \sum_{n=2}^{\infty} \frac{E^{(n)}}{N}$$

$$= 0.046 + \frac{B}{Nr_s^2} \int q\, d^3\mathbf{q} \int_{-\infty}^{+\infty} du \sum_{n=2}^{\infty} \frac{(-1)^n}{n} \left[\frac{C r_s\, Q_q(u)}{q^2} \right]^n. \tag{12.29}$$

In the various partial summations up to now we encountered sums of geometric series and square root series. In this partial sum, $\sum\limits_{n=2}^{\infty}$, over rings, we have a logarithmic series, which is easily summed to yield

$$\frac{E_{\text{correl.}}}{N} = 0{\cdot}046 + \frac{B}{Nr_s^2}\int q\,d^3q \int\limits_{-\infty}^{+\infty} du\left\{-\ln\left(1+\frac{Cr_s\,Q_q(u)}{q^2}\right)+\frac{Cr_s\,Q_q(u)}{q^2}\right\}$$

$$\text{for } \left[\frac{Cr_s\,Q_q(u)}{q^2}\right]^2 < 1, \quad (12.30)$$

which may be evaluated to give

$$\frac{E_{\text{correl.}}}{N} = 0{\cdot}0622\ln r_s - 0{\cdot}096 + \mathcal{O}(r_s),$$

where

$$\mathcal{O}(r_s) \to 0 \quad \text{as} \quad r_s \to 0. \quad (12.31)$$

Thus, the calculation is exact in the high-density limit.

The GB argument makes the usual assumption that the result of the partial summation is valid even in the small $\mathbf{q}$ region where the inequality in (12.30) shows the series diverges (see §3.3). This has since been validated, first by Sawada (Pines (1961), p. 201) who got the same result by a non-perturbative method, and then by Nozières and Pines (Pines (1961), p. 235) using a technique based on the calculation of the 'generalized dielectric constant' $\epsilon(\mathbf{k},\omega)$ (see (10.37)).

12.4 Brief view of Brueckner theory of nuclear matter

Nuclear matter was defined in §4.8. It was mentioned there that the binding energy per particle ($=$ ground state energy, if the energy zero is taken to be that of the non-interacting system) was about -16 MeV. We will now give a very short glimpse of how this may be calculated by the diagrammatic method.

To do such a calculation, it is necessary to have a more realistic inter-nucleon potential than the simple Yukawa interaction (4.80). High energy scattering experiments indicate that the true potential looks roughly like an infinitely hard repulsive core plus a short-range attractive tail, as shown in Fig. 12.1. Despite its rather violent appearance, the attractive part of the nuclear potential is 'weak', in a sense, since, if the nuclear matter is taken to have a density equal to that at the centre of heavy nuclei, then the mean separation of two nucleons is $\sim 1{\cdot}1 \times 10^{-13}$ cm or about three times the hard core radius. This means that nuclear matter acts as a low-density fermion gas, the hard core occupying only $\frac{1}{27}$ of the total volume.

The ground state energy may be calculated by means of the expansion in (12.1). As pointed out in §5.1, because of the hard core, V_{klmn}, and hence all the terms in the perturbation series, are infinite. Nevertheless, analogous to the electron gas case, it is possible to perform a partial sum over the most important types of diagrams and obtain a finite result. In the electron case,

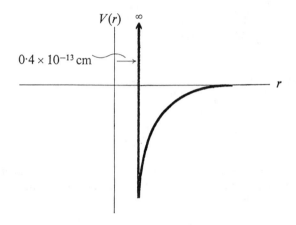

Fig. 12.1 *Form of Interaction Between Two Nucleons (Schematic).*
V = Potential Energy, and r = Internucleon Separation

the ring diagrams dominated because of high density. In this low-density nuclear case, the big contributors are the diagrams with only two hole lines, as discussed in §10.3. The approximate series for the energy involving just these graphs is

The technique for summing these 'ladder' diagrams becomes clear by writing out the first few in detail (see Table 12.1 and remember the factor of $\frac{1}{2}$ for symmetric graphs!)

$$\mathbf{i} \,\bigcirc\!\!\!\sim\!\!\!\sim\!\!\!\bigcirc\, \mathbf{j} \;=\; \tfrac{1}{2} \sum_{i,j<k_F} V_{ijij} \qquad\qquad (12.33)$$

$$\mathbf{i}\,\left(\!\!\begin{array}{c}\mathbf{n}\\ \mathbf{m}\end{array}\!\!\right)\,\mathbf{j} \;=\; \frac{1}{2} \sum_{\substack{i,j<k_F \\ m,n>k_F}} V_{ijmn} V_{mnij} \frac{1}{\epsilon_i + \epsilon_j - \epsilon_m - \epsilon_n} \qquad (12.34)$$

$$\mathbf{i}\,\left(\!\!\begin{array}{c}\mathbf{m}\;\mathbf{n}\\ \mathbf{p}\;\mathbf{q}\end{array}\!\!\right)\,\mathbf{j} \;=\; \frac{1}{2} \sum_{\substack{i,j<k_F \\ m,n,p,q>k_F}} V_{ijmn} V_{mnpq} V_{pqij} \times$$

$$\times \frac{1}{(\epsilon_i + \epsilon_j - \epsilon_m - \epsilon_n)(\epsilon_i + \epsilon_j - \epsilon_p - \epsilon_q)} \qquad (12.35)$$

and so on. The sum may be carried out with the aid of a frequency-independent K-matrix, similar to (but not identical with!) the frequency-dependent K-matrix in §10.3:

$$K_{mnij} \;=\; V_{mnij} + \sum_{p,l>k_F} V_{mnpl} K_{plij} \frac{1}{(\epsilon_i + \epsilon_j - \epsilon_p - \epsilon_l)}. \qquad (12.36)$$

This is the analogue of (10.21) used in the propagator case. Writing this equation for the special case $\mathbf{m}=\mathbf{i}$, $\mathbf{n}=\mathbf{j}$, and iterating yields

$$K_{ijij} \;=\; V_{ijij} + \sum_{p,l>k_F} V_{ijpl} V_{plij} \frac{1}{(\epsilon_i + \epsilon_j - \epsilon_p - \epsilon_l)}$$

$$+ \sum_{p,l,s,q>k_F} V_{ijpl} V_{plsq} V_{sqij} \frac{1}{(\epsilon_i + \epsilon_j - \epsilon_p - \epsilon_l)(\epsilon_i + \epsilon_j - \epsilon_s - \epsilon_q)} + \cdots.$$

$$(12.37)$$

Summing this over $i, j < k_F$ and comparing with (12.33, 34, 35) it is easily found that

$$\bigcirc\!\!\!\sim\!\!\!\sim\!\!\!\bigcirc \;+\; \left(\!\!\uparrow\downarrow\!\!\right) \;+\; \left(\!\!\uparrow\downarrow\!\!\right) \;+\; \cdots \;=\; \tfrac{1}{2} \sum_{i,j<k_F} K_{ijij} \quad (12.38)$$

In a similar fashion:

$$\text{(diagrams)} + \text{(diagrams)} + \text{(diagrams)} + \cdots = -\tfrac{1}{2} \sum_{i,j<k_F} K_{ijji} \quad (12.39)$$

so that

$$E_0 \approx W_0 + \tfrac{1}{2} \sum_{i,j<k_F} (K_{ijij} - K_{ijji}). \quad (12.40)$$

Thus, the ground state energy may be expressed in terms of the K-matrix. At first sight, this does not seem to be progress, since one would expect that the fact that V_{ijkl} in (12.36) is infinite implies that K_{ijkl} is also infinite. Remarkably enough, this is not true. It can be shown that even for a hard core potential, (12.36) may be solved, and it yields a finite K. Hence (12.40) is a well-behaved first approximation for the ground state energy.

A simple example of an infinite V with a finite K is the 'pairing interaction'

$$V_{mnij} = \chi \delta_{m,-n} \delta_{i,-j}, \quad \chi \to \infty, \quad (12.41)$$

which means the only interactions are between particles in oppositely directed momentum states. The χ is an infinite constant. In this case, it is easily seen that

$$K_{mnij} = \frac{\delta_{m,-n}\delta_{i,-j}}{\dfrac{1}{\chi} + \sum_{q>k_F} \dfrac{1}{2(\epsilon_q - \epsilon_i)}} = \text{finite}, \quad \text{as } \chi \to \infty \quad (12.42)$$

is the solution, which can be checked by just substituting (12.41, 42) into (12.36), remembering that $\epsilon_k = \epsilon_{-k} = k^2/2m$.

The method of extending the calculation to include higher-order non-ladder diagrams omitted from (12.32) becomes clear if we note that (12.40) looks like the ordinary first-order approximation to the energy

$$E_0 \approx W_0 + \text{(diagram)} + \text{(diagram)}$$
$$= W_0 + \tfrac{1}{2} \sum_{i,j<k_F} (V_{ijij} - V_{ijji}), \quad (12.43)$$

except that V_{ijkl} has been replaced by K_{ijkl}. Let us define a new 'effective interaction' equal to K_{ijkl}, by the new interaction diagram

$$\text{(diagram)} \equiv K_{ijkl}. \quad (12.44)$$

Then (12.40) may be written diagrammatically as

$$E_0 \approx W_0 \; + \; \text{⟨diagram⟩} \; + \; \text{⟨diagram⟩} \tag{12.45}$$

The higher-order diagrams may now be included in a fashion which guarantees they are all finite by just drawing - - - - - - instead of $\sim\!\!\sim\!\!\sim$ in the ordinary diagrams, thus

$$E_0 = W_0 \; + \; \text{⟨diagram⟩} + \text{⟨diagram⟩} + \text{⟨diagram⟩} + \text{⟨diagram⟩}$$

$$+ \; \text{⟨diagram⟩} + \text{⟨diagram⟩} + \dots . \tag{12.46}$$

(Note that there are no second-order diagrams like ⟨diagram⟩ or ⟨diagram⟩ since

these are already included in ⟨diagram⟩ and ⟨diagram⟩ .)

[Note: In actual calculations on nuclear matter, one employs the 'method of undefined single-particle energies' in which an arbitrary single-particle potential is added to the unperturbed H_0 and subtracted from the interaction. This is described in the article by Goldstone in Pines (1961), p. 109.]

Further reading

Raimes (1972), chaps. 7, 8, 9
Pines (1963), chap. 3.
Kittel (1963), chap. 6.
Thouless (1972).
Brout (1963).

Exercises

12.1 Verify that the last two graphs in (12.2) are anomalous.
12.2 Write out the expression for the first graph on the right side of (12.21).
12.3 Carry out the summation of the following set of energy diagrams:

$$E = \text{⟨diagram⟩} + \text{⟨diagram⟩} + \text{⟨diagram⟩} + \text{⟨diagram⟩} + \dots .$$

12.4 Verify (12.42).

Chapter 13

Collective Excitations and the Two-Particle Propagator

13.1 Introduction

Up to now, we have been mainly concerned with quasi-particle excitations in many-body systems. Now we turn to the second of the two types of elementary excitations introduced in the zeroth chapter, i.e., collective excitations. As pointed out there, collective excitations are the quanta associated with collective motions of the system as a whole, such as, for instance, phonons, which are the quanta of the sound wave. Like quasi particles, collective excitations have particle-like qualities but, unlike quasi particles, these qualities do not at all resemble those of the original particles of the system.

It should be noted here that, despite the fact that collective excitations cannot be described in terms of the 'bare particle plus cloud' picture used in the quasi-particle case, we often hear such expressions as 'dressed' plasmon or 'clothed' phonon. This is due to the fact that if the collective excitations are allowed to interact with one another (or with other elementary excitations in the system), then a given collective excitation may become surrounded by a cloud of other elementary excitations, thus giving rise to the 'dressed' or 'quasi' collective excitation.

Collective excitations may be handled by means of the 'density fluctuation' or 'polarization' propagator, F, which is a special case of the two-particle Green's function. Just as the quasi-particle energies and lifetimes were found from the poles of the single-particle propagator, so the collective excitation energies and lifetimes are determined from the poles of the polarization propagator. This new propagator, F, may be expanded in a diagram series which turns out to be just the sum over all the 'polarization parts' occurring in the effective interaction (10.49).

Calculation of F for a high-density electron gas in RPA (i.e., summing over just repeated 'pair bubbles') shows that F describes the collective excitation called the 'free plasmon'. When higher-order diagrams are included, the plasmons become 'dressed', and acquire a renormalized frequency dispersion law and a finite lifetime.

13.2 The two-particle Green's function propagator

In the classical case, the two-particle propagator gives the probability $P(\mathbf{r}_4 t_4, \mathbf{r}_3 t_3, \mathbf{r}_2 t_2, \mathbf{r}_1 t_1)$ that if one particle is introduced into the many-particle system at point $\mathbf{r}_1$ at time t_1, and another at $\mathbf{r}_3$ at time t_3, then one of the particles will be observed at $\mathbf{r}_2$ at later time t_2, and the other at $(\mathbf{r}_4, t_4)$. This can be evaluated as the sum of the probabilities for all the different ways this could happen, including the particles scattering off each other. Thus, in the case of the accelerator in Fig. 2.3, we could write this diagrammatically as

where the ------ line stands for the particles colliding with each other. If a scattering probability $\cdot P_s$ (here $P_s = 1$) is associated with ------, then the diagrams may be evaluated by methods similar to those used in the single-particle case.

On the left side of (13.1) we have used the conventional 'box with four tails' diagram for the two-particle propagator. Note that we should *not* associate free propagators with the four 'tails' sticking out of the square when translating the diagram into functions. To avoid confusion here, it helps to use a 'stretched skin' picture for the propagator (see appendix M, and Mattuck and Theumann (1971)), i.e., the tail-less diagram:

where a dot in the corner means 'line emerging' and no dot means 'line entering'.

The quantum mechanical two-particle propagator may be defined analogously as the probability amplitude $G_2(\mathbf{r}_4 t_4, \ldots, \mathbf{r}_1 t_1)$ that if one particle is introduced at $\mathbf{r}_1$ at time t_1 and another at $(\mathbf{r}_3, t_3)$, then a particle will be observed at $(\mathbf{r}_2, t_2)$ and another at $(\mathbf{r}_4, t_4)$. This G_2 may be expanded diagrammatically, analogous to the pinball case. Evidently, G_2 is the sum of the amplitudes for all possible virtual processes in which the two particles interact with the system and with each other. Thus, abbreviating $(\mathbf{r}_1, t_1) \equiv 1$, etc., we find equation (13.2). (Regarding the minus sign in front of the exchange diagrams in (13.2), see exercise 13.7.)

Just as in the case of the single-particle propagator, there are other possibilities corresponding to the other time orders. For example,

$$G_2(t_3 > t_4 > t_1 > t_2)$$

$$\tag{13.2}$$

is defined as the probability amplitude that if a particle is introduced at $(\mathbf{r}_1, t_1)$ and a particle removed (i.e., hole introduced) at $(\mathbf{r}_2, t_2)$, then there will be a hole observed at $(\mathbf{r}_3, t_3)$ and a particle at $(\mathbf{r}_4, t_4)$. This may be shown diagrammatically as follows

$$- iG_2(4, 3, 2, 1) \equiv \quad = \quad + \quad + \quad + \cdots$$

$$+ \quad - \quad + \cdots \quad + \quad + \cdots. \qquad (13.3)$$

This form of the two-particle propagator is called the 'particle–hole' propagator.

All possible pieces of the definition of G_2, corresponding to all possible time orders, are summarized in the closed expression:

$$G_2(4, 3, 2, 1) = -i\langle \Psi_0| T\{\psi(\mathbf{r}_4 t_4)\,\psi^\dagger(\mathbf{r}_3 t_3)\,\psi(\mathbf{r}_2 t_2)\,\psi^\dagger(\mathbf{r}_1 t_1)\} |\Psi_0\rangle \quad (13.4)$$

where $\psi^\dagger(\mathbf{r}, t)$, $\psi(\mathbf{r}, t)$ create and destroy a particle at point $\mathbf{r}$ at time t (see (7.83)), $|\Psi_0\rangle$ is the interacting ground state, and T is the time-ordering operator, all as described in §9.2.

It is possible to derive the analogue of Dyson's equation for $G_2(4, 3, 2, 1)$ and to use it for determining various properties of the system (see Thouless (1972) or Galitskii (1960)). We will discuss here only one special form of G_2, namely, the 'polarization propagator'.

13.3 Polarization ['density fluctuation'] propagator

It was mentioned in the introduction that collective excitations are. essentially regular variations in the density, i.e., 'density fluctuations' in the many-body medium. It seems plausible that such waves might be described by a propagator which propagates a density disturbance from one point to another, analogous to the way the single-particle propagator propagates a single particle. It is easy to get such a propagator from G_2 in (13.4) by letting $(\mathbf{r}_3, t_3) = (\mathbf{r}_4, t_4)$ and $(\mathbf{r}_1, t_1) = (\mathbf{r}_2, t_2)$. This yields the '*density fluctuation propagator*' (which is a special case of the particle-hole propagator):

$$F(3, 1) = (-i)\langle \Psi_0| T\{\psi^\dagger(3)\,\psi(3)\,\psi^\dagger(1)\,\psi(1)\} |\Psi_0\rangle. \qquad (13.5)$$

(By definition of T in (9.4), for equal times, the $\psi^\dagger(\mathbf{r}, t)$ must always stand to the left of $\psi(\mathbf{r}, t)$, so this has been put in explicitly.)

The physical significance of F may be seen from the fact that

$$\begin{aligned}
\rho(\mathbf{r}, t) &= \psi^\dagger(\mathbf{r}, t)\psi(\mathbf{r}, t) \\
&= e^{iHt}\psi^\dagger(\mathbf{r})e^{-iHt}e^{+iHt}\psi(\mathbf{r})e^{-iHt} \\
&= e^{+iHt}\psi^\dagger(\mathbf{r})\psi(\mathbf{r})e^{-iHt}
\end{aligned} \tag{13.6}$$

is just, by (7.84) and (9.3), the operator giving the density (or strictly speaking, number of particles) at the point $\mathbf{r}, t$. Hence, F may be rewritten

$$\begin{aligned}
F(\mathbf{r}_2 - \mathbf{r}_1, t_2 - t_1) &= -i\langle\Psi_0| T\{\rho(\mathbf{r}_2, t_2)\rho(\mathbf{r}_1, t_1)\}|\Psi_0\rangle \\
&= -i\langle\Psi_0| T\{\rho(\mathbf{r}_2, t_2)\rho^\dagger(\mathbf{r}_1, t_1)\}|\Psi_0\rangle,
\end{aligned} \tag{13.7}$$

where the last line follows from

$$\rho(\mathbf{r}, t) = \psi^\dagger(\mathbf{r}, t)\psi(\mathbf{r}, t) = [\psi^\dagger(\mathbf{r}, t)\psi(\mathbf{r}, t)]^\dagger = \rho^\dagger(\mathbf{r}, t). \tag{13.8}$$

Thus, F creates a 'density disturbance' at $(\mathbf{r}_1, t_1)$ and propagates it to $(\mathbf{r}_2, t_2)$. (We have assumed H is time independent, and the system is homogeneous (no external potential) so that F depends only on space and time differences.)

The diagrammatic expansion for F is gotten immediately from that of G_2 in (13.3) by just setting $(\mathbf{r}_3, t_3) = (\mathbf{r}_4, t_4)$ and $(\mathbf{r}_1, t_1) = (\mathbf{r}_2, t_2)$, i.e., by tying the loose ends together in each diagram, thus:

This is just the series of 'polarization parts' (see (10.50)), and shows why F is called the 'polarization propagator'.

If (13.9) is transformed to $(\mathbf{k}, \omega)$-space, then it is easily evaluated by means of the effective interaction equation (10.49), which yields

$$\frac{(\text{~~~~}) - (\text{~~~})}{(\text{~~~})^2} = \bigcirc + \bigcirc + \cdots = \bigcirc. \quad (13.10)$$

Hence, substituting (10.53) into (13.10), we find

$$\bigcirc = \frac{\bigcirc(\pi)}{1 - \bigcirc(\pi)} \quad (13.11)$$

or

$$F(\mathbf{k}, \omega) = \frac{\pi(\mathbf{k}, \omega)}{1 + V_k \pi(\mathbf{k}, \omega)} = \frac{\pi(\mathbf{k}, \omega)}{\epsilon(\mathbf{k}, \omega)} \quad (13.12)$$

where (10.55) has been used. Thus, the density fluctuation or 'polarization' propagator has been expressed in terms of the sum over irreducible polarization parts, π, and the related generalized dielectric constant, $\epsilon(\mathbf{k}, \omega)$.

13.4 Retarded polarization propagator and linear response

The time-ordered polarization propagator above is closely related to the 'retarded polarization propagator', defined by

$$F^R(\mathbf{r}_2 - \mathbf{r}_1, t_2 - t_1) = -i\theta(t_2 - t_1) \langle \Psi_0 | [\rho(\mathbf{r}_2, t_2), \rho(\mathbf{r}_1, t_1)]_+ | \Psi_0 \rangle. \quad (13.12')$$

This is the analogue of the retarded single particle propagator, G^R in (9.20″) just as F in (13.7) is the analogue of G in (9.1).

The physical meaning of F^R is the following: It can be shown (see Fetter and Walecka (1971), pp. 172–75) that if we have, for example, a system with charge e per particle, and we apply a small external scalar electric potential $\phi(\mathbf{k}, \omega)$ to the system, then the change in the density of the system caused by ϕ is given to first order by

$$\delta \langle \rho(\mathbf{k}, \omega) \rangle = F^R(\mathbf{k}, \omega) e\phi(\mathbf{k}, \omega) \quad (13.12'')$$

where $F^R(\mathbf{k}, \omega)$ is the Fourier transform of the retarded polarization propagator. That is, F^R gives the linear response of the system to a small perturbation.

Now, since $F^R(\mathbf{k}, \omega)$ and $F(\mathbf{k}, \omega)$ are functions of only one frequency variable, they have the same relation to each other as the corresponding single particle propagators (see appendix L, part E). Hence we may write (see appendix (L. 26))

$$\operatorname{Re} F(\mathbf{k}, \omega) = \operatorname{Re} F^R(\mathbf{k}, \omega)$$

$$\operatorname{Im} F(\mathbf{k}, \omega) = \operatorname{sgn}(\omega) \operatorname{Im} F^R(\mathbf{k}, \omega), \qquad (13.12''')$$

so that if we have calculated F diagrammatically, it is easy to find F^R from it. (Note that F^R itself cannot be calculated directly from diagrams.)

13.5 The collective excitation propagator

Since quasi particles resembled free particles, physical arguments led us to conclude (see chapter 3) that they should have a propagator like that for free particles, except that the energy dispersion law was different and there was a damping factor. Physical intuition is not quite as helpful in the case of collective excitations since they do not at all resemble the free particles. However, it is possible to guess at the general form of the collective excitation propagator by looking at it in a well-known case, i.e., the phonon.

In a lattice with purely harmonic interatomic forces as in (1.28), phonons describe exact eigenstates of the system. The phonon propagator appears in the reprint (chapter 16) just after Equation (16.41), p. 284. From this we guess that the general form of $F(\mathbf{k}, \omega)$ when the collective excitations describe exact eigenstates of the system is:

$$
\begin{aligned}
F_0(\mathbf{k}, \omega) &= \frac{B_k}{\omega^2 - \omega_k^2 + 2i\omega_k \delta} \\
&= \frac{(2\omega_k)^{-1} B_k}{\omega - \omega_k + i\delta} - \frac{(2\omega_k)^{-1} B_k}{\omega + \omega_k - i\delta}
\end{aligned}
\qquad (13.13)
$$

where B_k is independent of ω, and ω_k is the frequency dispersion law of the excitation. This is called the 'free propagator' for a collective excitation, hence the subscript '0'. It is the analogue of the free-particle propagator G_0 in (8.35), but, unlike G_0, has both positive and negative frequency parts. And, of course, ω_k is totally different from $\epsilon_k (= k^2/2m)$.

If the collective excitation is not an exact eigenstate of the system, i.e., if the free collective excitations are allowed to interact with each other (or with other elementary excitations) then we get something analogous to quasi particles. The interaction between excitations produces a renormalized frequency dispersion law ω_k' and a finite lifetime τ_k (fairly long, for the picture to hold) so that (13.13) is replaced by

$$F_{\text{dressed}}(\mathbf{k}, \omega) = \frac{B_k'}{\omega^2 - \omega_k'^2 + 2i\omega_k'/\tau_k} \quad \text{where } 1/\tau_k \ll \omega_k' \qquad (13.14)$$

similar to (8.37). This is the dressed collective excitation propagator.

According to (13.13) (or 13.14), the poles at $\omega = \omega_k$ yield the energy of the collective excitations. This can be simply interpreted with the aid of the retarded polarization propagator F^R described in §13.4. From (13.12″) it is easy to see that the F_0^R corresponding to F_0 in (13.13) is

$$F_0^R(\mathbf{k}, \omega) = \frac{(2\omega_k)^{-1} B_k}{\omega - \omega_k + i\delta} - \frac{(2\omega_k)^{-1} B_k}{\omega + \omega_k + i\delta}. \tag{13.14'}$$

When $\omega = \omega_k$, $F_0^R \to \infty$ which means that by the linear response equation (13.12″), there can be finite oscillating density fluctuations in the system without any external driving field. This is just like, e.g., resonant oscillations in an ideal condenser, or on a frictionless guitar string. These resonant density fluctuations are just the collective oscillations.

The general circumstances under which $F(\mathbf{k}, \omega)$ in (13.12) has the collective excitation form (13.14) may be found by expanding F about its poles in the same way as we did for $G(\mathbf{k}, \omega)$ in §11.2. First break up π into real + imaginary parts

$$F(\mathbf{k}, \omega) = \frac{\pi_R + i\pi_I}{1 + V_k \pi_R + i V_k \pi_I}. \tag{13.15}$$

Analogous to the quasi-particle case, (11.15), define Ω_k as the solution of

$$1 + V_k \pi_R(\mathbf{k}, \Omega_k) = 0 \tag{13.16}$$

and expand π_R about Ω_k, assuming that π_R is a slowly varying function of ω^2 rather than just ω (this will be true in the example we will consider in the next section):

$$\pi_R(\mathbf{k}, \omega) = \pi_R(\mathbf{k}, \Omega_k) + \left(\frac{\partial \pi_R}{\partial \omega^2}\right)_{\Omega_k} (\omega^2 - \Omega_k^2) + \cdots. \tag{13.17}$$

Substituting into $F(\mathbf{k}, \omega)$

$$F(\mathbf{k}, \omega) \approx \frac{2\Omega_k}{V_k} \times \frac{\pi(\mathbf{k}, \Omega_k)}{\left(\dfrac{\partial \pi_R}{\partial \omega}\right)_{\Omega_k}} \times \frac{1}{\omega^2 - \Omega_k^2 + 2i\Omega_k \left(\dfrac{\pi_I}{(\partial/\partial\omega)\,\pi_R}\right)_{\Omega_k}}. \tag{13.18}$$

This evidently has the damped collective excitation form (13.14) provided

$$\frac{1}{\tau_k} = \left(\frac{\pi_I}{(\partial/\partial\omega)\,\pi_R}\right)_{\Omega_k} \ll \Omega_k. \tag{13.19}$$

13.6 Plasmons and quasi plasmons

An easy place to look for collective excitations is in the high-density electron gas since, as shown in §10.4 and §12.3, the RPA (random phase approximation) applies. This means that the sum over all irreducible polarization parts may be replaced by just the first pair bubble diagram as in (10.36):

$$\left(\pi\right) \approx \left(\right) \equiv -i\pi_0(\mathbf{k}, \omega)$$

$$= -i(\pi_{0R} + i\pi_{0I}). \qquad (13.20)$$

This may be substituted for π in (13.11), (13.18), giving

$$F_{\text{RPA}}(\mathbf{k}, \omega) = \frac{2\Omega_k}{V_k} \times \frac{\pi_0(\mathbf{k}, \Omega_k)}{\left(\frac{\partial \pi_{0R}}{\partial \omega}\right)_{\Omega_k}} \times \frac{1}{\omega^2 - \Omega_k^2 + 2i\Omega_k\left(\frac{\pi_{0I}}{(\partial/\partial\omega)\pi_{0R}}\right)_{\Omega_k}}. \qquad (13.21\text{a})$$

We now evaluate (13.21a) using the frequency-dependent $\pi_0(\mathbf{k}, \omega)$ (for small $\mathbf{k}$) which appears in (10.77). In the high frequency limit, $\zeta\ (= m\omega/k_F k)$ in (10.77) is large, so we may use the expansion:

$$\ln\left(\frac{\zeta+1}{\zeta-1}\right) = \frac{2}{\zeta} + \frac{2}{3} \cdot \frac{1}{\zeta^3} + \frac{2}{5} \cdot \frac{1}{\zeta^5} + \cdots; \quad \zeta^2 > 1. \qquad (13.21\text{b})$$

Placing this in (10.77), using (10.70) and retaining only terms up to ζ^{-2} yields

$$\pi_0(\mathbf{k}, \omega) = -\frac{mk_F}{3\pi^2 \zeta^2} = -\left(\frac{k_F^3}{3\pi^2 m}\right)\frac{k^2}{\omega^2} = -\frac{nk^2}{m\omega^2}, \qquad (13.21\text{c})$$

where n = electron density, and we have used (9.44). The pole equation (13.16) is therefore (note that π_0 is pure real in this approximation):

$$1 - \frac{4\pi e^2}{k^2} \cdot \frac{nk^2}{m\Omega_k^2} = 0; \quad \Omega_k = \sqrt{\frac{4\pi e^2 n}{m}} \equiv \omega_p. \qquad (13.21\text{d})$$

Thus, in this approximation, Ω_k is independent of $\mathbf{k}$ and is equal to the so-called 'plasma frequency' which is the frequency of free oscillations in a classical electron gas (see Raimes (1961), chap. 10). In a better approximation Ω_k depends on $\mathbf{k}$ (see Exercise 13.5).

Using (13.21c) to work out the other quantities in (13.21a) yields for the polarization propagator

$$F_{\text{RPA}}(\mathbf{k}, \omega) = \frac{-nk^2/m}{\omega^2 - 4\pi e^2 n/m}. \qquad (13.21\text{e})$$

Comparing this with (13.13), (13.14) shows that we have here the propagator for a collective excitation of infinite lifetime, and having the plasma frequency (see exercise 13.5 for the **k**-dependence of the excitation frequency). This excitation is called the '*free plasmon*', and is the second of the two types of elementary excitations in the electron gas (the first was the quasi electron).

Let us represent the free or 'bare' plasmon propagator by a vertical dotted line; then its diagram expansion is, by (13.11, 20):

$$\text{(RPA)} \equiv \quad = \quad + \quad + \quad + \cdots \qquad (13.22)$$

If higher-order polarization parts are added to the series for ⬭, they represent interaction of the plasmon with other plasmons and quasi electrons, and convert the bare plasmon to the 'dressed' or 'quasi' plasmon with a finite lifetime. This is easily seen by including, for example, the ⬡ diagram in the polarization propagator expansion:

$$\approx \quad + \quad + \quad + \quad + \quad + \quad + \quad + \quad + \cdots . \qquad (13.23)$$

Regrouping diagrams and using (13.22) yields

$$\approx \left[\quad + \quad + \quad + \cdots \right]\left[\quad + \quad + \quad + \quad + \quad + \cdots \right]\left[\quad + \cdots \right] + \cdots$$

$$= \quad + \quad + \quad + \quad + \quad + \quad + \cdots \quad + \cdots \qquad (13.24)$$

$$= \quad + \quad + \quad + \cdots = \qquad (13.25)$$

where the last line is the conventional way of drawing this. Thus, the polarization part here acts like the irreducible self-energy parts in the single-electron propagator, and dresses the plasmon in the same way that the addition of self-energy parts dressed the electron.

13.7 Expressing the two-particle propagator in terms of the scattering amplitude

The amplitude for the scattering of two quasi particles was shown in (10.63). It is useful to note that by combining (10.63) with (13.2), the two-particle propagator may be expressed in terms of the scattering amplitude thus:

(13.26)

Further reading

Fetter and Walecka (1971), chap. 5.
Raimes (1961), chap. 10.
Pines (1963), chap. 3.
Schultz (1964), chap. 3, pp. 99–103.
Pines (1961), chap 2, pp. 34–42; chap. 3, pp. 55-65.

Exercises

13.1 Assume that the interaction is such that the most important scattering processes in (13.2) are those of the ladder type. Write out the series for such processes (ignore the 'crossed' ladder diagrams, for simplicity) and obtain an integral equation for G_2 in diagram form. Re-write the result in terms of the K-matrix (10.19), and compare with (13.26).

13.2 Sum the series for the plasmon propagator in RPA (13.22) and show that the result is just (13.11) with $\left(\pi \right) = \left\langle \right\rangle$.

13.3 Sum the clothed plasmon series:

 (a) Before regrouping (as in (13.23)).

 (b) After regrouping (see (13.24)).

 (c) Use the answer to Ex. 13.2 to prove the two results are the same.

13.4 Verify (13.14′).

13.5 Verify (13.21e).

13.6 Show that if the ζ^{-5} term of (13.21b) is included in the approximation for π_0, then the plasmon dispersion law becomes

$$\omega^2 = \omega_p^2 + \frac{3}{5} \frac{k_F^2}{m^2} k^2.$$

13.7 Use (13.4) to show why the exchange diagrams in (13.2) must have a minus sign. (Consider the zeroth order term in the perturbation expansion of (13.4).)

Chapter 14

Fermi Systems at Finite Temperature

14.1 Generalization of the $T=0$ case

The problems considered until now—excited states (i.e., elementary excitations) and ground state energy—are physically unrealistic in the sense that they assume the many-body system is at zero temperature. This is fine for nuclei, whose energies are large compared with ordinary thermal energies ($\sim \frac{1}{40}$ eV), and for solids at extremely low temperatures. However, for nuclei in hot plasmas and inside stars, or for solids in ordinary environments, the $T=0$ approximation may be a very bad one. At finite temperatures, the system will be statistically distributed over all of its excited levels. This means that the ground state 'average', $\langle \Psi_0 | \dots | \Psi_0 \rangle$, which was used to calculate the $T=0$ propagator, must now be replaced by an average over a grand canonical ensemble.

For one who has just mastered the complicated graphical art for the $T=0$ case, the sight of temperature rearing its ugly head might provoke the Archimedean response: 'Don't disturb my diagrams!'. However, the remarkable thing is that the diagrams are in fact not disturbed by the addition of temperature to the problem. The $T>0$ graphs are drawn precisely the same as the $T=0$ ones, the difference between the two cases lying solely in the dictionary used to translate the lines and wiggles into words. This makes it possible to get the whole finite temperature theory from the zero temperature one at no extra charge.

Take, for instance, the finite temperature single-particle propagator, $\mathcal{G}$. This can be used to find, among other things, the temperature dependence of the quasi-particle energy dispersion law, and of the bare particle momentum distribution function, etc. The diagram expansion for $\mathcal{G}$ is the same as in the $T=0$ case; the only modification required is to associate a statistical weighting factor with each directed line, change the time variable to imaginary time and subtract the chemical potential μ from the energy.

Another example is the grand partition function, Z, which is the key to the equilibrium thermodynamics of a many-body system. If we can find Z, everything—energy, entropy, pressure, etc.—can be calculated from it in a simple way. It will be shown that Z is just proportional to the finite temperature vacuum amplitude, and can be obtained from the same diagrams used to find the ground state energy E_0.

14.2 Statistical mechanics in occupation number formalism

Since occupation number formalism refers to a system in which the number of particles, N, is in general not fixed, we have to use a statistical method tailored to the variable N problem, e.g., the method of the *grand canonical ensemble*. This means that our system of particles is immersed in a reservoir composed of the same kind of particles held at fixed temperature T. The system can release particles or energy to the reservoir, or receive particles or energy from it.

Suppose the system Hamiltonian, H, is given by (7.51). Since N is variable, the eigenstates $|\Psi_i\rangle$ of H, will depend on N; let us call N_i the number of particles when the system is in $|\Psi_i\rangle$, and call the corresponding energy E_i. That is:

$$H|\Psi_i\rangle = E_i|\Psi_i\rangle, \qquad \sum_k c_k^\dagger c_k |\Psi_i\rangle = N_i|\Psi_i\rangle. \tag{14.1}$$

The probability that the system (regarded as a member of an ensemble) will be found in state $|\Psi_i\rangle$ is given by

$$\mathscr{P}_i = \frac{e^{-\beta[E_i - \mu N_i]}}{\sum_n e^{-\beta[E_n - \mu N_n]}} = \frac{\rho_i}{Z} \tag{14.2}$$

where $\mu =$ chemical potential, which is the energy required to remove one particle from the system, and $\beta = 1/kT$ ($k =$ Boltzmann factor). The denominator of (14.2), Z, is the '*grand partition function*', and the numerator, ρ_i, is the '*grand distribution function*'. It is convenient to write these quantities in terms of the *distribution operator*

$$\rho = e^{-\beta(H - \mu N)}, \tag{14.3}$$

(N stands for number operator) thus:

$$\rho_i = \langle \Psi_i | \rho | \Psi_i \rangle \tag{14.4}$$

$$Z = \operatorname{tr}\rho, \qquad \mathscr{P}_i = \frac{\langle \Psi_i | \rho | \Psi_i \rangle}{\operatorname{tr}\rho}. \tag{14.5}$$

In order to get the average value of any operator $\mathcal{O}$, we just calculate the weighted average:

$$\langle \mathcal{O} \rangle = \sum_i \langle \Psi_i | \mathcal{O} | \Psi_i \rangle \mathscr{P}_i = \frac{\operatorname{tr}\mathcal{O}\rho}{\operatorname{tr}\rho}. \tag{14.6}$$

An example of this is the average energy

$$\bar{E} = \langle H \rangle = \frac{\sum_i E_i e^{-\beta(E_i - \mu N_i)}}{\sum_i e^{-\beta(E_i - \mu N_i)}} = -\left(\frac{\partial}{\partial \beta} \ln Z\right)_{\beta\mu} \tag{14.7}$$

Equation (14.6) uses the exact many-body state vectors $|\Psi_i\rangle$. Since these are not known, in general, one makes use of the fact that the trace is invariant under change of representation, and transforms to a more comfortable basis, say $|\Phi_n\rangle$. The $|\Phi_n\rangle$ are usually chosen to be eigenstates of the non-interacting system. Then we have

$$\langle \mathcal{O} \rangle = \frac{\sum\limits_{i,p} \langle \Phi_i | \mathcal{O} | \Phi_p \rangle \langle \Phi_p | \rho | \Phi_i \rangle}{\sum\limits_{i} \langle \Phi_i | \rho | \Phi_i \rangle}. \qquad (14.8)$$

Let us apply the above to a rapid (although perhaps a bit weird-looking!) derivation of some old familiar results for a system of non-interacting fermions. The Hamiltonian and state vector are just

$$H_0 = \sum_k \epsilon_k c_k^\dagger c_k, \quad |\Phi_i\rangle = |n_1^i, n_2^i, \ldots, n_k^i, \ldots\rangle. \qquad (14.9)$$

The grand distribution function is

$$\begin{aligned}
\rho_{i0} &= \langle \Phi_i | e^{-\beta \sum_k (\epsilon_k - \mu) c_k^\dagger c_k} | \Phi_i \rangle \\
&= e^{-\beta \sum_k (\epsilon_k - \mu) n_k^i} = \prod_k e^{-\beta(\epsilon_k - \mu) n_k^i},
\end{aligned} \qquad (14.10)$$

where μ = chemical potential, and the subscript '0' denotes non-interacting system. The grand partition function may be written

$$Z_0 = \operatorname{tr} \rho_0 = \sum_i \rho_{i0} = \sum_{n_1^i, \ldots, n_k^i, \ldots} \prod_k e^{-\beta(\epsilon_k - \mu) n_k^i}. \qquad (14.11)$$

The first term in the sum has $\{n_1^i, \ldots, n_k^i, \ldots\} = \{0,0,0,\ldots\}$, the second term $= \{1,0,0,0,\ldots\}$, etc., so that

$$\begin{aligned}
Z_0 &= 1 + e^{-\beta(\epsilon_1 - \mu)} + e^{-\beta(\epsilon_2 - \mu)} + \cdots + e^{-\beta(\epsilon_1 - \mu)} \times e^{-\beta(\epsilon_2 - \mu)} + \cdots \\
&= \prod_k [1 + e^{-\beta(\epsilon_k - \mu)}]
\end{aligned} \qquad (14.12)$$

where we have used that $n_k = 0$ or 1.

From these results one can compute the average values of operators for the non-interacting system. For example, the average value of the number of particles in state k of the non-interacting system is (note that ρ, $c_k^\dagger c_k$ are diagonal in the $|\Phi_i\rangle$-basis, so $\langle \Phi_i | \rho | \Phi_j \rangle = 0$ for $i \neq j$, etc.):

$$\begin{aligned}
\langle n_k \rangle_0 = \langle c_k^\dagger c_k \rangle_0 = \sum_{\ldots n_k^i \ldots} &\langle n_1^i, \ldots, n_k^i, \ldots | c_k^\dagger c_k | n_1^i, \ldots, n_k^i, \ldots \rangle \times \\
&\times \prod_l \frac{e^{-\beta(\epsilon_l - \mu) n_l^i}}{[1 + e^{-\beta(\epsilon_l - \mu)}]}.
\end{aligned} \qquad (14.13)$$

The only terms contributing to the sum are those for which

$$\langle \ldots, n_k^i, \ldots | c_k^\dagger c_k | \ldots, n_k^i, \ldots \rangle = 1 \quad (\text{i.e., } n_k = 1),$$

so this becomes:

$$\langle n_k \rangle_0 = \frac{e^{-\beta(\epsilon_k - \mu)}}{1 + e^{-\beta(\epsilon_k - \mu)}} \times \sum_{\ldots n_k^i \ldots} \prod_{l \neq k} \frac{e^{-\beta(\epsilon_l - \mu) n_l^i}}{[1 + e^{-\beta(\epsilon_l - \mu)}]}. \tag{14.14}$$

The sum over products = 1 (by the same method used to prove (14.12)) leading to the final result

$$\langle n_k \rangle_0 = \langle c_k^\dagger c_k \rangle_0 = \frac{1}{e^{+\beta(\epsilon_k - \mu)} + 1} = f_k^- \tag{14.15}$$

which is just the well-known Fermi distribution function obtained by an elegant but rather unpleasant method. Another useful average value is

$$\langle c_k c_k^\dagger \rangle_0 = \langle 1 \rangle_0 - \langle c_k^\dagger c_k \rangle_0$$

$$= 1 - f_k^- = \frac{1}{e^{-\beta(\epsilon_k - \mu)} + 1} = f_k^+. \tag{14.16}$$

14.3 The finite temperature propagator

The zero temperature propagator for a mutually interacting N particle system, with no external potential was:

$$G(\mathbf{k}, t_2 - t_1) = -i \langle \Psi_0 | T\{c_k(t_2) c_k^\dagger(t_1)\} | \Psi_0 \rangle \tag{14.17}$$

which is evidently just the average or 'expectation' value of the operator $T\{c_k(t_2) c_k^\dagger(t_1)\}$ in the ground state. Hence the finite temperature propagator may be gotten simply by averaging this same operator over an ensemble of systems at temperature T by means of (14.6):

$$G^T(\mathbf{k}, t_2 - t_1) = -i \langle T\{c_k(t_2) c_k^\dagger(t_1)\} \rangle$$

$$= -i \frac{\text{tr } T\{c_k(t_2) c_k^\dagger(t_1)\} \rho}{\text{tr } \rho}. \tag{14.18}$$

Now our object is to evaluate this function and use it to find the system properties. It would be nice if we could somehow expand G^T diagrammatically like we did for G, and calculate it by partial summation. This cannot be done with G^T as it stands; however, it is possible to do it on a modified G^T called the '*imaginary time Green's function*' defined by (note that we drop the i-factor in order to agree with the convention in Fetter and Walecka (1971), p. 227 ff.)

$$\mathcal{G}(\mathbf{k}, \tau_2 - \tau_1) = -\langle T\{c_k(\tau_2) c_k^\dagger(\tau_1)\} \rangle,$$

where

$$\mathcal{O}(\tau) = e^{(H-\mu N)\tau}\, \mathcal{O}\, e^{-(H-\mu N)\tau} \tag{14.19}$$

and

$$0 < \tau_1, \tau_2 < \beta, \quad \tau \text{ real.}$$

Evidently, $\mathcal{G}$ is obtained from G^T by making the following replacements

$$\begin{aligned} H &\to H - \mu N \\ it &\to \tau \end{aligned} \tag{14.20}$$

Since τ is real, $t = -i\tau$ will be imaginary. Hence $\mathcal{G}$ is called the 'imaginary time propagator'. Note that T in (14.19) means that the operators are arranged so that τ is decreasing from left to right.

The reason why $\mathcal{G}$ can be expanded in the same diagram series as G (for $T=0$) is this: As shown in appendix B, the diagram series for G comes fundamentally from the time-dependent Schrödinger equation. Now the distribution operator ρ obeys the equation

$$\frac{\partial \rho}{\partial \beta} = -(H - \mu N)\rho \quad \text{(Bloch's equation)} \tag{14.21}$$

as can be verified by differentiating (14.3). If this is compared with the time-dependent Schrödinger equation, one sees the correspondence $\Psi \leftrightarrow \rho$, $H \leftrightarrow H - \mu N$, $it \leftrightarrow \beta$. This suggests that by making replacements (14.20) everywhere in sight, we can build up a finite temperature theory based on the Bloch equation in the same way the zero temperature theory was based on the Schrödinger equation. In particular, it turns out that if these replacements are made in G, the resulting $\mathcal{G}$ can be expanded in a perturbation series which is the nearly identical twin of the series for the zero temperature propagator. (See appendices, (D.14→19), (E.15→17).)

Despite the unphysical clang of 'imaginary time', it is not hard to see that it will be just about as easy to get physical information out of $\mathcal{G}$ as out of G^T. This is because, first of all, the only effect the $H \to H - \mu N$ replacement has, is to shift the single-particle energy by μ, since

$$H_0 - \mu N = \sum_k (\epsilon_k - \mu)\, c_k^\dagger c_k.$$

Secondly, although the result we get for $\mathcal{G}$ will be a function of τ, we can always get this result in terms of real time by replacing τ by it or, more properly, by analytically continuing the result back to the real t-axis.

The variables τ_1, τ_2 are restricted to the interval $(0, \beta)$ (so that $\tau_2 - \tau_1 = \tau$ is on $(-\beta, +\beta)$) because $\mathcal{G}(\mathbf{k}, \tau_2 - \tau_1)$ is guaranteed to converge on this interval

(Mills (1969), pp. 72–3). This can be seen by using (14.19). Considering the $\tau > 0$ case first, we have (let $\bar{H} \equiv H - \mu N$)

$$\mathscr{G}(\mathbf{k}, \tau) = \frac{-1}{Z} \operatorname{tr} \rho c_k(\tau) \, c_k^\dagger(0) = \frac{-1}{Z} \sum_i \langle \Psi_i | e^{-\beta \bar{H}} e^{+\tau \bar{H}} c_k \, e^{-\tau \bar{H}} c_k^\dagger | \Psi_i \rangle.$$

Inserting $\sum_j |\Psi_j\rangle\langle\Psi_j|$ at the appropriate point, this becomes (let $\bar{E}_i \equiv E_i - \mu N_i$):

$$\mathscr{G}(\mathbf{k}, \tau) = \frac{-1}{Z} \sum_{i,j} e^{(-\beta+\tau)\bar{E}_i} \langle \Psi_i | c_k | \Psi_j \rangle e^{-\tau \bar{E}_j} \langle \Psi_j | c_k^\dagger | \Psi_i \rangle. \qquad (14.22)$$

Since $\bar{E}_{i,j}$ can be arbitrarily large positive quantities, we can guarantee that this sum is finite if the exponents are negative, i.e., if $\tau < \beta$ and $\tau > 0$. A similar proof holds for the $\tau < 0$ case.

The imaginary time propagator obeys the so-called 'quasi-periodic boundary condition' on the interval $(-\beta, +\beta)$:

$$\mathscr{G}(\mathbf{k}, \tau) = -\mathscr{G}(\mathbf{k}, \tau+\beta) \quad \text{for} \quad -\beta < \tau < 0. \qquad (14.23)$$

This is proved with the aid of the theorem that $\operatorname{tr} AB = \operatorname{tr} BA$ as follows (for $-\beta < \tau < 0$):

$$\mathscr{G}(\mathbf{k}, \tau) = \frac{1}{Z} \operatorname{tr} e^{-\beta \bar{H}} c_k^\dagger(0) \, c_k(\tau)$$

$$= \frac{1}{Z} \operatorname{tr} c_k(\tau) \, e^{-\beta \bar{H}} c_k^\dagger(0)$$

$$= \frac{1}{Z} \operatorname{tr} e^{-\beta \bar{H}} \underbrace{e^{+\beta \bar{H}} c_k(\tau) \, e^{-\beta \bar{H}}}_{= \, c_k(\tau+\beta)} c_k^\dagger(0)$$

$$= -\mathscr{G}(\mathbf{k}, \tau+\beta). \qquad (14.24)$$

The restriction of τ to the interval $(-\beta, \beta)$ causes trouble when going over to the equivalent of $(\mathbf{k}, \omega)$ space, since the Fourier transform requires integrations from $\tau = -\infty \to +\infty$. The difficulty is eliminated by adding still another propagator to the menagerie—call it $\mathscr{G}_{\text{per}}$—which is just $\mathscr{G}$ periodically repeated from $-\infty$ to $+\infty$. This is given by the Fourier series:

$$\mathscr{G}_{\text{per}}(\mathbf{k}, \tau) = \frac{1}{\beta} \sum_{n=-\infty}^{+\infty} e^{-i\omega_n \tau} \mathscr{G}(\mathbf{k}, \omega_n) \qquad (14.25)$$

where

$$\mathscr{G}(\mathbf{k}, \omega_n) = \frac{1}{2} \int\limits_{-\beta}^{+\beta} d\tau \, e^{i\omega_n \tau} \mathscr{G}(\mathbf{k}, \tau) \tag{14.26}$$

$$\omega_n = \frac{\pi n}{\beta}, \quad n = 0, \pm 1, \pm 2, \dots. \tag{14.27}$$

The function $\mathscr{G}_{\mathrm{per}}(\mathbf{k}, \tau)$ is equal to $\mathscr{G}$ on the interval $(-\beta, +\beta)$ and repeats on $(\beta, 3\beta)$, $(3\beta, 5\beta)$, ..., $(-3\beta, -\beta)$, ..., etc. The Fourier transform of $\mathscr{G}_{\mathrm{per}}(\mathbf{k}, \tau)$ is evidently just $\mathscr{G}(\mathbf{k}, \omega_n)$.

Substituting (14.23) into (14.26) shows that even n terms equal zero, so that (14.27) is replaced by

$$\omega_n = \frac{(2n+1)\pi}{\beta} \tag{14.28}$$

(In the case of bosons, the odd n terms are eliminated.) Note that with the aid of (14.23) we can also show that (14.26) becomes

$$\mathscr{G}(\mathbf{k}, \omega_n) = \int\limits_{0}^{\beta} d\tau \, e^{i\omega_n \tau} \mathscr{G}(\mathbf{k}, \tau) \tag{14.28'}$$

Let us now look at the free imaginary time propagator. Using (14.19) and the fact that $c_k(\tau) = c_k \exp(-\tau(\epsilon_k - \mu))$ for the non-interacting system (this is just (F.7) with replacements (14.20)), and employing (14.15), (14.16) gives

$$\mathscr{G}_0(\mathbf{k}, \tau_2 - \tau_1) = -[\theta_{\tau_2 - \tau_1} f_k^+ - \theta_{\tau_1 - \tau_2} f_k^-] e^{-(\epsilon_k - \mu)(\tau_2 - \tau_1)}. \tag{14.29}$$

The physical meaning of (14.29) may be seen by comparing it with the ordinary $T=0$ free propagator in (8.34). Aside from the shift of energy zero by μ, and the $it \to \tau$ change, the essential difference is the replacement of $\theta_{\epsilon_k - \epsilon_F}$, $\theta_{\epsilon_F - \epsilon_k}$ by the statistical factors f_k^+, f_k^-. These factors smear out the θ-functions, as shown in Fig. 14.1. Interpreting a 'hole' now not as an empty state below k_F but in a more general fashion (see paragraph following (4.30)) as that part of $\mathscr{G}_0$ for $\tau_2 < \tau_1$, it is seen that

> For $T>0$ it is possible to have a hole and particle both in the
> same k-state. (14.30)

If $\mathscr{G}_0$ is now periodically repeated (from now on we drop the label 'per') by (14.25), then its Fourier transform is obtained by substituting (14.29) into (14.26), yielding

$$\mathscr{G}_0(\mathbf{k}, \omega_n) = \frac{1}{i\omega_n - \epsilon_k + \mu} \tag{14.31}$$

with ω_n as in (14.28). Because of the $i\omega_n$ appearing in it, $\mathscr{G}_0(\mathbf{k}, \omega_n)$ (or $\mathscr{G}(\mathbf{k}, \omega_n)$) is called an 'imaginary frequency propagator'.

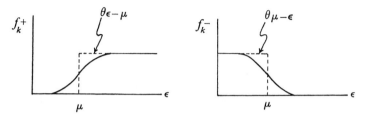

FIG. 14.1 *Statistical Factors for the Particle* (f^+) *and Hole* (f^-) *Parts of the Free Imaginary Time Propagator*

As discussed in appendix E, section 3, $\mathscr{G}$ may be expanded in a perturbation series which is the twin brother of the one for G in the $T=0$ case. From this it follows that $\mathscr{G}$ may be expressed in the same diagrams as G was:

$$\tag{14.32}$$

where now 'anomalous' diagrams (see after (9.32)) must be included because of (14.30). This series may be evaluated by means of the finite temperature dictionary in Table 14.1. (See appendix (D.13, 19) regarding missing i-factor in front of V.) Note that the imaginary frequency non-propagating lines in Table 14.1 require a convergence factor, $\exp(i\omega_n 0^+)$ analogous to that in the $T=0$ case, Table 9.1. The frequency sum:

$$\frac{1}{\beta} \sum_{n=-\infty}^{+\infty} \mathscr{G}_0(\mathbf{k}, \omega_n) e^{i\omega_n 0^+} = \frac{1}{\beta} \sum_{n=-\infty}^{+\infty} \frac{e^{i\omega_n 0^+}}{i\omega_n - \epsilon_k + \mu} = f_k^- \tag{14.32'}$$

may be carried out with the aid of (14.54).

Dyson's equation has the same graphical form as in (10.7), so that

$$\mathscr{G}(\mathbf{k}, \omega_n) = \frac{1}{i\omega_n - \epsilon_k + \mu - \Sigma(\mathbf{k}, \omega_n)}. \tag{14.33}$$

Table 14.1 *Diagram dictionary for interacting fermion system at finite temperature (no external potential)*

$(\mathbf{k},\tau)$-space — Diagram element	Factor	$(\mathbf{k},\omega_n)$-space — Diagram element	Factor
	$\mathscr{G}(\mathbf{k},\tau)$		$\mathscr{G}(\mathbf{k},\omega_n)$
	$\mathscr{G}_0(\mathbf{k},\tau) = -[\theta_\tau f_k^+ - \theta_{-\tau} f_k^-] \times e^{-(\epsilon_k-\mu)\tau}$		$\mathscr{G}_0(\mathbf{k},\omega_n) = +\dfrac{1}{i\omega_n - \epsilon_k + \mu}$, $\quad \omega_n = \dfrac{(2n+1)\pi}{\beta}$, $\quad \beta = 1/kT$
	$\mathscr{G}(\mathbf{k},0^-)$		$\mathscr{G}(\mathbf{k},\omega_n)\exp(i\omega_n 0^+)$
	$\mathscr{G}_0(\mathbf{k},0^-) = f_k^-$		$\mathscr{G}_0(\mathbf{k},\omega_n)\exp(i\omega_n 0^+)$ (so that: $\dfrac{1}{\beta}\sum\limits_{n=-\infty}^{+\infty}\mathscr{G}_0(\mathbf{k},\omega_n)\exp(i\omega_n 0^+) = f_k^-$)
	$-V_{klmn}$ or $-V_q$		$-V_{klmn}$ or $-V_q$
fermion loop	-1	fermion loop	-1
Intermediate $\mathbf{k},\tau$	$\sum\limits_k$ or $\int \dfrac{d^3k}{(2\pi)^3}$, $\int_0^\beta d\tau$	Intermediate $\mathbf{k},\omega_n$	$\sum\limits_k$ or $\int \dfrac{d^3k}{(2\pi)^3}$, $\dfrac{1}{\beta}\sum\limits^{+\infty}$

A simple example of this is the finite temperature Hartree–Fock approximation in which the irreducible self-energy, Σ, is given by (10.11). Evaluating this with the aid of Table 14.1 produces (multiply V_{kpkp} by 2 if spin is included)

$$\Sigma(\mathbf{k}, \omega_n) = \left| \text{\small\raisebox{0pt}{$\sim$}} \hspace{-2pt}\bigcirc \right.\, \mathbf{P}, \omega_i \; + \; \text{\small$\bigvee$}$$
$$\qquad\qquad\qquad\qquad\qquad \mathbf{P}, \omega_i$$

$$= \sum_p \frac{1}{\beta} \sum_{i=-\infty}^{+\infty} [(-1)(-V_{kpkp}) - V_{pkkp}] \mathcal{G}_0(\mathbf{p}, \omega_i)$$

$$= \sum_p (V_{kpkp} - V_{pkkp}) f_p^-. \tag{14.34}$$

Substituting this in (14.33) shows that we have the finite temperature analogue of the Hartree–Fock quasi particle, with energy

$$\epsilon_k' = \epsilon_k + \sum_{all\,\mathbf{p}} (V_{kpkp} - V_{pkkp})(e^{\beta(\epsilon_p - \mu)} + 1)^{-1}. \tag{14.35}$$

The effective field seen by the particle in $\mathbf{k}$ is modified by the fact that some of the particles causing this field are now above the Fermi surface on account of the finite temperature. This is reflected in the statistical factor on the right. Thus the quasi-particle energies have acquired a dependence on temperature through the $\beta = 1/kT$-factor—a good example of the rather bizarre-sounding concept of 'temperature-dependent energy levels' in quantum mechanics. (The true levels are not temperature dependent, of course.)

Note that when $T \to 0, f_p^- \to \theta(\mu - \epsilon_p)$. If now N is fixed during the calculation, so that $\mu = \epsilon_F$ (see §9.7), then $f_p^- = \theta(\epsilon_F - \epsilon_p)$ and (14.35) reduces to (4.78) as it should.

The above energy expression (14.35) may be made 'self-consistent' by replacing ϵ_p by ϵ_p' in the exponential on the right of (14.35), giving us a messy equation to be solved for ϵ_p'. Graphically, this means that for the irreducible self-energy we are using the first two terms of (11.3), i.e.:

$$\text{\small$\bigotimes\limits_{\Sigma}$} \; \approx \; \left| \text{\small\raisebox{0pt}{$\sim$}}\hspace{-2pt}\bigcirc \right. \; + \; \text{\small$\bigvee$} \tag{14.36}$$

14.4 The finite temperature vacuum amplitude

The zero temperature vacuum amplitude was defined by

$$R(t) = \langle \Phi_0 | U(t) | \Phi_0 \rangle e^{iW_0 t}. \tag{14.37}$$

For our present purposes we need the explicit expression for this in terms of the operator, $\tilde{U}$, as it appears in appendix (E.13), with $t_0 = 0$:

$$
\begin{aligned}
R(t) &= \langle \Phi_0 | \, \tilde{U}(t) \, | \Phi_0 \rangle \\
&= \langle \Phi_0 | \, e^{iH_0 t} e^{-iHt} \, | \Phi_0 \rangle
\end{aligned}
\tag{14.38}
$$

where appendix (D.1) has been used.

The finite temperature vacuum amplitude would then be given by

$$
R^T = \langle \tilde{U}(t) \rangle_0,
\tag{14.39}
$$

where $\langle \ \rangle_0$ means average over ensemble of *non*-interacting systems at temperature T. As was the case with G^T in (14.18), R^T is not very useful. A more convenient function is obtained by making the replacement (14.20), yielding

$$
\mathscr{R}(\beta) = \langle \tilde{U}(\beta) \rangle_0 = \langle e^{\beta(H_0 - \mu N)} e^{-\beta(H - \mu N)} \rangle_0.
\tag{14.40}
$$

We now notice that the grand partition function is proportional to $\mathscr{R}(\beta)$ since

$$
Z = \mathrm{tr}\, e^{-\beta(H - \mu N)} = \mathrm{tr}\, [e^{-\beta(H_0 - \mu N)}\, \tilde{U}(\beta)]
$$

$$
Z_0 = \mathrm{tr}\, e^{-\beta(H_0 - \mu N)}
\tag{14.41}
$$

so that

$$
Z/Z_0 = \langle \tilde{U}(\beta) \rangle_0 = \mathscr{R}(\beta).
\tag{14.42}
$$

Hence, by (14.7), we can easily get a formula for the average energy:

$$
\langle E \rangle = -\frac{\partial}{\partial \beta} \ln Z_0 - \frac{\partial}{\partial \beta} \ln \mathscr{R}(\beta),
\tag{14.43}
$$

which is evidently the $T \neq 0$ counterpart of the zero temperature theorem (5.4):

$$
E_0 = W_0 + i \left[\frac{d}{dt} \ln R(t) \right]_{t \to \infty(1 - i\eta)}.
\tag{14.44}
$$

As shown in appendix (E.19), $\mathscr{R}(\beta)$ may be expanded in a perturbation series which is identical in form to that for $R(t)$ in the $T = 0$ case. This means that the $T = 0$ diagram expansion still holds good for the $T > 0$ case:

$$
\tag{14.45}
$$

Note that it is necessary to include anomalous diagrams. Examples of the evaluation of these diagrams, using Table 14.1, are:

$$
\text{k} \,\bigcirc\!\!\sim\!\!\sim\!\!\bigcirc\, \text{l} \;=\; (-1)^2 \int_0^\beta d\tau \sum_{k,l} (-V_{klkl})(-f_k^-)(-f_l^-). \quad (14.46)
$$

$$
\text{k} \,\left(\!\!\sim\!\!\sim\!\!\right)\, \text{k} \;=\; (-1)^2 \sum_{k,l} (-V_{kllk})(-V_{lkkl}) f_k^+ (-f_k^-)(-f_l^-)^2
$$

$$
\times \int_0^\beta d\tau_1 \int_0^\beta d\tau_2 \, e^{-(\epsilon_k - \epsilon_k)(\tau_2 - \tau_1)} \theta_{\tau_2 - \tau_1}. \quad (14.47)
$$

The expansion for $\langle E \rangle$ may be derived from (14.45) with the aid of (14.43); again the linked cluster theorem holds, and we find

$$
\langle E \rangle = \langle E \rangle_0 + \bigcirc\!\!\sim\!\!\sim\!\!\bigcirc + \text{⬡} + \text{diagram} + \text{diagram} + \text{diagram} + \text{diagram} + \cdots.
$$
$$
(14.48)
$$

A new dictionary is required to evaluate these—the reader is referred to Bloch (1962) for details and examples.

[14.5 The pair-bubble at finite temperature]

We will now show the finite temperature machinery in full operation, using it to calculate the pair-bubble at $T \neq 0$. Recall that the $T=0$ pair-bubble's value appears in (10.66), and at the end of this section we will demonstrate that our present result reduces to (10.66) in the $T \to 0$ limit, as it should.

Let us first draw the pair-bubble and translate it with the aid of Table 14.1 (note that there is a factor (-1) difference between our pair-bubble and that in Fetter and Walecka (1971), pp. 271–75):

$$
\begin{array}{c} \mathbf{l+q,} \\ \omega_n + \omega_i \end{array} \bigcirc \begin{array}{c} \mathbf{l}, \omega_n \end{array} \equiv \Pi_0(\mathbf{q}, \omega_i)
$$

$$
= -2 \int \frac{d^3 l}{(2\pi)^3} \frac{1}{\beta} \sum_{n=-\infty}^{+\infty} \frac{1}{i\omega_n - (\epsilon_l - \mu)} \cdot \frac{1}{i\omega_n + i\omega_i - (\epsilon_{l+q} - \mu)}
$$
$$
(14.49)
$$

$(-1$ from fermion loop, factor of two for spin sum). The sum over n is evaluated as follows: First define the function $F(\omega)$ by

$$F(\omega) = \frac{1}{\omega - (\epsilon_l - \mu)} \cdot \frac{1}{\omega + i\omega_i - (\epsilon_{l+q} - \mu)}. \qquad (14.50)$$

Thus, $F(i\omega_n)$ is just the summand in (14.49). We now use Poisson's formula, (which is valid for any $F(\omega)$) which does not have poles on the imaginary axis), to convert the sum to a contour integral:

$$\sum_{n=-\infty}^{+\infty} F(i\omega_n) = \frac{-\beta}{2\pi i} \oint_C d\omega\, F(\omega) f(\omega), \qquad (14.51)$$

where $f(\omega)$ is the Fermi function $(= (\exp(\beta\omega) + 1)^{-1})$ and where the contour C surrounds the imaginary axis as shown in Fig. 14.2. Equation (14.51) follows immediately from the residue theorem when we note that $f(\omega)$ has poles all along the imaginary axis at $i\omega_n$, i.e.

$$f(\omega) = \frac{1}{e^{\beta\omega} + 1} = \infty \quad \text{when} \quad \omega = \frac{\pi i}{\beta}(2n+1), \quad n = 0, \pm 1, \pm 2, \ldots. \quad (14.52)$$

If we now look at contour C' in Fig. 14.2, we see that the integral around C' will be equal to the sum of the residues from the poles on the imaginary axis $(=$ integral around $C)$ plus the residues from the poles of $F(\omega)$ at $\omega = \epsilon_l - \mu$ and

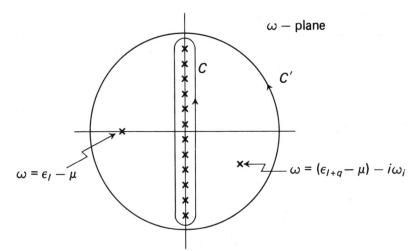

Fig. 14.2 *Contours used in the Finite Temperature Calculation*

$\omega = \epsilon_{l+q} - \mu - i\omega_i$. Since the integral around C' goes to zero when $C' \to \infty$ we have:

$$\oint_{C'} = 0 = \oint_C d\omega\, F(\omega) f(\omega) + 2\pi i \sum \text{Residues of } F(\omega) f(\omega)$$
$$\text{at poles of } F(\omega). \qquad (14.53)$$

Utilizing (14.51) then gives us a simple formula for evaluating sums like (14.49):

$$\frac{1}{\beta} \sum_{n=-\infty}^{+\infty} F(i\omega_n) = \text{sum of the residues of } F(\omega) f(\omega)$$
$$\text{at the poles of } F(\omega). \qquad (14.54)$$

Applying (14.54) to (14.49) we have

$$\text{Res}\, F(\omega) f(\omega) \quad \text{at} \quad (\omega = \epsilon_l - \mu) = \frac{f(\epsilon_l - \mu)}{\epsilon_l - \mu + i\omega_i - (\epsilon_{l+q} - \mu)} \qquad (14.55)$$

$$\text{Res}\, F(\omega) f(\omega) \quad \text{at} \quad (\omega = \epsilon_{l+q} - \mu - i\omega_i) = \frac{f(\epsilon_{l+q} - \mu - i\omega_i)}{\epsilon_{l+q} - \mu - i\omega_i - (\epsilon_l - \mu)}. \qquad (14.56)$$

Now, since by dictionary Table 14.1 both ω_n and $\omega_n + \omega_i$ are proportional to odd integers, it follows that ω_i is proportional to an even integer. Hence

$$f(\epsilon_{l+q} - \mu - i\omega_i) = f(\epsilon_l - \mu). \qquad (14.57)$$

Substituting this in (14.56), and using (14.56, 55, 54, 49) yields for the finite T pair-bubble (let $f_l \equiv f(\epsilon_l - \mu)$):

$$\Pi_0(\mathbf{q}, \omega_i) = -2 \int \frac{d^3 l}{(2\pi)^3} \frac{f_l - f_{l+q}}{i\omega_i - \epsilon_{l+q} + \epsilon_l}. \qquad (14.58)$$

This can be brought into a more convenient form by adding and subtracting $f_l f_{l+q}$ in the numerator, yielding:

$$\Pi_0(\mathbf{q}, \omega_i) = 2 \int \frac{d^3 l}{(2\pi)^3} \frac{f_{l+q}(1 - f_l) - f_l(1 - f_{l+q})}{i\omega_i - \epsilon_{l+q} + \epsilon_l}. \qquad (14.59)$$

Substituting $\mathbf{l} + \mathbf{q} \to -\mathbf{l}$ in the first term and noting that $f_{-k} = f_k$ and $\epsilon_{-k} = \epsilon_k$, we find

$$\Pi_0(\mathbf{q}, \omega_i) = 2 \int \frac{d^3 l}{(2\pi)^3} f_l(1 - f_{l+q}) \left[\frac{1}{i\omega_i - \epsilon_l + \epsilon_{l+q}} - \frac{1}{i\omega_i - \epsilon_{l+q} + \epsilon_l} \right]. \qquad (14.60)$$

Expression (14.60) is the imaginary frequency pair-bubble. But, since experiments are done at real frequencies, we want the real frequency version of this. This can be obtained by the method in appendix L, since the pair-bubble is actually a polarization propagator and depends on only a single frequency (see appendix L, part E). Thus, using (L.51), (L.39), and (L.40), we find for the real frequency finite T pair-bubble:

$$\pi_0^T(\mathbf{q}, \omega) = f^+(\omega) \Pi_0(\mathbf{q}, \omega + i\delta) + f^-(\omega) \Pi_0(\mathbf{q}, \omega - i\delta) \qquad (14.61)$$

where we have analytically continued $i\omega_l$ to $\omega + i\delta$ and $\omega - i\delta$.

Finally, let us take the $T \to 0$ limit of this and compare it with $\pi_0(\mathbf{q}, \omega)$ in (10.66). We have

$$\pi_0^{T=0}(\mathbf{q}, \omega) = 2 \int_{\substack{\epsilon_l < \mu \\ \epsilon_{l+q} > \mu}} \frac{d^3 l}{(2\pi)^3} \left\{ \theta_\omega \left[\frac{1}{\omega - \epsilon_l + \epsilon_{l+q} + i\delta} - \frac{1}{\omega + \epsilon_l - \epsilon_{l+q} + i\delta} \right] \right.$$

$$\left. + \theta_{-\omega} \left[\frac{1}{\omega - \epsilon_l + \epsilon_{l+q} - i\delta} - \frac{1}{\omega + \epsilon_l - \epsilon_{l+q} - i\delta} \right] \right\}. \qquad (14.62)$$

Using (3.76″) to find the real and imaginary parts of this, we find that it may be written in the form

$$\pi_0^{T=0}(\mathbf{q}, \omega) = 2 \int_{\substack{\epsilon_l < \mu \\ \epsilon_{l+q} > \mu}} \frac{d^3 l}{(2\pi)^3} \left\{ \frac{1}{\omega - \epsilon_l + \epsilon_{l+q} - i\delta} - \frac{1}{\omega - \epsilon_{l+q} + \epsilon_l + i\delta} \right\} \qquad (14.63)$$

(since (14.63) has the same real and imaginary parts as (14.62)). This result is for fixed chemical potential, μ. If we fix N instead, then $\mu = \epsilon_F$ and (14.63) becomes just (10.66). Note that there are no μ's appearing in the denominators since they have cancelled.

Further reading

Fetter and Walecka (1971), chaps. 7, 8, 9.
Schultz (1964), chap. 5.
Thouless (1972).
Luttinger (1960a).
Abrikosov (1965), chap. 3.
Bloch (1962).

(The non-perturbative method of finding the finite temperature propagator makes use of the differential equation for the propagator. This method is described briefly in appendix M, and in detail in Ter Haar (1962), Parry (1964), Kadanoff (1962).)

Exercises

14.1 Express the average particle number, $\bar{N}$, in terms of the grand partition function.

14.2 Use the result of Ex. 14.1 to find an equation for the chemical potential μ in terms of the average particle number (given to be N_0) in a non-interacting Fermi system at temperature T.

14.3 Verify (14.29).

14.4 Verify (14.31).

14.5 Translate into finite temperature functions (in $\mathbf{k}, \omega_n$-space) the fourth diagram on the right of (14.32).

14.6 Verify (14.51) by calculating the residues.

14.7 Show $\oint_{C'} = 0$ for F as in (14.50).

14.8 Show that (14.62) and (14.63) have the same real and imaginary parts.

14.9 Verify (14.32′).

Chapter 15

Diagram Methods in Superconductivity

15.1 Introduction

It is well known that below a certain critical temperature, $T_c \sim 1$–$10°\text{K}$, a large number of metals and alloys undergo a transition to a new phase called the superconducting state (see Rikayzen (1965), chap. I, or Tinkham (1962) for a review). The spectacular physical properties of this phase, such as zero electrical resistance and perfect diamagnetism, have made it one of the great centres of interest in modern many-body physics.

The intense interest in superconductivity is not confined merely to metals. For one thing, Bohr, Mottelson, and Pines (1958) have demonstrated that even–even nuclei show characteristics of the superconducting phase (in particular, an energy gap—see below). This is also true of nuclear matter (Fetter and Walecka (1971), p. 383 ff.). There is evidence that the interior of neutron stars is superconducting (Baym (1969)) . In addition, a great flurry of activity has been stirred up by the discovery that liquid He^3 becomes a neutral superconductor at temperatures below $2\cdot7 \times 10^{-3}\,°\text{K}$ (Osheroff (1972), Anderson and Brinkman (1973), Leggett (1973)). And finally, much attention has been devoted to the possibility of developing revolutionary high temperature superconductors using organic materials and metal–dielectric combinations (Ginzburg (1968)). So superconductivity is a real grab-bag, filled with goodies for all kinds of physicists.

The mechanism responsible for superconductivity is now known to be the effective attractive interaction between two electrons due to the exchange of 'virtual' phonons. As a result of this attraction, the electrons become '*paired*' in states of opposite momentum and spin. That is, the super-conducting ground state is a superposition of just non-interacting states like, for example, $|0_{k_1\uparrow},\ 0_{-k_1\downarrow},\ 1_{k_2\uparrow},\ 1_{-k_2\downarrow},\ \ldots\rangle$, in which the single-particle states, $\phi_{+k\uparrow}$, $\phi_{-k\downarrow}$, are either both occupied or both empty. On account of the pairing, the elementary excitations of the system ('bogolons') turn out to have an energy dispersion law of the form $E_k = [(k^2/2m^* - \epsilon_F)^2 + \varDelta^2]^{\frac{1}{2}}$ instead of the usual $E_k = k^2/2m^* - \epsilon_F$. Thus, a minimum 'gap energy', $\varDelta$, is required to excite the system; this makes the superconductor immune to the sort of scattering which gives rise to electrical resistance in the normal case.

In the nuclear case, pairing is caused by attractive forces originating in the exchange of mesons between two nucleons. In He^3 the attractive forces are due to exchange of 'spin fluctuations' or 'paramagnons' (critically damped spin wave quanta found in nearly-ferromagnetic materials) between two He^3 atoms, with parallel spins and p-state pairing (Layzer and Fay (1974)). In metal–dielectric combinations, and in organic materials, pairing forces are expected to originate via the creation of particle–hole excitations (i.e., excitons).

There are several ways of getting ground and excited energies, E_0 and E_k, for a superconductor. The original Bardeen, Cooper, Schrieffer (BCS) derivation, which will be briefly reviewed, involves a variational solution of the Schrödinger equation for a gas of electrons with mutual instantaneous attractive interactions. (The actual interaction is retarded.) It is also possible to get E_0, E_k by a diagrammatic method, and it is this method which we will concentrate on here.

The diagram method for superconductors is not the one we have been using up to now. We shall first show that powerful though they may be, the usual graphical methods do not have enough muscle to handle the superconducting case. In fact, we'll see that because of the pairing, the ordinary graphical perturbation expansion breaks down completely in the case of a superconductor.

There are several equivalent ways out of this tragedy. The one which we will describe involves a *matrix propagator*, $\mathbb{G}(\mathbf{k}, \omega)$ which has the pairing already built into it. This has the form:

$$\mathbb{G} = \begin{pmatrix} G(\mathbf{k}\uparrow) & F \\ F^\dagger & -G(-\mathbf{k}\downarrow) \end{pmatrix}, \tag{15.1}$$

where G is the ordinary propagator and F is the so-called *anomalous propagator*, which gives the probability amplitude for the creation or destruction of a pair of particles in the system. The matrix $\mathbb{G}$ can be expanded in diagrams just like G, and can be used to find the bogolon energy, E_k.

The advantage of the diagram method in superconductivity is that it can take into account the retarded nature of the electron–phonon–electron interaction. This allows it to tackle problems which are inaccessible to the ordinary BCS method and provides the basis for a much more general formulation of the theory of superconductivity.

15.2 Hamiltonian for coupled electron–phonon system

In the examples up to now, our Utopian model metal was a box containing a smeared out static positive background (representing the positive ions) against which the electrons moved. For our present purposes, this is much too idealized. In order to explain superconductivity, we need to also take account of the fact that the positive ions which form the crystal lattice can vibrate about their equilibrium positions. That is, it is necessary to put the lattice vibration quanta or *phonons* into our box.

The Hamiltonian for a system of non-interacting phonons is (see ($\mathscr{A}$.41)):

$$H = \sum_q \Omega_q [B_q^\dagger B_q + \tfrac{1}{2}], \quad (\hbar = 1) \tag{15.2}$$

where Ω_q is the frequency of the phonon of momentum **q** and polarization direction λ ($q \equiv \mathbf{q}, \lambda$), and $B_q^\dagger, B_q$ are the phonon creation and destruction operators. Phonons have what might be roughly described as one 'longitudinal polarization' mode of vibration with frequency Ω_q^L and two 'transverse' modes, with frequencies Ω_q^T, $\Omega_q^{T'}$. If we use a model in which a lattice of positive ions, interacting by Coulomb forces, vibrates against a static

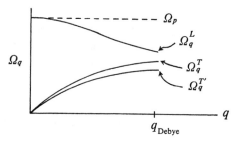

Fig. 15.1 *Frequency Dispersion Law for Longitudinal and Transverse Modes of Bare Phonons*

negative charge background representing the electrons (this is not the same as the simple 'spring coupling' model of Fig. $\mathscr{A}$.1 because of the long range of the Coulomb force!), and neglect coupling between ions and electrons, we find that the modes have the appearance shown in Fig. 15.1, where $q_{\text{Debye}} \sim 1$ Å^{-1}.

The frequency

$$\Omega_p = \sqrt{(4\pi Z^2 e^2 N/M)} \tag{15.3}$$

where Ze = ion charge, N = ions/cm^3 and M = mass of ion, is the ion 'plasma' frequency (cf. (13.21d)). This is the natural frequency of oscillation of the ions if they are smeared out to form a continuous jelly vibrating against a fixed uniform negative charge background ('jellium' approximation). The fact that $\Omega_q^L \to \Omega_p$ as $\mathbf{q} \to 0$ is physically incorrect; we should have $\Omega_q^L \to 0$, as $\mathbf{q} \to 0$. This odd behaviour is due to the fact that interaction of ions with electrons has been neglected; when this is taken into account, the longitudinal phonons become 'dressed' with an electron cloud, and their frequency becomes renormalized so that $\Omega_q \propto q$ for small **q** (see appendix J).

Let us now allow the electrons and phonons in the box to interact. Then the Hamiltonian may be written (neglecting phonon–phonon interaction)

$$H = H_{\substack{\text{free} \\ \text{electron}}} + H_{\text{Coulomb}} + H_{\substack{\text{free} \\ \text{phonon}}} + H_{\substack{\text{electron--} \\ \text{phonon}}}, \tag{15.4}$$

where

$$H_{\substack{\text{free} \\ \text{electron}}} = \sum_{k,\sigma} \epsilon_k c^\dagger_{k,\sigma} c_{k,\sigma} \tag{15.5}$$

$$H_{\text{Coulomb}} = \tfrac{1}{2} \sum_{\substack{k,k',K \\ \sigma,\sigma'}} V_K c^\dagger_{k'-K,\sigma'} c^\dagger_{k+K,\sigma} c_{k,\sigma} c_{k'\sigma'} \quad \text{(see (7.71))} \tag{15.6}$$

$$H_{\substack{\text{free} \\ \text{phonon}}} = \sum_q \Omega_q (B^\dagger_q B_q + \tfrac{1}{2}) \tag{15.7}$$

$$H_{\substack{\text{electron--} \\ \text{phonon}}} = \sum_{\substack{k,k',q \\ \sigma}} g_q [B_q + B^\dagger_{-q}] c^\dagger_{k',\sigma} c_{k,\sigma}. \tag{15.8}$$

For the derivation of (15.8), the reader is referred to Schultz (1964), p. 107 ff., or Schrieffer (1964a), p. 89 ff. We have written it in jellium approximation, where only longitudinal phonons couple to electrons and g_q is

$$g_q = \frac{4\pi e^2}{q} \Bigg/ \sqrt{\left(\frac{Z^2 N}{2\Omega_p M}\right)} \tag{15.9}$$

with Ω_p given by (15.3). Equation (15.8) describes a process in which an electron scatters from state $\mathbf{k} \to \mathbf{k}'$ with the emission or absorption of a phonon of wavenumber $\mathbf{q}$.

15.3 Short review of BCS theory (see Rikayzen (1965) chap. 4, for details)

(a) The BCS Hamiltonian

It would be idle for us to hope that if we simply stare at Hamiltonian (15.4) long enough, it will reveal the secret of superconductivity. One must look to nature for clues. Some of the important clues coming from experiment (see Lynton (1962), p. 109 ff.) are: (1) Superconductivity occurs in an enormous variety of metals and alloys. (2) The binding energy of the superconducting phase is very tiny—around 10^{-8} eV/atom. (3) The phenomenon appears to come from electrons in a thin shell around the Fermi surface. The shell thickness is $\sim kT_c$, where T_c=superconducting transition temperature (~ 1–$10°$K). These are called the 'superconducting electrons'. (4) There are strong correlations between the motions of electrons lying within a 'coherence length' ($\sim 10^{-4}$ cm) of each other. (5) The isotope effect, showing $T_c \propto 1/\sqrt{}$(lattice ion mass), indicates that the electron–phonon interaction is of great importance in superconductivity.

Clue (1) tells us that the details of metal structure play no decisive role in superconductivity. This means that in first approximation we can neglect the periodic lattice potential and all the resulting distortions (like 'monsters' and 'dog's bones') in the Fermi surface. Now we have seen that the interaction of electrons via Coulomb repulsion produces only normal quasiparticle behaviour. Therefore we look to clue (5) for a possible interaction of electrons via phonons which could give rise to the new sort of correlated state described in clue (4). Physically, such an interaction could come from the fact that (a): one electron pulls in the $(+)$ ions in its vicinity, thus deforming the lattice, creating phonons (this is the electron–phonon interaction of (15.8)), and (b): another electron is influenced by this deformation, i.e., absorbs a phonon. Because it is of second order, this interaction doesn't appear explicitly in (15.4). However, Fröhlich (1952) showed that it can be made visible by performing a complicated canonical transformation on H (see Kittel (1963), p. 151, or Schrieffer (1959)), which produces

$$H_{\text{transf.}} = H_{\substack{\text{quasi}\\ \text{electron}}} + H_{\substack{\text{shielded}\\ \text{Coulomb}}} + H_{\substack{\text{electron–}\\ \text{phonon–}\\ \text{electron}}} + H_{\substack{\text{dressed}\\ \text{phonon}}} + \cdots, \qquad (15.10)$$

where

$$H_{\substack{\text{quasi}\\ \text{electron}}} = \sum_{k,\sigma} \epsilon_k c^{\dagger}_{k,\sigma} c_{k,\sigma}, \quad \left(\epsilon_k = \frac{\hbar^2 k^2}{2m^*} - \epsilon_F \right) \qquad (15.11)$$

$$H_{\substack{\text{dressed}\\ \text{phonon}}} = \sum_k \hbar\omega_k (B^{\dagger}_k B_k + \tfrac{1}{2})$$

$$H_{\substack{\text{shielded Coul.}\\ \text{+el.–phon.–el.}}} = \sum_{\substack{k,k',q\\ \sigma,\sigma'}} \mathscr{V}_{kq} c^{\dagger}_{k'-q,\sigma'} c^{\dagger}_{k+q,\sigma} c_{k\sigma} c_{k'\sigma'}, \qquad (15.12)$$

where the 'BCS interaction', $\mathscr{V}_{kq}$, is given by

$$\mathscr{V}_{kq} \equiv V_{k+q,\,k'-q,\,k,\,k'} = \underbrace{\frac{4\pi e^2}{q^2 + \lambda^2}}_{\substack{\text{shielded}\\ \text{Coulomb}\\ \text{interaction}}} + \underbrace{\frac{2\hbar\omega_q |M_q|^2}{(\epsilon_k - \epsilon_{k+q})^2 - (\hbar\omega_q)^2}}_{\substack{\text{shielded electron–phonon}\\ \text{electron interaction}\\ \text{(Fröhlich interaction)}}}. \qquad (15.13)$$

The plasmon term has been dropped, since it will not concern us. In (15.11) it is assumed that the electron energy has been renormalized by including the first-order contributions from the interaction (15.12); i.e., the new ϵ_k is the quasi electron energy in Hartree–Fock approximation. We assume, for simplicity, that this can be written in terms of an effective mass, m^*. Note that energy is measured relative to the Fermi energy. That is, we are using the modified H of (9.46) for a system in contact with a reservoir. The M_q is proportional to the shielded electron–phonon coupling (see appendix J), and ω_k is the longitudinal phonon frequency, renormalized by interaction with electrons (appendix J).

Let us examine the BCS interaction, $\mathscr{V}_{kq}$, more closely. The average value of $\hbar\omega_q$ is ~ 0.025 eV, while the average value of $(\epsilon_k - \epsilon_{k+q})$ for electrons in the thin shell about the Fermi surface (see clue (3)) is $\sim 4 \times 10^{-4}$ eV. Hence on the average, the Fröhlich term in $\mathscr{V}_{kq}$ is negative (attractive). Now in a metal which is always normal, the electron–phonon coupling, M_q, is small, so the Coulomb term predominates and the BCS interaction is positive (repulsive). However, in a superconductor (i.e., a metal which is capable of entering the superconducting state when the temperature is low enough), M_q is large, so the Fröhlich term dominates and the BCS interaction is negative. This means that in a superconductor there is an effective attractive interaction between all electrons located in a thin shell about the Fermi surface. (Note that this attraction is always present in a superconductor regardless of whether or not it is actually in the superconducting state.)

(b) Pairing and the reduced Hamiltonian

The effect of such an attractive interaction was analysed by Cooper. He showed that two electrons in a superposition of two-particle states, $|1_{k\uparrow}, 1_{-k\downarrow}\rangle$, ($\uparrow$, $\downarrow$ means spin up, down), where $|\mathbf{k}| > k_F$, with an attractive interaction between them, would form a bound state, no matter how weak the interaction. The bound state had a wave function $\sim 10^{-4}$ cm wide, and an energy lower than that when the electrons were non-interacting. This indicated that in the presence of attractive interactions, the Fermi sea was unstable, and that it was energetically favourable for the system to form some sort of correlated state in which every $\mathbf{k}\uparrow$ electron was 'paired off' with its mate in $-\mathbf{k}\downarrow$ ('Cooper pair'). (See Fetter and Walecka (1971), p. 320 ff.)

The 'pairing' concept is clear enough for two electrons, but what does it mean for N electrons? BCS assumed that it meant that the superconducting ground state Ψ_0 was not just the usual linear combination of all wave functions $|n_1, n_2, \ldots, n_i, \ldots\rangle$ for the non-interacting system, but rather was composed only of wave functions in which particles occurred in Cooper pairs. That is,

$$|\Psi_0\rangle = \sum_{\ldots n_{ki\uparrow}, n_{-ki\downarrow}, \ldots} A_{\ldots n_{ki\uparrow}, n_{-ki\downarrow}, \ldots} |\ldots n_{ki\uparrow}, n_{-ki\downarrow}, \ldots\rangle \qquad (15.14)$$

where $n_{ki\uparrow} = n_{-ki\downarrow}$ for all k_i. This may be written in abbreviated form:

$$|\Psi_0\rangle = \sum_{\ldots N_i \ldots} A_{\ldots N_i \ldots} |N_1 \ldots N_i \ldots\rangle, \quad N_i = 0, 1 \qquad (15.15)$$

where $|N_1, \ldots, N_i, \ldots\rangle$ means: N_1 pairs in pair state $\mathbf{k}_1\uparrow$, $-\mathbf{k}_1\downarrow$, N_2 in $\mathbf{k}_2\uparrow$, $-\mathbf{k}_2\downarrow$, etc.

With $|\Psi_0\rangle$ restricted in this way, it is found that the only terms in H (15.12) which have non-vanishing matrix elements between two $|N_1, \ldots, N_i, \ldots\rangle$ are those in which the $c_{k\sigma}^\dagger$, $c_{k\sigma}$ operators also occur in pairs; this enables us to drop all other terms and write (15.11, 12) in 'reduced' or 'paired' form:

$$H_{\text{red}} = 2 \sum_k \epsilon_k b_k^\dagger b_k - \sum_{k \neq k'} V_{kk'} b_{k'}^\dagger b_k \quad (V_{kk'} \equiv V_{-k', k', -k, k}) \quad (15.16)$$

where the pairing operators (these are neither fermion nor boson operators—see exercise 15.8):

$$b_k^\dagger = c_{k\uparrow}^\dagger c_{-k\downarrow}^\dagger; \quad b_k = c_{-k\downarrow} c_{k\uparrow} \quad (15.17)$$

create and destroy Cooper pairs when they operate on $|N_1, \ldots, N_i, \ldots\rangle$. Note that the justification for the form of the first term in (15.16) is that it yields the same result as (15.11) when it operates on a paired wave function. Observe that in (15.16) we have incorporated the $V_{kk'}$ $(k = k')$ term into the unperturbed part of H_{red}, so that ϵ_k is in reality $\epsilon_k + \frac{1}{2} V_{kk}$ (we just write ϵ_k for brevity.)

(c) The two-particle superconductor

The superconductor ground state wave function $|\Psi_0\rangle$ will satisfy the Schrödinger equation

$$H_{\text{red}} |\Psi_0\rangle = E_0 |\Psi_0\rangle. \quad (15.18)$$

Before stating the BCS solution to this equation, it is a good idea to look at a trivial 'two-particle superconductor' to see how the negative interaction term in H_{red} leads to a correlated ground state (clue (4)) with an energy slightly lower than that of the normal state (clue (2)). Suppose there are two electrons and six states, all having energy ϵ_0, thus:

$$\overline{\quad\quad} \; \overline{\quad\quad} \; \overline{\quad\quad} \; \overline{\quad\quad} \; \overline{\quad\quad} \; \overline{\quad\quad} \quad \epsilon_0$$
$$\mathbf{k}_1\uparrow \quad -\mathbf{k}_1\downarrow \quad \mathbf{k}_2\uparrow \quad -\mathbf{k}_2\downarrow \quad \mathbf{k}_3\uparrow \quad -\mathbf{k}_3\downarrow \quad\quad (15.19)$$

Assume $V_{kk'} = V$ (constant). Then

$$H_{\text{red}} = 2\epsilon_0 \sum_k b_k^\dagger b_k - V \sum_{kk'} b_{k'}^\dagger b_k. \quad (15.20)$$

The solution for the non-interacting $(V = 0)$ case is (use pairing notation in (15.15)):

$$|\Phi_1\rangle = |1_{k_1}, 0, 0\rangle, \quad |\Phi_2\rangle = |0, 1_{k_2}, 0\rangle, \quad |\Phi_3\rangle = |0, 0, 1_{k_3}\rangle$$
$$E_1 = 2\epsilon_0, \quad E_2 = 2\epsilon_0, \quad E_3 = 2\epsilon_0. \quad (15.21)$$

In the 'superconducting' case we have by (15.15):

$$|\Psi\rangle = A_1 |1_{k_1}, 0, 0\rangle + A_2 |0, 1_{k_2}, 0\rangle + A_3 |0, 0, 1_{k_3}\rangle. \quad (15.22)$$

It is easiest to solve the problem in matrix form using the normal states as basis. Thus

$$\langle \Phi_i | 2\epsilon_0 \sum_k b_k^\dagger b_k | \Phi_j \rangle = 2\epsilon_0 \begin{pmatrix} 1 & 0 & 0 \\ 0 & 1 & 0 \\ 0 & 0 & 1 \end{pmatrix} \qquad (15.23)$$

$$\langle \Phi_i | -V \sum_{kk'} b_{k'}^\dagger b_k | \Phi_i \rangle = -V \begin{pmatrix} 1 & 1 & 1 \\ 1 & 1 & 1 \\ 1 & 1 & 1 \end{pmatrix} \qquad (15.24)$$

In matrix form (15.18) is:

$$\left[2\epsilon_0 \begin{pmatrix} 1 & 0 & 0 \\ 0 & 1 & 0 \\ 0 & 0 & 1 \end{pmatrix} - V \begin{pmatrix} 1 & 1 & 1 \\ 1 & 1 & 1 \\ 1 & 1 & 1 \end{pmatrix} \right] \begin{pmatrix} A_1 \\ A_2 \\ A_3 \end{pmatrix} = E \begin{pmatrix} A_1 \\ A_2 \\ A_3 \end{pmatrix} \qquad (15.25)$$

which has the eigen-solutions

$$E = 2\epsilon_0, \; 2\epsilon_0, \; 2\epsilon_0 - 3V. \qquad (15.26)$$

The last solution is seen to be $3V$ lower than the normal state energy—this is the 'superconducting' state. The corresponding state vector is:

$$|\Psi\rangle = \frac{1}{\sqrt{3}} [|1_{k_1}, 0, 0\rangle + |0, 1_{k_2}, 0\rangle + |0, 0, 1_{k_3}\rangle]. \qquad (15.27)$$

We see that this is a state in which the electrons are highly correlated—first, they are always paired, and second they are in a special $1:1:1$ mixture of the three possible paired states.

(d) BCS solution for ground state energy (see Ex. 15.9)

In solving (15.18) for the real N-particle superconductor, BCS assumed that the interaction $V_{kk'}$ in (15.16) had the simple form

$$V_{kk'} = V \text{ (constant), for } \mathbf{k}, \mathbf{k'} \text{ in a shell about the Fermi surface}$$
$$\text{such that } (-\hbar\omega_c) < \epsilon_k, \epsilon_{k'} < +\hbar\omega_c$$
$$= 0 \text{ otherwise.} \qquad (15.28)$$

The $\hbar\omega_c$ was an average phonon energy $\sim 10^{-2}$ eV. They showed that to a good approximation (exact in the $N\rightarrow\infty$ limit) the ground state wave function was given by (see Rikayzen (1965), chapter 4, for details)

$$|\Psi_0\rangle = \prod_k (u_k + v_k b_k^\dagger) |0\rangle, \text{ where } u_k = \sqrt{(1 - v_k^2)} \qquad (15.29)$$

For example, if there were only two pair states, then:

$$|\Psi_0\rangle = u_1 u_2 |0_1, 0_2\rangle + u_1 v_2 |0_1, 1_2\rangle + v_1 u_2 |1_1, 0_2\rangle + v_1 v_2 |1_1, 1_2\rangle. \qquad (15.29')$$

The quantity v_k is the probability amplitude that the pair state $k\uparrow,\ -k\downarrow$ is occupied. It is given by

$$v_k^2 = \frac{1}{2}\left[1 - \frac{\epsilon_k}{E_k}\right] \tag{15.30}$$

$$E_k = \sqrt{(\epsilon_k^2 + \Delta_k^2)} \tag{15.31}$$

where

$$\Delta_k = \sum_{k'} V_{kk'}\, u_{k'} v_{k'} \tag{15.31'}$$

$$\Delta_k \ (\text{for small } V) = \begin{cases} \Delta = 2\hbar\omega_c \exp\left[-\dfrac{1}{N(0)V}\right] \ \text{for } \mathbf{k} \text{ in shell} \\ 0 \ \text{for } \mathbf{k} \text{ outside of shell} \end{cases} \tag{15.32}$$

where $N(0)$ is the density of states at the Fermi surface.

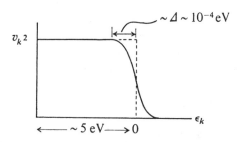

Fig. 15.2 *Probability that the Bare Single-particle State of Energy ϵ_k is Occupied in the Superconducting Ground State*

The quantity v_k^2 is plotted in Fig. 15.2. Since v_k^2 is just the probability that the state $\mathbf{k}\uparrow$ (or $-\mathbf{k}\downarrow$) is occupied, we see that there is no discontinuity at the Fermi surface, i.e., the Fermi surface does not exist in a superconductor (cf. chapter 11).

It should be noted that (15.29) is a mixture of states with 0, 2, 4, 6, ... $N-2$, N, $N+2$, $N+4$, ... particles, so that N is not a good quantum number here. This is due to the fact that we are using the modified H of (9.46) for a system in contact with a reservoir, so the particle number is variable. However, using (15.30, 31, 32) it can be shown that the contribution to $|\Psi_0\rangle$ is extremely large from components of (15.29) containing just around N particles, and very small from other components, so this causes no trouble.

For the ground state energy in the small V limit, BCS obtained

$$E_0 = E_{0(\text{normal})} - 2N(0)\,(\hbar\omega_c)^2 \exp\left[-\frac{2}{N(0)V}\right]. \tag{15.33}$$

Thus the energy is lowered by the attractive interaction.

(e) Solution for the excited states: bogolons

The excited states of (15.18) may be obtained by diagonalizing H_{red} by means of the 'Bogoliubov–Valatin' canonical transformation. This is defined by

$$\alpha_k^\dagger = u_k c_{k\uparrow}^\dagger - v_k c_{-k\downarrow}, \quad \beta_k^\dagger = u_k c_{-k\downarrow}^\dagger + v_k c_{k\uparrow}, \tag{15.34}$$

$$c_{k\uparrow} = u_k \alpha_k + v_k \beta_k^\dagger, \quad c_{-k\downarrow} = u_k \beta_k - v_k \alpha_k^\dagger. \tag{15.35}$$

Substituting this into H_{red} (with the first term replaced by (15.11)) and using (15.30, 31) yields, after some labour (see Fetter and Walecka (1971), p. 326 ff.)

$$H'_{\text{red}} = E_0 + \sum_k E_k(\alpha_k^\dagger \alpha_k + \beta_k^\dagger \beta_k) + \text{small terms}. \tag{15.36}$$

Comparing this with (1.43, 44) shows that we have here a set of nearly independent elementary excitations. They are called '*Bogoliubov quasi particles*' or '*bogolons*' (see §1.3), and have number operators $\alpha_k^\dagger \alpha_k$, $\beta_k^\dagger \beta_k$ and energy given by (15.31):

$$E_k = \sqrt{(\epsilon_k^2 + \Delta_k^2)} = \sqrt{\left[\left(\frac{k^2}{2m^*} - \epsilon_F\right)^2 + \Delta_k^2\right]}. \tag{15.37}$$

It is important to understand the significance of $\alpha_k^\dagger$, $\beta_k^\dagger$. By (15.34), $\alpha_k^\dagger$ creates bogolons of momentum $+\mathbf{k}$, spin↑ (since $c_{-k\downarrow}$ subtracts $-\mathbf{k}\downarrow$ from the system, which is the same as adding $\mathbf{k}\uparrow$). Similarly, $\beta_k^\dagger$ creates bogolons with $-\mathbf{k}$, spin↓. The eigenstate corresponding to (15.37) is the 1-bogolon state:

$$|1_{k\uparrow}\rangle = \alpha_k^\dagger |\Psi_0\rangle \tag{15.38}$$

or

$$|1_{-k\downarrow}\rangle = \beta_k^\dagger |\Psi_0\rangle. \tag{15.39}$$

Thus, the superconducting ground state, $|\Psi_0\rangle$, acts as the 'vacuum' for bogolons. (See Schrieffer (1964a), p. 44 ff., for further discussion.)

The bogolon dispersion law has an '*energy gap*' at the Fermi surface equal to Δ, as shown in Fig. 15.3. This curve comes from (15.37), with Δ_k as in (15.32). The k_1, k_2 are the k-values for the inner and outer radii of the shell about the Fermi surface described in (15.28). The discontinuities at k_1, k_2 come from the approximation in (15.28). They should not be regarded as real physical effects! Note that in the case of the normal metal ($\Delta_k = 0$) we have $E_k = |\epsilon_k| = |k^2/2m^* - \epsilon_F|$, where the absolute value sign is used since below k_F removal of an electron with $\epsilon_k < 0$, adds energy to the system.

Figure 15.3 shows that whereas in a normal metal it takes zero energy to excite a quasi electron at the Fermi surface, in a superconductor it requires a minimum energy Δ ($\Delta \sim kT_c \sim 10^{-4}$ eV) to excite a bogolon. It is this fact which accounts for the remarkable stability of the superconducting state against the scattering of electrons which causes resistance in normal metals.

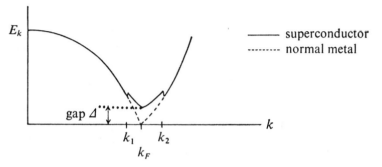

Fig. 15.3 *Energy Dispersion Law for Bogolons (Greatly Exaggerated)*

15.4 Breakdown of the perturbation expansion in a superconductor

In chapter 10, we saw that in a normal Fermi system it is possible to get the quasi particle energy dispersion law from the single-particle propagator. Let us therefore try to get the energies of the quasi particles in a superconductor (bogolons) by the propagator method. We expect that the propagator will look something like

$$G(\mathbf{k}, \omega) \propto \frac{1}{\omega - \sqrt{\epsilon_k^2 + \Delta_k^2} + i\delta} \qquad (15.40)$$

since this has poles which yield the bogolon energy dispersion law (15.37).

The expansion for the propagator will presumably be given by the usual one (9.40), where the interaction is now the BCS interaction of (15.13) (which is assumed to be attractive, since we are dealing with the super-conducting case). However, when we attempt to carry out the calculation of G this way, an unfortunate thing happens, e.g., we find that there are classes of diagrams in the propagator expansion (9.40) (or equivalently, in the proper self-energy expansion (10.8)), which produce an *unstable* result when summed over. These are the classes in which a pair of particles with momenta equal in magnitude but opposite in direction (i.e., a Cooper pair) multiply scatter against each other. An example of such a class is the bracketed term in

$$(15.41)$$

where $\sim\!\!\!\sim\!\!\!\sim$ stands for the BCS interaction, $\mathscr{V}_{kq}$. This is just a ladder sum and may be evaluated by means of the K-matrix equation (10.20) and equation (10.22) with total momentum $\mathbf{q}=0$. For simplicity, assume $V_{pp'}=Vu_pu_{p'}$, where $u_p=1$ for $|\epsilon_p|<\omega_c$ and $u_p=0$ for $|\epsilon_p|>\omega_c$ (ϵ_p relative to ϵ_F). Following the same procedure as in Problem 10.7 yields

$$K(\omega) = \frac{Vu_pu_{-p}}{1+N(0)\,V\left\{\tfrac{1}{2}\ln\left|\dfrac{(2\omega_c)^2-\omega^2}{\omega^2}\right|+i\,\dfrac{\pi}{2}\theta_{2\omega_c-\omega}\,\theta_{\omega+2\omega_c}\right\}}. \quad (15.42)$$

We now make use of some results in appendix L. Since this $K(\omega)$ is obtained directly from summing diagrams, it is the 'time-ordered' K analogous to the time-ordered single-particle propagator $G(\mathbf{k},\omega)$ (see appendix L, part A, and part E). We can easily construct the corresponding 'retarded' K-matrix,

$$K^R(\omega) = \frac{Vu_pu_{-p}}{1+N(0)\,V\left\{\tfrac{1}{2}\ln\left|\dfrac{(2\omega_c)^2-\omega^2}{\omega^2}\right|+i\,\mathrm{sgn}\,(\omega)\,\dfrac{\pi}{2}\theta_{2\omega_c-\omega}\,\theta_{\omega+2\omega_c}\right\}} \quad (15.42')$$

This obeys the same relations as G^R in (L.26).

We now regard $K^R(\omega)$ as a function of the complex variable ω and examine its behaviour in the upper half-plane. Assuming $\omega\ll\omega_c$ and using the fact that $\ln(x+iy)=\ln(x^2+y^2)^{\frac{1}{2}}+i\tan^{-1}(y/x)$ we find that when $V<0$, $K^R(\omega)$ has a pole at the point

$$\omega_{\text{pole}} = +i2\omega_c\,e^{-1/N(0)V}, \quad (15.43)$$

i.e., a pole in the upper half-plane. (There is no pole if $V>0$.) But this violates the fact that the retarded propagator must be analytic in the upper half-plane (appendix L, part C). Hence for negative V, the perturbation series leading to (15.42) must be invalid. That is, the ordinary perturbation series is invalid in a superconductor. (Note: this statement is strictly true only in the case of an infinitely large superconductor—see Mattuck and Johansson (1968), Appendix D.)

It is important to note that when V has its value at the transition point, i.e., $V=0$, then $\omega_{\text{pole}}=0$. That is, when V approaches 0 from above, the pole first appears at $\omega=0$, then moves into the upper half-plane when V becomes a finite negative quantity.

The pole (15.43) in the upper half-plane indicates that the normal system is unstable, and will undergo a phase transition to the superconducting state (see §17.8 and Pines (63), p. 288 ff.). Physically, the appearance of this pole may be interpreted as follows: The retarded K-matrix for Cooper pairs is the probability amplitude for a Cooper pair coming out if we put a Cooper pair in. That is, it is the 'response' of the system to an applied 'pair field'. For $V>0$, there is no pole and K^R is finite. But when V first becomes negative, there is a pole in K^R at $\omega=0$, i.e., K^R becomes infinite. This means that pairs come out even if we don't put any pairs in, i.e., the system spontaneously manufactures Cooper pairs, which means a transition to the superconducting state.

The above argument is generalized to finite temperature in §15.7, where it is used to find the transition temperature of a superconductor.

Actually, it was unnecessary to go through all these contortions—we could have seen this result immediately from the remark after Fig. 15.2. That is, there is no Fermi surface here, so by §11.3, the perturbation series cannot hold. Another way of seeing this is to expand the ground state energy (15.33) into a series:

$$E_0 = E_{0(\text{normal})} - 2N(0)(\hbar\omega_c)^2\left[1 - \frac{2}{N(0)V} + \frac{1}{2!}\left(\frac{2}{N(0)V}\right)^2 + \cdots\right]. \quad (15.44)$$

A true perturbation series should have the form

$$E = a_0 + a_1 V + a_2 V^2 + \cdots, \quad (15.45)$$

so each term containing V goes to zero as $V \to 0$. But the terms in (15.44) each go to ∞ as $V \to 0$. (Of course, by (15.33) their sum is finite!) That is, E_0 is a 'non-analytic' function of V, and cannot be expanded in a perturbation series.

The difficulty here may be expressed in the general statement that the perturbation series (which we have been using throughout this book) is valid only when the perturbed state is qualitatively similar to (or 'has the same symmetry as') the unperturbed state. This means that whenever a system undergoes a *change of phase*—like gas $\to$ liquid, liquid $\to$ solid, paramagnet $\to$ ferromagnet, or normal metal $\to$ superconductor—since the two states involved are qualitatively dissimilar, the perturbation series breaks down. (For a general discussion of the theory of phase transitions, see chapter 17.)

These ideas can be illustrated with a simple classical case. Consider the marble on the frictionless circular track in Fig. 15.4. The stable position of the marble, $\theta(F/mg)$, under the influence of applied force F, and its own weight, mg, may be developed in a perturbation series in powers of F:

$$\theta = \left(\frac{F}{mg}\right) - \frac{1}{3}\left(\frac{F}{mg}\right)^3 + \frac{1}{5}\left(\frac{F}{mg}\right)^5 - \cdots$$

$$\left(= \tan^{-1}\frac{F}{mg} \right). \tag{15.46}$$

This series produces mathematically valid results in the region $|F/mg| < 1$ or $-\frac{1}{4}\pi < \theta < +\frac{1}{4}\pi$. But physically, it obviously breaks down when $|\theta| > \theta_0$, where the marble undergoes a 'phase change', falling off the track and rolling into corner A or B. That is, even if carried to infinite order (i.e., use $\tan^{-1} F/mg$ formula), the series produces the wrong result.

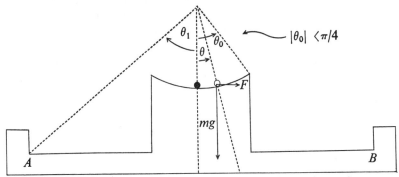

Fig. 15.4 *Classical System Illustrating Breakdown of Perturbation Series when a Phase Transition Occurs*

The mathematical reason why the diagram perturbation series for G breaks down is easily seen from appendix E.12. There it is pointed out that the series is invalid if

$$\langle \Phi_0 | \Psi_0 \rangle = 0. \tag{15.47}$$

This is just the statement that the ground state of the interacting system is orthogonal to that of the non-interacting one, i.e., that they have different symmetries.

15.5 A brief look at Nambu formalism

A clue regarding how to fix up the perturbation series so it will yield the superconducting state, comes from observing that BCS were able to obtain this state only by introducing the assumption that the superconducting wave function was paired (15.14, 15). If this assumption could somehow be squeezed into the perturbation formalism, it might be possible to get the bogolon propagator (15.40).

The way to do this (Gorkov, Nambu—see Schrieffer (1964b) and (1964a), p. 169. Alternative methods are in Pines (1961), p. 98, Johansson (1966)) is to introduce new '*anomalous propagators*', F, F', which destroy or create a Cooper pair in the superconducting ground state. These are defined by:

$$F(\mathbf{k}, t) = -i\langle \Psi_0| T\{c_{k\uparrow}(t)c_{-k\downarrow}(0)\} |\Psi_0\rangle, \qquad (15.48)$$

$$F'(\mathbf{k}, t) = -i\langle \Psi_0| T\{c^\dagger_{-k\downarrow}(t)c^\dagger_{k\uparrow}(0)\} |\Psi_0\rangle. \qquad (15.49)$$

The state $|\Psi_0\rangle$ here is the ground state of the superconductor, and is considered to be of the type (15.29), i.e., the number of particles is not sharp, but is peaked about the average value N (see after (15.32), and Schrieffer (1964a), p. 172). Physically, F' is the probability amplitude that if the $\mathbf{k}\uparrow$ member of a Cooper pair is added to the system at time 0 and the $-\mathbf{k}\downarrow$ member is added at time t, then the system will still be in its ground state at time t.

There are two equivalent ways of fitting these new propagators into the formalism. The first requires introducing new diagrams for F, F'. (See Abrikosov *et al.*, (1965).) The second, due to Nambu, uses the same old diagrams as before, but combines F, F' and G into a *matrix propagator*. (The generalization of the matrix propagator method to arbitrary fermion phase transitions is discussed in chapter 17.) We will discuss just the Nambu method, very briefly.

The matrix propagator, $\mathbb{G}$, is defined by

$$\begin{aligned}\mathbb{G}(\mathbf{k}, t) \; &= \begin{pmatrix} G_{11} & G_{12} \\ G_{21} & G_{22} \end{pmatrix} = \begin{pmatrix} G(\mathbf{k}\uparrow, t) & F(\mathbf{k}, t) \\ F'(\mathbf{k}, t) & -G(-\mathbf{k}\downarrow, -t) \end{pmatrix} \\ &= -i\begin{pmatrix} \langle \Psi_0| T\{c_{k\uparrow}(t)c^\dagger_{k\uparrow}(0)\} |\Psi_0\rangle & \langle \Psi_0| T\{c_{k\uparrow}(t)c_{-k\downarrow}(0)\} |\Psi_0\rangle \\ \langle \Psi_0| T\{c^\dagger_{-k\downarrow}(t)c^\dagger_{k\uparrow}(0)\} |\Psi_0\rangle & \langle \Psi_0| T\{c^\dagger_{-k\downarrow}(t)c_{-k\downarrow}(0)\} |\Psi_0\rangle \end{pmatrix}. \end{aligned}$$

$$(15.50)$$

The free propagator is obtained by replacing Ψ_0 by Φ_0, the ground state of the non-interacting system. This yields, after transformation to $(\mathbf{k}, \omega)$-space:

$$\mathbb{G}_0(\mathbf{k}, \omega) \; = \begin{pmatrix} \dfrac{1}{\omega - \epsilon_k + i\,\omega\delta} & 0 \\ 0 & \dfrac{1}{\omega + \epsilon_k + i\,\omega\delta} \end{pmatrix}. \qquad (15.51)$$

It is easy to get the perturbation expansion of $\mathbb{G}$ if everything is first expressed in terms of the two-component fermion spinor operators

$$\Psi_k = \begin{pmatrix} \Psi_1(\mathbf{k}) \\ \Psi_2(\mathbf{k}) \end{pmatrix} \equiv \begin{pmatrix} c_{k\uparrow} \\ c^\dagger_{-k\downarrow} \end{pmatrix}, \quad \Psi^\dagger_k = (c^\dagger_{k\uparrow}, c_{-k\downarrow}). \qquad (15.52)$$

In terms of these the matrix propagator assumes the compact form:

$$G_{ij}(\mathbf{k},t) = -i\langle \Psi_0| T\{\Psi_i(\mathbf{k},t)\, \Psi_j^\dagger(\mathbf{k},0)\}|\Psi_0\rangle, \qquad (15.53)$$

while the BCS Hamiltonian (15.10) becomes

$$H = \sum_k \epsilon_k\, \Psi_k^\dagger\, \tau_3\, \Psi_k + \tfrac{1}{2} \sum_{k,k',q} \mathscr{V}_{kq}(\Psi_{k+q}^\dagger \tau_3\, \Psi_k)(\Psi_{k'-q}^\dagger \tau_3\, \Psi_{k'}). \quad (15.54)$$

The τ_i's here are the Pauli matrices

$$\tau_1 = \begin{pmatrix} 0 & 1 \\ 1 & 0 \end{pmatrix}, \quad \tau_2 = \begin{pmatrix} 0 & -i \\ i & 0 \end{pmatrix}, \quad \tau_3 = \begin{pmatrix} 1 & 0 \\ 0 & -1 \end{pmatrix}, \quad \mathbb{1} = \begin{pmatrix} 1 & 0 \\ 0 & 1 \end{pmatrix}. \quad (15.55)$$

(Note: When (15.52) is substituted in (15.54) and the matrices are multiplied out, we get (15.10) plus some extra terms. These are compensated by choice of $\mathbb{G}_0$. See Schrieffer (1964a), pp. 174–5.) Thus, in terms of the spinor operators, $\mathbb{G}$ and H have the same form as the ordinary G and H for a normal system. Hence $\mathbb{G}$ may be expanded employing the same diagrams as for G:

$$\mathbb{G} \equiv \qquad = \qquad + \qquad + \qquad + \qquad + \cdots, \qquad (15.56)$$

but with a new dictionary in which matrices are associated with the various lines as shown in Table 15.1 (p. 270). (Observe that since all free propagators are diagonal, (15.56) would not include any anomalous processes if we had free propagators in the self-energy parts. It is necessary to use full propagators instead, i.e., self-consistent renormalization to obtain the superconducting phase.)

For example, the lowest-order diagram is

$$\equiv \left[i\mathbb{G}_0(\mathbf{k},\omega) \right]^2 \times \int \frac{d^3k'\, d\omega'}{(2\pi)^4}(-i)\mathscr{V}_{k',k'-k}\,\tau_3\, i\mathbb{G}(\mathbf{k}',\omega') e^{i\omega 0^+}\, \tau_3.$$

$$(15.57)$$

Evidently, all the partial summation tricks hold here also. The Dyson equation is

$$= \frac{1}{^{-1} - } \quad ; \qquad = \qquad + \qquad + \qquad + \qquad + \cdots,$$

$$(15.58)$$

Table 15.1 *Diagram dictionary for a superconductor*
(Nambu formalism)

Diagram	Function
$\uparrow\uparrow$ or $\uparrow\uparrow$	$i\mathbb{G}(\mathbf{k},\omega)$ $\begin{bmatrix} \times \exp(i\omega 0^+),\text{ for} \\ \text{non-propagating lines} \end{bmatrix}$
$\uparrow$ or $\downarrow$	$i\mathbb{G}_0(\mathbf{k},\omega) = i\begin{pmatrix} \dfrac{1}{\omega-\epsilon_k+i\omega\delta} & 0 \\ 0 & \dfrac{1}{\omega+\epsilon_k+i\omega\delta} \end{pmatrix}$
$\mathbf{k+q}$ $\qquad$ $\mathbf{k'-q}$ $\mathbf{k}\quad\mathbf{q}\quad\mathbf{k'}$ BCS interaction	Each vertex: τ_3 Wiggle: $-i\mathscr{V}_{kq}$
fermion loop Ex:	$(-1)\times$ Trace of the product of $\mathbb{G}$'s and τ_3's forming the loop

and the interactions and propagators may be renormalized, yielding as before just (10.60), (11.3) or (11.5), with all diagram elements now evaluated by means of the Nambu dictionary, Table 15.1.

The Nambu formalism yields bogolon excitations in the lowest order of self-consistent renormalization, i.e., using the first two terms of (15.58). That is, we must solve simultaneously the two matrix equations:

$$\uparrow\uparrow \;=\; \frac{1}{\uparrow^{-1} - \bigotimes} \;;\quad \bigotimes \;\approx\; \rightimes + \multimap\bigcirc \qquad (15.59)$$

Translated, with the aid of Table 15.1, this becomes:

$$\mathbb{G}(\mathbf{k},\omega) = \frac{1}{\omega\mathbb{1} - \epsilon_k\,\tau_3 - \sum(\mathbf{k},\omega)} \;, \qquad (15.60)$$

where we note that $\sum$ involves the dressed propagator $\mathbb{G}$, in contrast to (15.57). When (15.60) is solved using the BCS interaction, it is found that

$$G_{11} \equiv G(\mathbf{k}, \omega) = \frac{u_k^2}{\omega - \sqrt{\epsilon_k^2 + \Delta_k^2} + i\delta} + \frac{v_k^2}{\omega + \sqrt{\epsilon_k^2 + \Delta_k^2} - i\delta}, \quad (15.61)$$

where u_k, v_k, Δ_k are just as in the BCS solution (15.29, 30, 32). Thus, by using the Nambu method, we have obtained a propagator with first term of the bogolon form (15.40).

The physical meaning of the two terms in (15.61) is easily seen by performing the Bogoliubov canonical transformation directly on the propagator. Using (15.35) yields

$$\begin{aligned}
G(\mathbf{k}, t) &= -i\langle \Psi_0 | T\{c_{k\uparrow}(t) c_{k\uparrow}^\dagger(0)\} | \Psi_0 \rangle \\
&= -iu_k^2 \langle \Psi_0 | T\{\alpha_k(t) \alpha_k^\dagger(0)\} | \Psi_0 \rangle + \\
&\quad -iv_k^2 \langle \Psi_0 | T\{\beta_k^\dagger(t) \beta_k(0)\} | \Psi_0 \rangle.
\end{aligned} \quad (15.62)$$

(The $\alpha_k \beta_k$ terms $= 0$.) The $\alpha_k(t)$, $\beta_k(t)$, etc., are in Heisenberg picture like the c_k's in (9.3). Using (15.38, 39) and taking the Fourier transform produces just (15.61), showing that the first term of (15.61) describes α-bogolons while the second describes β-bogolons (see after (15.37)).

15.6 Treatment of retardation effects by Nambu formalism

The BCS theory, as well as the Nambu calculation we have just described, suffer from the defect that they treat the Fröhlich interaction as static. In reality, because it takes time for the phonon to travel from one electron to another, this interaction is retarded (i.e., time dependent). The only way of handling this retardation properly is to treat the phonons field-theoretically from the very beginning. This means that instead of taking the canonically transformed H in (15.10) as our starting Hamiltonian, it is necessary to go all the way back to the H in (15.4), which contains the electron–phonon interaction explicitly. Then, in order to calculate the propagator, it is necessary to introduce special diagrams for the free phonon, and for the electron–phonon interaction.

This programme is carried out for a normal system of interacting electrons and phonons in appendix J. It is shown that the calculation of the electron propagator is just like that in the no-phonon case, except that the Coulomb interaction is replaced by a *combined interaction* which is the sum of the Coulomb and retarded Fröhlich interactions:

$$ \hspace{8cm} (15.63) $$

combined   Coulomb   Fröhlich
interaction   interaction   interaction

The Fröhlich interaction diagram shows a virtual phonon (⨋⨋⨋⨋⨋⨋) being sent out by one electron and absorbed by another. The diagrams are drawn in $(\mathbf{k}, t)$-space to show the retardation.

The BCS interaction is just a static approximation to the combined interaction, as shown in appendix J. Hence, in a rough way, we may say that when the combined interaction is predominantly positive, the system acts normally, while if it is predominantly negative, the system is a superconductor.

The calculation of the electron propagator in the superconducting system, taking retardation into account, may be carried out simply by replacing the BCS interaction by the combined interaction of (15.63), in the Nambu dictionary, Table 15.1. The calculation then proceeds just as in §15.5.

Such a computation using the retarded interaction has been carried out by Schrieffer et al. (Schrieffer (1964a), pp. 180–93), and it yields a frequency-dependent energy gap which shows good agreement with the results of tunnelling experiments in lead.

15.7 Transition temperature of a superconductor

It is easy to generalize the discussion of the Cooper pair K-matrix in §15.4 to finite T, and use the result to find the transition temperature. The finite T K-matrix or scattering amplitude is (remember that energies here are relative to the Fermi energy, $\mu = \epsilon_F$ so μ doesn't appear explicitly, and note that $\epsilon_p = \epsilon_{-p}$, so $u_p u_{-p} = u_p$):

$$
\mathscr{K}(i\omega_m) = V u_p \left\{ 1 - (-V) \int \frac{d^3\mathbf{p}}{(2\pi)^3} u_p \frac{1}{\beta} \sum_{n=-\infty}^{+\infty} \left(\frac{1}{i\omega_n + i\omega_m - \epsilon_p} \right) \left(\frac{1}{-i\omega_n - \epsilon_p} \right) \right\}^{-1}.
$$

$$(15.64)$$

Evaluating the frequency sum by the same method as in §14.5 yields

$$
\mathscr{K}(i\omega_m) = V u_p \left\{ 1 + V \int \frac{d^3\mathbf{p}}{(2\pi)^3} u_p \frac{\tanh(\beta\epsilon_p/2)}{2\epsilon_p - i\omega_m} \right\}^{-1}.
$$

$$(15.65)$$

Analytically continuing $i\omega_m \to \omega + i\delta$ to get the retarded K-matrix (see appendix L.39), changing to an integral over ϵ_p by introducing $N(0)$, the density of states in the vicinity of the Fermi surface, then using (3.76) to find the real and imaginary parts, we obtain:

$$
K^R(\omega) = \mathscr{K}(\omega + i\delta)
$$
$$
= V u_p \left\{ 1 + V N(0) \dot{P} \int_{-\omega_c}^{+\omega_c} d\epsilon_p \frac{\tanh(\beta\epsilon_p/2)}{2\epsilon_p - \omega} + i V N(0) \pi \tanh\left(\frac{\beta\omega}{4} \right) \right\}^{-1}.
$$

$$(15.66)$$

Now we saw that at $T=0$, the illegal pole of K^R in the upper half-plane first reared its ugly head when V reached its critical value, $V_c=0$. At this point, $\omega=0$ (see (15.43)). Therefore, let us look at $K^R(\omega=0)$. In the limit $k_B T \ll \omega_c$ we get, after approximately evaluating the integral,

$$K^R(0) \approx V u_p \left\{ 1 + V N(0) \ln \left(\frac{\omega_c}{2k_B T} \right) \right\}^{-1}. \tag{15.67}$$

This diverges when the bracketed quantity is zero, i.e., for $V < 0$ and

$$T \equiv T_c = \frac{\omega_c}{2k_B} e^{-1/N(0)V}, \tag{15.68}$$

which is the BCS expression for the transition temperature of a superconductor.

Further reading

Rickayzen (1965).
Schrieffer (1964a, b).
Beliaev (1958).
Pines (1961), chap. 3, p. 91.
Abrikosov (1965), chap. 7 (Gorkov formalism).
Thouless (1972).

Electron–phonon interaction

Appendix J.
Pines (1963), chap. 5.
Schultz (1964), chap. 4.
Pines (1961), chap. 3, p. 82.
Schrieffer (1964a, b).

Exercises

15.1 Prove that $2 \sum_k \epsilon_k b_k^\dagger b_k$ and $\sum_{k\sigma} \epsilon_k c_{k\sigma}^\dagger c_{k\sigma}$ give the same result when operating on a paired wave function of form (15.14).

15.2 Show with the aid of (7.32) that the bogolon operators in (15.34) obey fermion commutation relations (for example, prove that $[\alpha_k, \alpha_{k'}^\dagger] = \delta_{kk'}$).

15.3 What is the energy of a superconductor in the state $\alpha_{k_2}^\dagger \beta_{k_2}^\dagger \alpha_{k_1}^\dagger |\Psi_0\rangle$?

15.4 Verify that (15.53) is equivalent to (15.50).

15.5 Verify that (except for an infinite constant term) the first term of (15.54) is equal to the quasi-electron Hamiltonian (15.11).

15.6 Write out the expression for the bubble diagram in (15.56) using Nambu formalism.

15.7 Carry out the Bogoliubov transformation of the propagator (15.62) in detail.

15.8 Show that the pairing operators in (15.17) obey the following commutation rules:

(a) $[b_k, b_{k'}^\dagger]_- = (1 - c_{k\uparrow}^\dagger c_{k\uparrow} - c_{-k\downarrow}^\dagger c_{-k\downarrow}) \delta_{kk'}$.

(b) $[b_k, b_{k'}]_- = 0 = [b_k^\dagger, b_{k'}^\dagger]_-$

(c) $[b_k, b_{k'}]_+ = 2b_k b_{k'}(1 - \delta_{kk'})$.

15.9 (a) Using H_{red} in (15.16) and $|\Psi_0\rangle$ in (15.29), show that

$$\langle \Psi_0 | H_{red} | \Psi_0 \rangle = 2 \sum_k \epsilon_k V_k^2 - \sum_{\substack{k,k' \\ k \neq k'}} V_{kk'} u_k u_{k'} V_k V_{k'}$$

where the contribution from V_{kk} is incorporated into ϵ_k. (Hint: to get first term, write $|\Psi_0\rangle = u_k |\phi_{k0}\rangle + v_k |\phi_{k1}\rangle$ where ϕ_{k0} means no pair in k while ϕ_{k1} means there is a pair in k.) Use a similar method to get second term.

(b) Carry out the variation $\delta\langle \Psi_0 | H_{red} | \Psi_0 \rangle = 0$ and obtain (15.30), (15.31).

(c) Show that Δ_k satisfies: $\Delta_{k'} = \sum_k \dfrac{V_{kk'}}{2\epsilon_k} \Delta_k$.

(d) Solve (c) for $V_{kk'}$ as in (15.28) and derive (15.32, 33).

15.10 Derive (15.42), (15.42'), (15.43).

15.11 Verify (15.64)→(15.68).

15.12 Show that the anomalous propagator F in (15.48) is finite in the superconducting state, and zero in the normal (e.g., non-interacting) state. (Take $t = 0^+$.)

Chapter 16

Phonons from a Many-Body Viewpoint (Reprint)

In appendix $\mathscr{A}$ we find the phonon ground state energy and frequency dispersion law by performing the canonical transformation ($\mathscr{A}$.29, 30), ($\mathscr{A}$.36) on the lattice Hamiltonian ($\mathscr{A}$.28). In the reprint on the following pages, it is shown how the same result may be obtained diagrammatically. This provides an extremely simple example of the application of the graphical technique to a boson system.

First the Hamiltonian ($\mathscr{A}$.28) is re-written as the sum of (1) an unperturbed part describing a very primitive sort of collective excitation called the *Einstein phonon*, and (2) a perturbation describing the strong interaction between Einstein phonons. On account of the interaction, the Einstein phonon becomes surrounded by a cloud of other Einstein phonons; this dresses it and converts it into an ordinary phonon. Thus we have

bare Einstein phonon	+	cloud of other Einstein phonons	=	quasi Einstein phonon
			=	ordinary phonon

Hence the ordinary phonon may be interpreted as a quasi particle in the 'quasi collective excitation' sense. This is similar to (13.24) where we had the dressed or quasi plasmon:

bare plasmon	+	cloud of electrons and other plasmons	=	quasi plasmon

or appendix J where we find the dressed or quasi phonon:

bare phonon	+	electron cloud	=	quasi phonon.

*Phonons from a Many-Body Viewpoint**

The phonon is considered as a quasi particle consisting of an Einstein phonon dressed by interaction with other Einstein phonons. By exact summation of an infinite series of self-energy graphs, the Einstein phonon propagator is evaluated, and its poles yield the well-known phonon dispersion law. An alternative derivation shows the propagator to be also valid in regions of divergence. The vacuum fluctuation diagrams—all of the ring type—are easily summed to get the ground state energy. The unusual simplicity of the graphs encountered make the arguments here an ideal illustration of the application of the field theoretical technique to the many-body problem.

* Reprinted from ANNALS OF PHYSICS. Volume 27, No. 2, April 1964
Copyright © by Academic Press Inc.   *Printed in U.S.A.*

16.1 Introduction

In the usual field-theoretic treatments (Migdal (1958)), the phonon is regarded as a bare particle which is converted into a clothed (i.e., quasi) particle by collisions with other phonons or with electrons. This tends to obscure the fact that the phonon itself is a quasi particle of a particularly simple sort, consisting of a bare Einstein phonon dressed by interaction with other Einstein phonons. Our object here is to develop this point of view in some detail, showing how one can derive the phonon dispersion law and ground state energy by exact summations over Einstein phonon self-energy and ring diagrams. The argument is probably the simplest existing example of the application of quantum field theory to a many-body problem, and constitutes a pedagogically ideal illustration of the qualities which made the graphical method famous: its power to do perturbation theory to infinite order (thus enabling it to cope with strong couplings beyond the reach of ordinary perturbation procedures), its highly systematic and so-called 'automatic' character, its vivid pictorial appeal, and its remarkable talent for producing results valid outside their region of convergence.

The Hamiltonian for the model system is written as the sum of an unperturbed part representing a set of independent constant-frequency lattice oscillators (Einstein phonons) and a strong coupling term. After reviewing the standard treatment by canonical transformation, an Einstein phonon propagator is introduced in two different forms:

$$G(k,t) = -i\langle\psi_0| T\{b_k(t) b_k^\dagger(0)\} |\psi_0\rangle,$$

and

$$D(k,t) = -i\langle\psi_0| T\{[b_{-k}(t) + b_k^\dagger(t)][b_k(0) + b_{-k}^\dagger(0)]\} |\psi_0\rangle,$$

where $b_k^\dagger$, b_k are the Einstein phonon creation and destruction operators. These are evaluated by summing self-energy diagrams exactly to all orders; the poles of the resultant clothed propagators yield the renormalized Einstein phonon frequencies, i.e., the well-known phonon dispersion law. The D-form of the propagator is shown to be easier to handle mathematically (since it eliminates the need for summing over backward-going graphs), but more difficult to interpret physically. Despite the fact that the propagator appears unusable in certain regions due to divergence difficulties, an alternative derivation shows it to be valid everywhere. In the last section, the ground state energy is calculated by summing exactly over an infinite set of ring diagrams which turns out to be the series for $\sqrt{(1-x)}$.

It is to be emphasized that although we believe the viewpoint taken here is new, the end results are *not* new, and the procedures used to obtain them are *not* simpler than the usual canonical transformation technique. The field-theoretic route is followed first for the purpose of showing how the phonon may be fitted nicely into the quasi-particle picture of matter which has been emerging over the last ten years (Ter Haar (1959–60)), and second, in order to exhibit an extraordinarily simple model system where the many-body methodology is transparent and the solutions exact.

16.2 Hamiltonian for coupled Einstein phonons

If we have a linear chain of N atoms, with interatomic distance d, and harmonic coupling constant $\frac{1}{2}m\omega_0^2$, its Hamiltonian may be written

$$
\begin{aligned}
H &= \sum_{l=1}^{N} \frac{p_l^2}{2m} + \frac{1}{4}m\omega_0^2 \sum_{l=1}^{N} (u_{l+1}-u_l)^2 \\
&= \sum_{l=1}^{N} \left[\frac{p_l^2}{2m} + \frac{1}{2}m\omega_0^2 u_l^2 \right] - \frac{1}{2}m\omega_0^2 \sum_{l=1}^{N} u_l u_{l+1},
\end{aligned}
\tag{16.1}
$$

where p_l, u_l are momentum and displacement operators for the atom on site l, and end effects have been neglected. This may be expressed in occupation number formalism by first introducing the operators $b_l^\dagger$, b_l, which respectively create or destroy a unit of excitation at the lth oscillator, by means of

$$
u_l = \sqrt{\frac{1}{2m\omega_0}}(b_l + b_l^\dagger)
$$

$$
p_l = \frac{1}{i}\sqrt{\frac{m\omega_0}{2}}(b_l - b_l^\dagger)
\tag{16.2}
$$

(we have set $\hbar = 1$); the result is

$$
H = \omega_0 \sum_l (b_l^\dagger b_l + \tfrac{1}{2}) - \frac{\omega_0}{4} \sum_l (b_l + b_l^\dagger)(b_{l+1} + b_{l+1}^\dagger).
\tag{16.3}
$$

Secondly, we make the Fourier transformation

$$
b_l = (1/\sqrt{N}) \sum_k b_k e^{+ikld}
$$

$$
b_l^\dagger = (1/\sqrt{N}) \sum_k b_k^\dagger e^{-ikld}
\tag{16.4}
$$

converting H to

$$
H = \omega_0 \sum_k (b_k^\dagger b_k + \tfrac{1}{2}) - \frac{\omega_0}{4} \sum_k e^{-ikd}(b_k + b_{-k}^\dagger)(b_{-k} + b_k^\dagger).
\tag{16.5}
$$

Making use of the fact that the b_k's obey boson commutation rules

$$
[b_k, b_{k'}^\dagger]_- = \delta_{kk'}; \quad [b_k, b_{k'}]_- = [b_k^\dagger, b_{k'}^\dagger]_- = 0,
\tag{16.6}
$$

the last term may be changed to a summation over $k > 0$ and we obtain the desired starting form for the Hamiltonian (see (5) in §7.7):

$$H = \omega_0 \sum_k (b_k^\dagger b_k + \tfrac{1}{2}) - \frac{\omega_0}{2} \sum_{k>0} \cos kd (b_k + b_{-k}^\dagger)(b_{-k} + b_k^\dagger) = H_0 + H_1. \quad (16.7)$$

The unperturbed term, H_0, is the Hamiltonian for a set of oscillators extending through the whole lattice, having number operator $b_k^\dagger b_k$, and a common frequency ω_0; these are the bare Einstein phonons. They interact via the perturbing term, H_1, become clothed and acquire a k-dependent frequency, thus being transformed into ordinary phonons. The problem is to find the new frequency dispersion law ω_k, and ground state energy E_0.

Since H_1 is as large as H_0, perturbation theory cannot be used here, at least not in its ordinary form. The standard textbook solution (Ziman (1962), chap. 1) therefore employs a canonical transformation of the Hamiltonian. Defining the new operators

$$\phi_k = b_k + b_{-k}^\dagger$$
$$\pi_k = i(b_k - b_{-k}^\dagger), \quad (16.8)$$

H becomes

$$H = \frac{\omega_0}{2} \sum_{k>0} \{\pi_k^\dagger \pi_k + (1 - \cos kd)\phi_k^\dagger \phi_k - 2\}. \quad (16.9)$$

Introducing the boson operators $a_k^\dagger$, a_k by

$$\phi_k = \sqrt{\omega_0/\omega_k}(a_k + a_{-k}^\dagger)$$
$$\pi_k = i\sqrt{\omega_k/\omega_0}(a_k^\dagger - a_{-k}), \quad \omega_k = \omega_0\sqrt{(1 - \cos kd)}, \quad (16.10)$$

yields

$$H = \sum_k \omega_k(a_k^\dagger a_k + \tfrac{1}{2}). \quad (16.11)$$

This is clearly the Hamiltonian for a set of quasi particles with number operators $a_k^\dagger a_k$ and dispersion law and ground state energy

$$\omega_k = \omega_0\sqrt{(1 - \cos kd)}, \quad E = \sum_k \frac{\omega_k}{2}, \quad (16.12)$$

which are the well-known relations for phonons.

It will now be shown how this exact result may be obtained by the graphical techniques of many-body theory. The dispersion law is derived in §§16.3, 16.4, and the ground state energy in §16.6.

16.3 Definition of Einstein phonon propagator

The field-theoretic method for finding the excitations of a many-body system consists in solving for the Green's function propagator; the poles of this propagator yield directly the energies of the excited states (Galitski and Migdal (1958)). The propagator for the Einstein phonon may be defined in one of two ways. The simplest, from the standpoint of physical interpretation, is

$$G(k,t) = -i\langle\psi_0| T\{b_k(t) b_k^\dagger(0)\} |\psi_0\rangle \qquad (16.13)$$

where $b_k(t)$ is the operator b_k in the Heisenberg picture:

$$b_k(t) = e^{+iHt} b_k e^{-iHt}. \qquad (16.14)$$

The symbol T is the Wick time-ordering operator for bosons, and has the property

$$T\{A(t')B(t)\} = A(t')B(t) \quad t' > t$$
$$= B(t)A(t') \quad t' < t. \qquad (16.15)$$

The ket vector $|\psi_0\rangle$ is the exact ground state or 'interacting vacuum' for the system of interacting Einstein phonons. This propagator has the physical significance of being the probability amplitude that if an Einstein phonon in the state k is introduced into the interacting vacuum at time $t=0$, and 'propagates' in the system, we will observe an Einstein phonon in state k at time $t=t$ ($t>0$). In the case of zero interaction, $G(k,t)$ assumes the free propagator form

$$G_0(k,t) = -i\{\theta_t\langle 0| b_k(t) b_k^\dagger(0) |0\rangle + \theta_{-t}\langle 0| b_k^\dagger(0) b_k(t) |0\rangle\}$$
$$= -i\theta_t e^{-i\omega_0 t} \qquad (16.16)$$

where $|0\rangle$ is the ket for the non-interacting vacuum (i.e., no bare Einstein phonons present). We shall also need an auxiliary propagator which represents propagation 'backwards' in time for the non-interacting case, e.g.,

$$G_0^- = -i\langle 0| T\{b_k^\dagger(t) b_k(0)\} |0\rangle_{H_1=0} = -i\theta_{-t} e^{+i\omega_0 t}. \qquad (16.17)$$

It is most convenient to work with the Fourier transforms of the propagator, given by

$$G_0(k,\omega) = \int_{-\infty}^{+\infty} dt\, e^{i\omega t} G_0(k,t) = +\frac{1}{\omega - \omega_0 + i\delta} \qquad (16.18)$$

$$G_0^-(k,\omega) = \int_{-\infty}^{+\infty} dt\, e^{i\omega t} G_0^-(k,t) = -\frac{1}{\omega + \omega_0 - i\delta} \qquad (16.19)$$

where δ is a positive infinitesimal used to remove terms oscillating at $t=\infty$.

An alternative propagator, harder to interpret physically, but easier to work with mathematically is

$$D(k,t) = -i\langle\psi_0| T\{\phi_k(t)\phi_k^\dagger(0)\} |\psi_0\rangle. \qquad (16.20)$$

The ϕ_k's here are defined by

$$\phi_k = b_k + b^\dagger_{-k}; \quad \phi^\dagger_k = b^\dagger_k + b_{-k} = \phi_{-k}, \qquad (16.21)$$

and are proportional to the Fourier transform of the displacement operator:

$$u_k = \sum_l u_l e^{-ikld} = \sqrt{\frac{1}{2m\omega_0}} (b_k + b^\dagger_{-k}) \qquad (16.22)$$

where (16.2) has been used. Although difficult to picture in terms of Einstein phonons, $D(k,t)$ may be viewed as propagating a 'displacement wave' in the interacting vacuum. The free propagator is

$$D_0(k,t) = -i[\theta_t e^{-i\omega_0 t} + \theta_{-t} e^{+i\omega_0 t}] \qquad (16.23)$$

which has the Fourier transform

$$D_0(k,\omega) = \frac{1}{\omega - \omega_0 + i\delta} - \frac{1}{\omega + \omega_0 - i\delta} = \frac{2\omega_0}{\omega^2 - \omega_0^2 + 2i\delta\omega_0}. \qquad (16.24)$$

It is seen that D_0 includes propagation both forwards and backwards in time; this makes it somewhat simpler to use.

16.4 Evaluation of propagator by exact summation of graphs

Propagators may be evaluated either by solving the differential equation they obey (see appendix M) or by expressing them as a perturbation series with the aid of graphs and summing over selected sets of graphs to infinite order. The latter technique has the advantage of being systematic and to a high degree automatic, and is the one we shall consider now. It will be seen that in the present case the summation may be carried out over all graphs and is exact.

The derivation of the dispersion law using the $G(k,t)$ form of the propagator is the most instructive from the physical standpoint. This begins with the Dyson perturbation expansion

$$G(k,t) = \lim_{\substack{\mathrm{Im}\, T_1, -T_2 \\ \to \infty}} \frac{\left\{ (-i) \sum_{n=0}^{\infty} (-1)^n/n! \times \int_{T_1}^{T_2} dt_1 \ldots \int_{T_1}^{T_2} \times dt_n \langle 0| T\{H_1(t_1) \ldots H_1(t_n) b_k(t) b^\dagger_k(0)\} |0\rangle \right\}}{\left\{ \sum_{n=0}^{\infty} (-1)^n/n! \times \int_{T_1}^{T_2} dt_1 \ldots \int_{T_1}^{T_2} \times dt_n \langle 0| T\{H_1(t_1) \ldots H_1(t_n)\} |0\rangle \right\}}, \qquad (16.25)$$

where $b_k(t)$ is now in the *interaction* picture:

$$b_k(t) = e^{iH_0 t} b_k e^{-iH_0 t} \tag{16.26}$$

and

$$H_1(t) = e^{iH_0 t} H_1 e^{-iH_0 t} = \sum_{p>0} V_p[b_p(t) + b_{-p}^\dagger(t)][b_{-p}(t) + b_p^\dagger(t)] \tag{16.27}$$

$$V_p = -\frac{\omega_0}{2}\cos pd. \tag{16.28}$$

With the conventions that $b^\dagger$ (b) are represented by lines above (below) interaction vertices, with $+p$ ($-p$) lines directed upward (downward), the four products in $H_1(t)$ may be represented graphically as shown in Fig. 16.1. These show the different types of interactions between Einstein phonons: (1) two Einstein phonons, one in state p, the other in $-p$, collide and annihilate each other at time t, (2) an Einstein phonon in state p is annihilated at time t

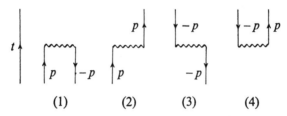

(1)   (2)   (3)   (4)

Fig. 16.1 *Graphs for Interactions Between Einstein Phonons*

with the simultaneous creation of a new Einstein phonon in state p, etc. According to the field-theory recipe for associating diagrams with matrix elements, the expansion, (16.25), may be shown to be the sum of all possible sequences of such interactions, with a bare Einstein phonon entering and leaving, as shown in Fig. 16.2. These self-energy diagrams reveal the Einstein phonon dressed in a cloud of other Einstein phonons having the same or opposite momentum. (Note that *unlike* the electron propagator, twisting the interaction wiggles through 180 degrees does *not* produce a new diagram.) They may be evaluated by Fourier transforming to (k, ω)-space and applying the rules:

(1) Factor of $iG_0(k, \omega)$ for the forward propagating $(+k)$ lines.
(2) Factor of $iG_0^-(k, \omega)$ for the backward propagating $(-k)$ lines. (16.29)
(3) Factor of $-iV_k$ for each interaction wiggle.

The summation is most instructively carried out with the aid of the Dyson equation. Define an irreducible self-energy diagram for the Einstein phonon as one which *cannot* be drawn as two parts connected by a positive k-line;

thus graphs (2) and (4) in Fig. 16.2 are irreducible, while (3) and (5) are reducible. Then, by summing over all reducible self-energy diagrams, it is easily shown that $G(k,t)$ satisfies Dyson's equation as drawn in Fig. 16.3. In Fig. 16.3, Σ is the sum over all irreducible self-energy parts as shown.

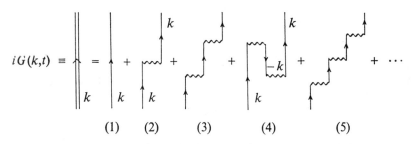

Fig. 16.2 *Expansion of Einstein Phonon Propagator*

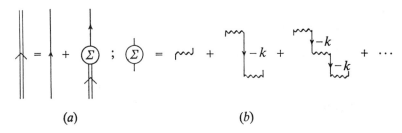

Fig. 16.3 (*a*) *Dyson Equation and* (*b*) *Irreducible Self-energy Expansion*

This is easily evaluated by means of the rules of (16.29), plus (16.19) (dropping the infinitesimals for convenience):

$$-i\Sigma = -iV_k[1+(-iV_k)(iG_0^-)+(-iV_k)^2(iG_0^-)^2+\cdots]$$
$$= \frac{-iV_k}{1-V_kG_0} = \frac{-iV_k(\omega+\omega_0)}{V_k+\omega+\omega_0}. \tag{16.30}$$

The Dyson equation is

$$iG(k,\omega) = iG_0+(iG_0)(-i\Sigma)(iG), \tag{16.31}$$

from which one readily obtains, with the aid of (16.18, 28 and 30),

$$G(k,\omega) = \frac{\omega+\omega_0(1-\frac{1}{2}\cos kd)}{\omega^2-\omega_0^2(1-\cos kd)}. \tag{16.32}$$

The renormalized Einstein phonon frequencies are given by the poles of $G(k, \omega)$; these occur at

$$\omega = \omega_0 \sqrt{(1 - \cos kd)} \tag{16.33}$$

which is identical with the phonon dispersion law, (16.12).

The derivation in terms of the $D(k, t)$ form of the propagator is mathematically simpler but lacks the charm of having direct physical interpretation. This alternative propagator may be expanded by means of (16.25) with the

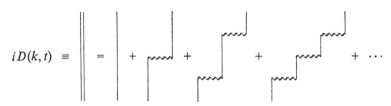

$$iD(k, t) \equiv \quad = \quad + \quad + \quad + \quad + \cdots$$

Fig. 16.4 *Expansion of Alternative Einstein Phonon Propagator*

substitution of $\phi_k(t)\phi_k^\dagger(0)$ for $b_k(t)b_k^\dagger(0)$, and H_1 written in terms of the ϕ's thus:

$$H_1(t) = \sum_{p>0} V_p \phi_p(t) \phi_p^\dagger(t). \tag{16.34}$$

The diagrammatic series for this case appears in Fig. 16.4. The rules are

(1) Factor of $iD_0(k, \omega)$ for each line. $\qquad\qquad$ (16.35)
(2) Factor of $-iV_k$ for each wiggle.

This yields in (k, ω)-space

$$iD(k, \omega) = iD_0[1 + iD_0(-iV_k) + (iD_0)^2(-iV_k)^2 + \cdots] = \frac{iD_0}{1 - D_0 V_k}, \tag{16.36}$$

from which

$$D(k, \omega) = \frac{2\omega_0}{\omega^2 - \omega_0^2[1 - \cos kd]}, \tag{16.37}$$

which produces again the correct phonon dispersion law.

16.5 Question of convergence

Throughout the above arguments, we have ignored the fact that the series summations converge only for |summand| < 1. This condition restricts the ω-range in which the above propagators are valid. Thus, the $G(k, \omega)$ form holds only for ω satisfying

$$(a)\ |V_k G_0^-| < 1, \quad (b)\ |G_0 \Sigma| < 1 \tag{16.38}$$

where (a) comes from (16.30) and (b) comes from the summation which yields the Dyson equation, (16.31), while $D(k, \omega)$ is true only for ω such that

$$|V_k D_0| < 1, \quad \text{i.e.,} \quad |\omega_0^2 \cos kd| < |\omega^2 - \omega_0^2|. \qquad (16.39)$$

This is typical of the convergence troubles which plague many-body theory. The usual household remedy is to assume without proof that the propagator may be analytically continued to all values of ω; it is often possible to justify this by using non-perturbative procedures to get the same result. No general validation is yet available (Katz (1962)).

In the present case, it is easily shown that the results are good *outside* the regions of convergence by direct canonical transformation of the propagator itself. Consider just $D(k, \omega)$ for simplicity. Let $\mathcal{O}$ be the operator transforming from ordinary phonon representation to Einstein phonon representation. If $|\gamma\rangle$ is the ground state in ordinary phonon representation, this means

$$\mathcal{O}|\gamma\rangle = |\psi_0\rangle. \qquad (16.40)$$

Substituting into (16.20) and using (16.10) yields

$$\begin{aligned} D(k, t) &= -i\langle\gamma|\, \mathcal{O}^{-1} T\{\phi_k(t)\, \phi_k^\dagger(0)\}\, \mathcal{O}\,|\gamma\rangle \\ &= -i\langle\gamma|\, T\{\mathcal{O}^{-1}\phi_k(t)\, \mathcal{O}\mathcal{O}^{-1}\phi_k^\dagger(0)\, \mathcal{O}\}\,|\gamma\rangle \\ &= -i\langle\gamma|\, T\{\sqrt{\omega_0/\omega_k}\,[a_k(t) + a_{-k}^\dagger(t)]\sqrt{\omega_0/\omega_k}\,[a_k^\dagger(0) + a_{-k}(0)]\}\,|\gamma\rangle \\ &= -i(\omega_0/\omega_k)\{\theta_t e^{-i\omega_k t} + \theta_{-t} e^{+i\omega_k t}\}, \qquad (16.41) \end{aligned}$$

from which

$$D(k, \omega) = \frac{2\omega_0}{\omega^2 - \omega_k^2 + 2i\delta\omega_k}.$$

This is just (16.37) but *without* any restrictions on ω, thus showing that in this transparent case, and perhaps in more opaque ones, the field-theoretic method is vindicated.

16.6 Ground state energy

The ground state energy of an interacting many-body system may be found from the theorem

$$E = E_0 + i \lim_{\text{Im}\, t \to -\infty} \frac{d}{dt} \ln \langle\psi_0|\, \tilde{U}(t)\,|\psi_0\rangle, \qquad (16.42)$$

where E_0 is the ground state energy of the non-interacting system, and $\tilde{U}(t)$ is the scattering operator, given by

$$\tilde{U}(t) = e^{+iH_0 t} e^{-iHt}. \qquad (16.43)$$

The term $\langle \psi_0| \tilde{U}(t) |\psi_0 \rangle$ is the interacting vacuum–vacuum scattering matrix, which is obtained from the expansion:

$$\langle \psi_0| \tilde{U}(t) |\psi_0 \rangle = \sum_{n=0}^{\infty} \frac{(-i)^n}{n!} \int_0^t dt_1 \ldots \int_0^t dt_n \langle 0| T\{H_1(t_1) \ldots H_1(t_n)\} |0 \rangle.$$

$$(16.44)$$

This may be expressed graphically as the sum over all closed, topologically distinct 'vacuum fluctuation' diagrams; its logarithm is just the sum over the linked diagrams as shown in Fig. 16.5. These ring diagrams are similar to those encountered in the random phase approximation, but much simpler. They may be calculated using the rules of either (16.29) or (16.35) (both give

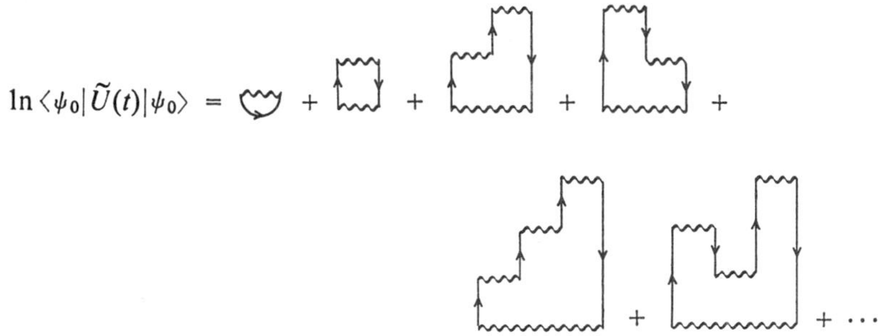

Fig. 16.5 *Graphs for Ground State Energy*

the same result). After evaluating each diagram and taking the derivative and limit according to (16.42), it is found that each distinct graph gives a contribution to the energy obtained by the following rules:

(1) Factor of V_k for each interaction wiggle.
(2) In each of the $n-1$ time intervals between successive interactions on the nth order graph, draw a horizontal line. Then have a factor of $-\omega_0^{-1}$ for each intersection of these horizontal lines with a vertical line of the graph.
(3) Sum over all k.   (16.45)

Carrying out the calculation according to these rules, we find

$$\sum \text{linked graphs} = \sum_{k>0} \left\{ V_k - V_k^2 \left[\frac{1}{2\omega_0} \right] + V_k^3 \left[\frac{1}{(2\omega_0)^2} + \frac{1}{(2\omega_0)^2} \right] \right.$$
$$\left. - V_k^4 \left[2 \times \frac{1}{(2\omega_0)^3} + 2 \times \frac{1}{(2\omega_0)^2} \frac{1}{(4\omega_0)} + 2 \times \frac{1}{(2\omega_0)^3} \right] + \cdots \right\}.$$

$$(16.46)$$

From (16.5), we have

$$E_0 = \sum_k \frac{\omega_0}{2} = \sum_{k>0} \omega_0. \tag{16.47}$$

Combining (16.28, 46 and 47) yields the final result

$$E = E_0 + \sum \text{ linked graphs}$$

$$= \sum_{k>0} \omega_0 \{1 - \tfrac{1}{2}\cos kd - \tfrac{1}{8}\cos^2 kd - \tfrac{1}{16}\cos^3 kd - \tfrac{1}{16}\cdot\tfrac{5}{8}\cos^4 kd \ldots\} \tag{16.48}$$

$$= \sum_{k>0} \omega_0 \sqrt{(1 - \cos kd)} = \sum_k \frac{\omega_k}{2}$$

in agreement with (16.12).

It is interesting to note that, unlike the case with the summation for the propagator, there are no convergence difficulties here, since $\cos kd$ is always < 1. This should be contrasted with the corresponding ground state summation in random phase approximation, where one is forced to assume that the result of summing a logarithmic series can be continued into divergent regions (Gell-Mann (1957)).

ACKNOWLEDGEMENT

The author wishes to thank Prof. H. Højgaard Jensen for the many stimulating conversations which motivated this work and for his careful reading and criticism of the manuscript.

Chapter 17

Quantum Field Theory of Phase Transitions in Fermi Systems

17.1 Introduction

It is everyday experience that under certain circumstances a peaceful-looking system of interacting particles in its 'normal' phase, will suddenly become unstable and undergo a dramatic transition to a 'condensed' phase with radically new properties. A familiar example of such a 'phase transition' is the change from gas to liquid when the temperature is decreased below a certain point, or the pressure is increased. Other examples are the gas $\rightarrow$ solid transition, paramagnet $\rightarrow$ ferromagnet, ordinary conductor $\rightarrow$ superconductor, paramagnet $\rightarrow$ spin-density-wave antiferromagnet (see first two columns of Fig. 17.1). The intriguing thing about phase transitions is the emergence of properties in the condensed phase which are completely different from the properties of the normal phase. Thus solids are hard while gases are 'soft', ferromagnets pick up iron nails but paramagnets do not, superconductors have zero resistance while ordinary conductors have finite resistance, etc.

The perturbation expansion of the propagator which we have used throughout most of this book unfortunately breaks down in condensed systems. We saw an example of this in the case of superconductivity, §15.4. In this chapter we are going to show that the *matrix propagator* used to describe the superconducting phase in §15.5 can be extended to provide a systematic, intuitively appealing framework for a theory of fermion phase transitions in general. The material here is a condensed version of the review article by Mattuck and Johansson (1968), hereafter abbreviated MJ.

To introduce the main ideas, let us begin with a simple example—the transition from the paramagnetic to the ferromagnetic state in a metal with non-localized electrons, like Ni. This is illustrated schematically in the first row of Fig. 17.1. Above a certain critical temperature, T_c, the system is in the normal, paramagnetic phase. There is *short-range order* (SRO), meaning that nearby spins tend to line up parallel to one another, but no *long-range order* (LRO), i.e., no tendency of spins separated by macroscopic distances to align themselves parallel. If we now reduce the temperature below T_c, all spins, even those of macroscopic separation, spontaneously start to line up parallel to one another. This emergence of long-range order marks the transition to the ferromagnetic state.

Normal phase	Condensed phase	Broken symmetry	Long-range order parameter
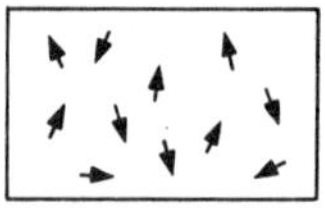 Paramagnet	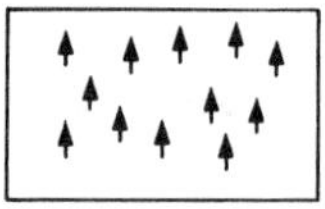 Ferromagnet	Rotation	Magnetization
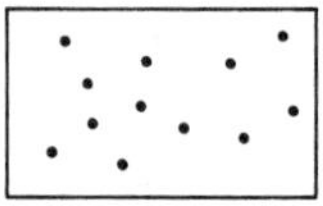 Gas	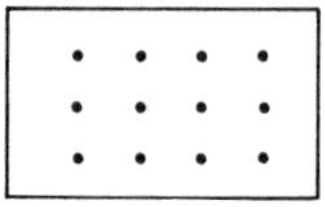 Solid	Translation rotation	ρ_K, amplitude of Kth Fourier component of density, where κ = reciprocal lattice vector
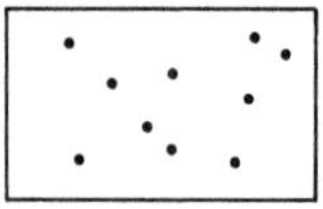 Normal conductor	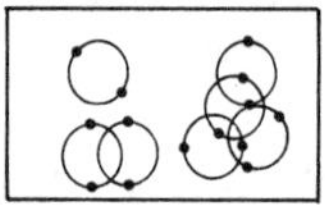 Superconductor	Global gauge symmetry	Pair creation amplitude ($\propto$ energy gap in Type I superconductor)
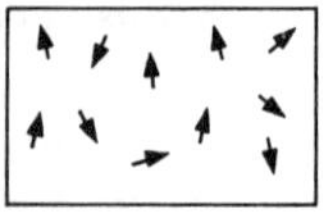 Paramagnet	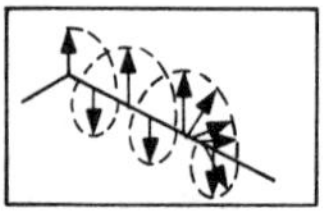 Spin-density wave antiferromagnet	Rotation translation	S_Q, amplitude of Qth Fourier component of spin density

Fig. 17.1. *Examples of Phase Transitions in Fermi Systems (Highly Schematic)*

By 'spontaneously' in the above description, we mean 'with no applied magnetic field'. If there were such a field, even the spins in a normal system would line up parallel to it, and there would be LRO. However, this is not referred to as a phase transition, since the LRO disappears as soon as the field is turned off and is therefore not an intrinsic property of the electron system.

The measure of long-range spin order is the 'relative magnetization' (= actual magnetization divided by maximum possible magnetization) given by

$$\mathbf{M} = \frac{\mu_B \langle \mathbf{S} \rangle}{N\mu_B/2} = \frac{2}{N} \langle \mathbf{S} \rangle \tag{17.1}$$

where N = number of electrons in unit volume, and $\langle S \rangle$ is the average spin per unit volume. The quantity M is called the '*long-range order parameter*' for the system.

Besides long-range order, the ferromagnet shows another characteristic feature of condensed phases, i.e., '*broken symmetry*'. The idea here is that in the normal paramagnetic phase, since spins of macroscopic separation are oriented randomly with respect to one another, the system as a whole is invariant under the group of all possible rotations in space. But in a ferromagnet, the spin system as a whole points in a definite direction in space. Hence the system is now invariant only under that subgroup of the rotation group consisting of rotation about the direction of spin orientation. Thus we have '*broken symmetry*' under rotation.

Of course if we are dealing with a perfectly isotropic system, all directions of spin orientation are equally likely so on the average there is no broken symmetry. True broken symmetry can arise only if there is a very tiny magnetic field present—the '*source field*'—which fixes the direction of spin alignment. In practice there are always such small fields around; thus the Earth's magnetic field may play the source field role.

A striking example of a phase transition quite similar to the paramagnet–ferromagnet transition is the 'staring crowd' phenomenon shown in Fig. 17.2. In a crowd in the 'normal' state, people gaze in random directions. However, it is often observed that if one person starts staring at, say, an empty second-story window, a large number of people in the crowd will start staring at the window, despite the fact that nothing of interest is happening there (i.e., no external field present)! This is the 'condensed' phase. Again, the broken symmetry is rotational invariance, while the LRO parameter is the 'starization' or number of starers divided by the total number of people.

The ideas here are general: a phase transition is a sudden change from a normal state with only SRO and perfect symmetry, to a condensed state with LRO and broken symmetry. Examples of broken symmetry and LRO parameter are shown in the last two columns of Fig. 17.1. (Regarding the superconducting case, a 'global gauge transformation' is a change of the phase of the single-particle operators:

$$c_{k\sigma}^{\dagger} \to c_{k\sigma}^{\dagger} \exp(i\phi), \quad c_{k\sigma} \to c_{k\sigma} \exp(-i\phi). \qquad (17.1')$$

The wave function of a normal system is invariant under this transformation, but the BCS wave function is not. Hence there is broken symmetry under global gauge transformation. See MJ for further discussion of this and the other cases.)

The parameters which determine whether the system is in a normal or condensed state are the temperature T, and density ρ (or alternatively, the coupling constant λ for the interaction between particles). A single fermion

(b)

(a)

Fig. 17.2. 'Staring Crowd' Analogue of Phase Transition. (a) 'Normal' Phase. (b) 'Condensed' Phase

system can usually exist in several possible condensed phases depending on T and ρ or λ. Figure 17.3 shows a *hypothetical* Fermi system of this sort. The points on the boundary between the normal and condensed phases are called '*transition points*'.

In addition to the nature of the broken symmetry and LRO, there is another important characteristic of a phase transition, i.e., whether it is first order, second order or some higher order. In the *second-order* transition, the LRO parameter changes continuously in going from the normal to the condensed

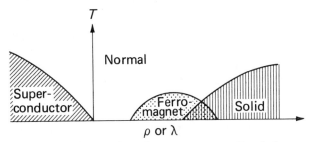

Fig. 17.3. *Phases of Hypothetical One-component Fermi System. T=
Temperature, ρ= Density, λ= Coupling Constant*

state, but has a discontinuity in its first derivative. In a *first-order* transition, the LRO parameter itself changes discontinuously at the transition point.

17.2 Qualitative theory of phase transitions

We now ask: Why do phase transitions occur? There is evidently no difficulty in understanding how there can be long-range order in a system, when this order is brought about by the presence of a finite external field. Thus, we saw that in an external magnetic field, a paramagnetic system will become magnetized parallel to the field. Or, if there is a fire in the third-storey window, it is obvious why the crowd will be staring at it. But in a condensed system, there is LRO in the *absence* of a finite external field. How is this possible?

As Brout (1965a) points out, the answer to this in all known cases is that *there is an internal field produced by the interactions between the particles of the system itself.* This field depends on interaction strength (or density) and temperature. In the normal phase, the field is limited in range and causes local, or short-range, order. But *in a condensed phase, the internal field extends throughout the whole system and produces long-range order.*

Consider for instance a set of spins coupled by strong negative spin–spin interaction (due to exchange), and for simplicity, assume the temperature is zero. If we look at some arbitrary spin—'spin A'—we see that another nearby spin sitting in the field of A will tend to line up parallel to A. This creates an even stronger field in which a third spin will align itself, etc., until all the spins are lined up pointing in the same direction (ferromagnetic state). If we now examine the final situation, we see that any one spin is lined up parallel to a long-range internal field due to all the other spins. In the case of localized spins, this is called (in lowest-order approximation) the 'Weiss field' or 'molecular field'. Notice that this internal field has the same aligning effect on spins as an external magnetic field has. We will call it the 'spin-aligning field'. However it is not a real magnetic field because it cannot bend the orbits of charged particles.

(The 'staring crowd' phenomenon may be explained in the same way. One person stares at the empty window. A second sees him staring, and this inter-action between the two people (not between the person and the window!) causes the second to stare also, etc., until finally the whole crowd is staring. Thus we have any one person polarized parallel to an internal 'staring field' due to all the others.)

In Table 17.1 is a list of the internal fields associated with various condensed phases.

Table 17.1   *Internal fields for various condensed phases*

Condensed phase	Internal field
Ferromagnet	Spin-aligning field
Solid	Periodic field
Superconductor	Pairing field
Spin-density wave antiferromagnet	Spiral spin-aligning field

The above ideas may be expressed mathematically in the following schematic way: Let us assume that the system has in it, in the final situation, some internal long-range field or fields which we will collectively call F. The corresponding LRO parameters will be collectively called $\mathcal{O}$. The $\mathcal{O}$ will be a function of F which depends in general on λ, the interparticle interaction strength, and T, the temperature:

$$\mathcal{O} = \mathcal{O}_{\lambda T}(F) \qquad (17.2a)$$

(for example, the relative magnetization, $\mathcal{M}$, is a function of the internal spin-aligning field, $\mathcal{H}$). The field F, in turn, is a function of $\mathcal{O}$:

$$F = F_{\lambda T}(\mathcal{O}) \qquad (17.2b)$$

(thus, with localized spins, $\mathcal{H} = \gamma \mathcal{M}$). These two equations may be combined into:

$$\mathcal{O} = \mathcal{O}_{\lambda T}(F_{\lambda T}(\mathcal{O})), \qquad (17.2\text{c})$$

which is to be solved for the order parameter $\mathcal{O}$. (In the magnetic case, this is $\mathcal{M} = \mathcal{M}(\gamma \mathcal{M})$, which is just the Weiss equation.) Note that in general (17.2c) is extremely complicated, and one must find $\mathcal{O}$ by a *self-consistent* method. That is, we assume a non-zero value for $\mathcal{O}$, find F from (17.2b), substitute in (17.2a) to get a new value of $\mathcal{O}$, put this in (17.2b), etc., until the value of $\mathcal{O}$ stops changing appreciably and the calculation is 'self-consistent'. If the self-consistent value of $\mathcal{O}$ turns out to be non-zero, then the system is in the condensed phase. Equations (17.2a, b, c), relating long-range order and internal field, are the expressions of the basic mechanism underlying all phase transitions.

17.3 Anomalous propagators and the breakdown of the perturbation series in the condensed phase

In chapter 9 and the appendices, we saw how to develop the single-particle propagator for normal systems in a perturbation expansion. But, as pointed out in the chapter on superconductivity, the normal perturbation series breaks down in the superconducting phase. This is quite general, and in fact the normal perturbation series breaks down in *any* condensed phase. A particularly simple example of why this breakdown occurs is the case of ferromagnetism. In the normal phase of a system with no external field, because of momentum conservation, the propagating particle always emerges in the same linear and spin momentum state that it entered. But in the ferromagnetic phase, because of the internal spin-aligning field, F, there is a possibility that the particle spin may be flipped as shown in Fig. 17.4. (Of course, it may also propagate through the internal field without flipping.) This means that in the ferromagnetic phase, in addition to the ordinary propagator, we can have 'anomalous' propagators which flip a spin (for simplicity we consider only the $T = 0$ case):

$$G(\mathbf{k}\downarrow, \mathbf{k}\uparrow, t' - t) = -i\langle \Psi_0 | T\{c_{k\downarrow}(t') c_{k\uparrow}^{\dagger}(t)\} | \Psi_0 \rangle$$

and

$$G(\mathbf{k}\uparrow, \mathbf{k}\downarrow, t' - t) = -i\langle \Psi_0 | T\{c_{k\uparrow}(t') c_{k\downarrow}^{\dagger}(t)\} | \Psi_0 \rangle. \qquad (17.3)$$

These propagators are zero in the normal phase.

If we now examine the normal perturbation expansion of Σ in (10.8), or its self-consistent version in (11.3), we see that the only propagators in it are normal ones. Hence the normal expansion is invalid in the ferromagnetic case because it does not include the anomalous spin-flip processes which are characteristic for the ferromagnetic phase. Note that this means that the propagator expansion in (9.40) and the expansion for Σ in (10.8) or (11.3) are invalid in the ferromagnetic case, even if they are summed exactly to infinite order.

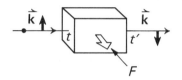

Fig. 17.4. *Anomalous Propagation in the Ferromagnetic Phase. The Internal Spin-Aligning Field, F, is Assumed to be in the x Direction*

In the above discussion, it is assumed (see Fig. 17.4) that the internal field F is in the x direction. If, on the other hand, F happened to be in the z direction, no spin flips could occur (since our test particles are polarized along z) and all propagation would apparently be normal. Hence we might conclude that the propagator was not a completely adequate tool for probing the structure of the ferromagnetic state. Actually this is not true, because there is a difference between propagation of spin up and spin down particles, which is caused by F_z, and this difference may itself be considered anomalous propagation. That is

$$\Delta G = G(\mathbf{k}\uparrow, \mathbf{k}\uparrow, t'-t) - G(\mathbf{k}\downarrow, \mathbf{k}\downarrow, t'-t) \qquad (17.4)$$

is zero in the normal phase since the system is completely symmetric, but finite in the ferromagnetic phase, if $F_z \neq 0$. For example, in the Hartree–Fock approximation, we find simply that (assume $t' > t$ and $k > k_F$):

$$\Delta G_{HF} = -i\{\exp[-i(\epsilon_k - \mu - g\beta F_z)(t'-t)] - \exp[-i(\epsilon_k - \mu + g\beta F_z)(t'-t)]\},$$

where $g\beta F_z$ is the energy of the spin in the internal spin-aligning field. (This will be discussed further in §17.5.)

The above argument for the breakdown of the perturbation series in the condensed phase is true in general, provided we always assume from the beginning, as we have been doing throughout, that our system is infinitely large. (If we start with a finite system, then there are some cases, such as superconductivity, in which it is possible to obtain the condensed phase from the ordinary perturbation series. This is discussed in MJ, Appendix C.) In each condensed phase, in addition to the normal process, there is an anomalous process (or processes) which can take place because of the long-range internal field, with a corresponding propagator (or propagators) (Bogoliubov (1960)). In Table 17.2 is a list of anomalous propagators for various phases (for each propagator listed there is also a Hermitian conjugate which has been omitted).

Table 17.2 *Anomalous propagators for various condensed phases*

Condensed phase	Anomalous propagator
Ferromagnetic	$G_{\text{fer}} = -i\langle\Psi_0|T\{c_{k\downarrow}(t')\,c_{k\uparrow}^{\dagger}(t)\}|\Psi_0\rangle$
Solid	$G_{\text{sol}} = -i\langle\Psi_0|T\{c_{k+\kappa',\sigma}(t')\,c_{k+\kappa,\sigma}^{\dagger}(t)\}|\Psi_0\rangle$
Superconductor	$G_{\text{sup}} = -i\langle\Psi_0|T\{c_{-k\downarrow}^{\dagger}(t')\,c_{k\uparrow}^{\dagger}(t)\}|\Psi_0\rangle$
Spin-density wave antiferromagnet	$G_{\text{SDW}} = -i\langle\Psi_0|T\{c_{k+Q',\sigma'}(t')\,c_{k+Q,\sigma}(t)\}|\Psi_0\rangle$

In the case of the solid, the propagating particle can pick up momentum $\varkappa'-\varkappa$, the difference between any two reciprocal lattice vectors, from Bragg scattering by the periodic internal field. In the superconductor, the anomalous process is not really a propagation, but rather the amplitude for the creation of a pair of particles in opposite momentum and spin states. In the spin-density wave case, a particle picks up momentum $\mathbf{Q}-\mathbf{Q}'$ from scattering against the periodic structure of the spiral internal field, and has its spin changed from σ to σ' by the spin-aligning character of the internal field.

The anomalous propagators have the valuable property that the LRO parameters may be obtained directly from them. For example, in the ferromagnetic case, the LRO parameter is, for magnetization in the x direction,

$$M = \frac{2}{N}\langle S_x\rangle = \frac{1}{N}\sum_k \langle\Psi_0|\,c_{k\uparrow}^{\dagger}c_{k\downarrow}+c_{k\downarrow}^{\dagger}c_{k\uparrow}|\Psi_0\rangle. \qquad (17.5)$$

where we have expressed the spin operator S_x in second quantized form (see exercise 7.11). Evidently we can get this from the anomalous propagators (17.3) as follows:

$$M = \frac{-i}{N}\sum_k [G(\mathbf{k}\downarrow,\mathbf{k}\uparrow,0^-)+G(\mathbf{k}\uparrow,\mathbf{k}\downarrow,0^-)]. \qquad (17.6)$$

If the magnetization is in the z direction, we have

$$M = \frac{2}{N}\langle S_z\rangle = \frac{1}{N}\sum_k \langle\Psi_0|c_{k\uparrow}^{\dagger}c_{k\uparrow}-c_{k\downarrow}^{\dagger}c_{k\downarrow}|\Psi_0\rangle$$

$$= \frac{-i}{N}\sum_k [G(\mathbf{k}\uparrow,\mathbf{k}\uparrow,0^-)-G(\mathbf{k}\downarrow,\mathbf{k}\downarrow,0^-)]. \qquad (17.7)$$

showing how $\varDelta G$ in (17.4) acts as an anomalous propagator.

In a superconductor, the order parameter is the pair creation amplitude ($\propto$ energy gap in a Type I superconductor) given by

$$\Delta = \sum_k \langle \Psi_0 | c_{k\uparrow}^\dagger c_{-k\downarrow}^\dagger | \Psi_0 \rangle$$

$$= -i \sum_k F'(\mathbf{k}, 0^-), \qquad (17.8)$$

where F' is the anomalous pair propagator in (15.49). Similar relations exist for the other condensed phases (see MJ, p. 530).

The fact that some sort of anomalous propagation exists in all condensed phases, and that this anomalous propagation is not taken into account in the normal expansion, is the physical reason for the breakdown of the normal perturbation expansion in the condensed system. There is also a more mathematical way of getting this result. This is based on the fact that for the normal perturbation expansion of the propagator to be valid, the interacting ground state $|\Psi_0\rangle$ must have the same 'structure' as the non-interacting ground state $|\Phi_0\rangle$, i.e. it must overlap $|\Phi_0\rangle$ so that (see appendix E.12, C.6):

$$\langle \Psi_0 | \Phi_0 \rangle \neq 0 \quad \textit{(non-orthogonality condition)}. \qquad (17.9)$$

Since at infinite volume the condensed phase always has a different structure than the normal phase, it follows that (17.9) is not satisfied, so that the expansion in (10.8) or (11.3) is not valid. For example, in the ferromagnetic state, there are, say, more spins up than down, so that we have (in HF approximation):

$$|\Psi_0\rangle = |1_{k_1\uparrow} 1_{k_2\uparrow}, \ldots, 1_{k_M\uparrow}; 1_{k_1\downarrow} 1_{k_2\downarrow}, \ldots, 1_{k_P\downarrow} 000, \ldots\rangle, \quad \text{where } M > P. \quad (17.10)$$

This is clearly orthogonal to the non-interacting $|\Phi_0\rangle$, which has $M = P$.

17.4 The generalized matrix propagator

In order to construct a perturbation theory valid for the condensed state, it is necessary to modify the self-consistent (see below for reason why self-consistent form is required) self-energy expansion in (11.3) so that it includes the appropriate anomalous propagators. We will show that the method of doing this is to replace each normal propagator in (11.3) by a *matrix propagator* which is a simple generalization of those of Nambu (1960) and Rajagopal (1964a, b, 1966). The diagonal elements of this generalized matrix propagator are normal propagators, while the off-diagonal elements are anomalous. A crude plausibility argument analogous to that used in the superconducting case §15.5 for this will be given here—a more rigorous demonstration is in MJ, §6.

For simplicity, consider the ferromagnetic case first. The matrix propagator here is (see (17.3)) defined by:

$$\mathbf{G}_{fer}(\mathbf{k}, t'-t) = -i \begin{pmatrix} \langle \Psi_0 | T\{c_{k\uparrow}(t')\, c_{k\uparrow}^\dagger(t)\} | \Psi_0 \rangle & \langle \Psi_0 | T\{c_{k\uparrow}(t')\, c_{k\downarrow}^\dagger(t)\} | \Psi_0 \rangle \\ \langle \Psi_0 | T\{c_{k\downarrow}(t')\, c_{k\uparrow}^\dagger(t)\} | \Psi_0 \rangle & \langle \Psi_0 | T\{c_{k\downarrow}(t')\, c_{k\downarrow}^\dagger(t)\} | \Psi_0 \rangle \end{pmatrix}.$$
(17.11)

In the non-interacting system, this becomes the bare matrix propagator, which in $\mathbf{k}, \omega$ space is:

$$\mathbf{G}_{0_{fer}}(\mathbf{k}, \omega) = -i \begin{bmatrix} \dfrac{1}{\omega - (\epsilon_k - \mu) + i\omega\delta} & 0 \\ 0 & \dfrac{1}{\omega - (\epsilon_k - \mu) + i\omega\delta} \end{bmatrix}.$$
(17.12)

It is possible to write $\mathbf{G}_{fer}$ in the same form as the ordinary normal propagator by introducing the spinor operators:

$$\gamma_k(t) = \begin{pmatrix} c_{k\uparrow}(t) \\ c_{k\downarrow}(t) \end{pmatrix}; \quad \gamma_k^\dagger(t) = (c_{k\uparrow}^\dagger(t), c_{k\downarrow}^\dagger(t)),$$
(17.13)

from which

$$\mathbf{G}_{fer}(\mathbf{k}, t'-t) = -i\langle \Psi_0 | T\{\gamma_k(t')\, \gamma_k^\dagger(t)\} | \Psi_0 \rangle.$$
(17.14)

(Note that when $t' < t$, we must change $\gamma_k^\dagger(t)$ to a column vector, and $\gamma_k(t')$ to a row vector.). Furthermore, the Hamiltonian (7.51) (including spin and assuming a spin-independent interaction of form (7.70)) has the same form when written out in terms of these γ_k's that it has when written out using the ordinary c_k's. Thus we find:

$$H = \sum_k (\epsilon_k - \mu)\gamma_k^\dagger \gamma_k + \tfrac{1}{2} \sum_{klmn} V_{klmn}(\gamma_l^\dagger \gamma_n)(\gamma_k^\dagger \gamma_m) + \text{constant}.$$
(17.15)

Hence, since both $\mathbf{G}$ and H have the same form in terms of the γ_k's as the ordinary G and H have in terms of the c_k's, it is plausible that one can use the same perturbation expansion as before, the only difference being that matrix propagators are to be associated with the directed lines, instead of ordinary propagators.

Thus we find that the Dyson equation (11.8), with self-energy Σ in the self-consistent form (11.3), may be written out in terms of matrix propagators as follows:

$$\mathbf{G}_{fer}(\mathbf{k}, \omega) = \frac{1}{\mathbf{G}_{0_{fer}}^{-1}(\mathbf{k}, \omega) - \mathbf{\Sigma}(\mathbf{G}_{fer})}.$$
(17.16)

As pointed out just after (10.11), in the normal system, Σ is the generalized local effective field which produces the short-range order (i.e., the 'cloud') around the propagating bare particle. By analogy, $\mathbf{\Sigma}$ here is the potential of the internal spin-aligning field in the ferromagnetic system, the potential which establishes the long-range order. Further, according to §17.3, the anomalous propagators in $\mathbf{G}$ directly give the LRO parameter. Hence (17.16) is just the self-consistent equation for the order parameter which was described schematically in (17.2c).

Note that it is absolutely necessary to use the perturbation series in the self-consistent form, (11.3). If the ordinary form, (10.8), were used instead, we would find that because $\mathbf{G}_0$ in (17.12) is diagonal, all diagrams in $\boldsymbol{\Sigma}$ would be diagonal. Hence $\mathbf{G}$ itself would be diagonal and therefore incapable of describing the condensed state. This difficulty disappears in the self-consistent theory, since only clothed propagators appear in the self-energy term. (Note that in the ferromagnetic case with magnetization in the z direction, $\mathbf{G}, \Sigma$ are always diagonal, but $G_{\uparrow\uparrow} \neq G_{\downarrow\downarrow}$ and $\Sigma_{\uparrow\uparrow} \neq \Sigma_{\downarrow\downarrow}$).

Matrix propagators for the other phases are constructed in the same way as in the ferromagnetic case, using for off-diagonal elements the anomalous propagators in Table 17.2. The superconducting case yields (omit Ψ_0, T, t', t for brevity only!):

$$\mathbf{G}_{\text{sup}}(\mathbf{k}, t'-t) = -i \begin{pmatrix} \langle c_{k\uparrow} c_{k\uparrow}^\dagger \rangle & \langle c_{k\uparrow} c_{-k\downarrow} \rangle \\ \langle c_{-k\downarrow}^\dagger c_{k\uparrow}^\dagger \rangle & \langle c_{-k\downarrow}^\dagger c_{-k\downarrow} \rangle \end{pmatrix}. \tag{17.17}$$

The solid gives a matrix propagator with an infinite number of rows and columns. For simplicity, consider a one-dimensional lattice with spacing d and first reciprocal lattice vector $\kappa_0 = 2\pi/d$. Then, suppressing spin index, etc., for brevity, we find:

$$\mathbf{G}_{\text{sol}}(\mathbf{k}, t'-t) = -i \begin{bmatrix} \langle c_k c_k^\dagger \rangle & \langle c_k c_{k+\kappa_0}^\dagger \rangle & \langle c_k c_{k+2\kappa_0}^\dagger \rangle \dots \\ \langle c_{k+\kappa_0} c_k^\dagger \rangle & \langle c_{k+\kappa_0} c_{k+\kappa_0}^\dagger \rangle & \dots\dots\dots\dots \\ \langle c_{k+2\kappa_0} c_k^\dagger \rangle & \dots\dots\dots\dots\dots\dots \\ \vdots \end{bmatrix} \tag{17.18}$$

(For very rough calculations, one could neglect all higher order harmonics, and replace this by a 2×2 matrix involving just $\mathbf{k}$ and $\mathbf{k} + \mathbf{\kappa}_0$.) The spin-density wave phase also produces an infinite matrix propagator:

$$\mathbf{G}_{\text{SDW}}(\mathbf{k}, t'-t) = -i \begin{bmatrix} \langle c_{k\uparrow} c_{k\uparrow}^\dagger \rangle & \langle c_{k\uparrow} c_{k+Q_0\downarrow}^\dagger \rangle & \langle c_{k\uparrow} c_{k+Q_0\uparrow}^\dagger \rangle \dots\dots\dots \\ \langle c_{k+Q_0\downarrow} c_{k\uparrow}^\dagger \rangle & \langle c_{k+Q_0\downarrow} c_{k+Q_0\downarrow}^\dagger \rangle & \langle c_{k+Q_0\downarrow} c_{k+Q_0\uparrow}^\dagger \rangle \dots \\ \langle c_{k+Q_0\uparrow} c_{k\uparrow}^\dagger \rangle & \dots\dots\dots\dots\dots\dots\dots\dots \\ \langle c_{k+2Q_0\downarrow} c_{k\uparrow}^\dagger \rangle & \dots\dots\dots\dots\dots\dots\dots\dots \\ \vdots \end{bmatrix}$$

$$\tag{17.19}$$

This is usually replaced by a 2×2 matrix which neglects all the higher-order terms (Rajagopal (1964a)).

The argument in (17.13)–(17.16) is easily generalized to any matrix propagator $\mathbf{G}$ and we find the general self-consistent Dyson equation:

$$\mathbf{G}(\mathbf{k}, \omega) = \frac{1}{\mathbf{G}_0^{-1}(\mathbf{k}, \omega) - \Sigma(\mathbf{G})}. \tag{17.20}$$

Another way of deriving (17.20) is based on the idea of turning on the source field so that the new unperturbed 'vacuum' state, $|\Phi_0\rangle$ is no longer orthogonal to the condensed state $|\Psi_0\rangle$ in (17.9). This more rigorous technique is described in detail in MJ, §6.

17.5 Application to ferromagnetic phase in system with δ-function interaction

As a very simple example of how the above formalism works, we will calculate the magnetization in HF approximation for a system with δ-function interaction, i.e., with

$$H = \sum_{k\sigma} \epsilon_k c_{k\sigma}^\dagger c_{k\sigma} + V \sum_{\substack{klmn \\ \sigma\sigma'}} c_{l\sigma'}^\dagger c_{k\sigma}^\dagger c_{m\sigma} c_{n\sigma'}. \tag{17.21}$$

This will turn out to yield the well-known Stoner result.

In self-consistent HF approximation, the matrix propagator equation is:

$$\tag{17.22}$$

To make things easy, we will assume that the internal magnetization field is in the z direction so that $G_{\uparrow\downarrow} = G_{\downarrow\uparrow} = 0$. The diagram dictionary for this case is in Table 17.3; see (9.39) regarding convergence factor $\exp(i\omega 0^+)$ in Table 17.3. See around (9.47') regarding $i\omega\delta$ in free propagator. Regarding the trace rule, see MJ around Fig. 21. Applying these rules to (17.22) yields:

$$= -iV \begin{pmatrix} N_\uparrow + N_\downarrow & 0 \\ 0 & N_\uparrow + N_\downarrow \end{pmatrix} \tag{17.23}$$

$$= +iV \begin{pmatrix} N_\uparrow & 0 \\ 0 & N_\downarrow \end{pmatrix}, \tag{17.24}$$

Table 17.3　*Diagram dictionary for the ferromagnetic phase at $T = 0$*
(magnetization in z direction)

Diagram element	Factor
$\mathbf{k}, \omega$	$i\mathbf{G}_{\text{fer}}(\mathbf{k}, \omega) = i\begin{pmatrix} G_{\uparrow\uparrow}(\mathbf{k}, \omega) & 0 \\ 0 & G_{\downarrow\downarrow}(\mathbf{k}, \omega) \end{pmatrix}$
$\mathbf{k}, \omega$　　　 $\mathbf{k}, \omega$	$i\mathbf{G}_{\text{fer}}(\mathbf{k}, \omega)\exp(i\omega 0^+)$
$\mathbf{k}, \omega$	$i\mathbf{G}_{0_{\text{fer}}}(\mathbf{k}, \omega) =$ $i\begin{pmatrix} \dfrac{1}{\omega - (\epsilon_k - \mu) + i\omega\delta} & 0 \\ 0 & \dfrac{1}{\omega - (\epsilon_k - \mu) + i\omega\delta} \end{pmatrix}$
$\begin{array}{ll}\mathbf{k}+\mathbf{q},\sigma & \mathbf{k}'-\mathbf{q},\sigma' \\ \mathbf{k},\sigma \quad \mathbf{q} & \mathbf{k}',\sigma'\end{array}$	$-iV_{k+q,\,k'-q,\,k,\,k'} \times \begin{pmatrix} 1 & 0 \\ 0 & 1 \end{pmatrix}$
fermion loop	$(-1) \times$ trace over product of $\mathbf{G}$'s forming the loop
intermediate momentum, $\mathbf{k}$ intermediate frequency, ω	$\displaystyle\int \frac{d^3\mathbf{k}}{(2\pi)^3},\quad \int \frac{d\omega}{2\pi}$

where we have used:

$$\langle \hat{n}_{l\sigma} \rangle = -i \int \frac{d\omega}{2\pi} G_{\sigma\sigma}(\mathbf{l}, \omega) e^{-i\omega 0^-} \tag{17.25}$$

and

$$N_\sigma = \int \frac{d^3\mathbf{l}}{(2\pi)^3} \langle \hat{n}_{l\sigma} \rangle. \tag{17.26}$$

Placing these results in (17.22) yields

$$\begin{pmatrix} G_{\uparrow\uparrow} & 0 \\ 0 & G_{\downarrow\downarrow} \end{pmatrix} = \begin{pmatrix} \dfrac{i}{\omega - (\epsilon_k - \mu) - VN_\downarrow + i\omega\delta} & 0 \\ 0 & \dfrac{i}{\omega - (\epsilon_k - \mu) - VN_\uparrow + i\omega\delta} \end{pmatrix} \tag{17.27}$$

whence

$$G_{\sigma\sigma}(\mathbf{k}, \omega) = \frac{1}{\omega - (\epsilon_k - \mu) - VN_{-\sigma} + i\omega\delta}. \qquad (17.28)$$

We can get an equation for N_σ by placing (17.28) in (17.25), (17.26). Doing the resulting integral by contours yields:

$$N_\sigma = \int_{(\epsilon_k = 0)}^{(\epsilon_k = \mu - VN_{-\sigma})} \frac{d^3\mathbf{k}}{(2\pi)^3} = \int_0^{\mu - VN_{-\sigma}} d\epsilon\, \rho(\epsilon) = \frac{3N}{4\epsilon_F^{\frac{3}{2}}} \int_0^{\mu - VN_{-\sigma}} d\epsilon\, \sqrt{\epsilon}$$

$$= \frac{N}{2\epsilon_F^{\frac{3}{2}}} (\mu - VN_{-\sigma})^{\frac{3}{2}}, \qquad (17.29)$$

where $\rho(\epsilon)$ is the density of energy levels. Now we express this in terms of the relative magnetization given by:

$$M = \frac{N_\uparrow - N_\downarrow}{N_\uparrow + N_\downarrow} = \frac{N_\uparrow - N_\downarrow}{N}. \qquad (17.30)$$

Using

$$N = N_\uparrow + N_\downarrow, \qquad (17.31)$$

and

$$N_\uparrow = \tfrac{1}{2}(MN + N), \quad N_\downarrow = \tfrac{1}{2}(N - MN), \qquad (17.32)$$

we find the magnetization equation:

$$M = \frac{2S}{N} = \frac{1}{2\epsilon_F^{\frac{3}{2}}} \{(\mu - \tfrac{1}{2}NV + \tfrac{1}{2}VNM)^{\frac{3}{2}} - (\mu - \tfrac{1}{2}NV - \tfrac{1}{2}VNM)^{\frac{3}{2}}\} \qquad (17.33)$$

where ϵ_F is equal to $(3\pi^2 N)^{\frac{2}{3}}/2m$, the Fermi energy of the non-interacting system. Thus, (17.33) is the self-consistent equation for the LRO parameter, M.

The value of μ is determined by the condition that the total number of particles is N. Using (17.29)–(17.32) yields:

$$N = \frac{N}{2\epsilon_F^{\frac{3}{2}}} \{(\mu - \tfrac{1}{2}NV + \tfrac{1}{2}NVM)^{\frac{3}{2}} + (\mu - \tfrac{1}{2}NV - \tfrac{1}{2}NVM)^{\frac{3}{2}}\}. \qquad (17.34)$$

Adding and subtracting (17.33), (17.34) in order to eliminate μ gives:

$$\frac{NV}{2\epsilon_F} = \frac{1}{2M} \{(1 + M)^{\frac{2}{3}} - (1 - M)^{\frac{2}{3}}\}. \qquad (17.35)$$

This is the well-known Stoner equation. The magnetization is plotted as a function of $NV/2\epsilon_F$ in Fig. 17.5. For fixed V, we can regard this curve as a function of the particle density, N. (In MJ, N is fixed and the coupling V is allowed to vary.)

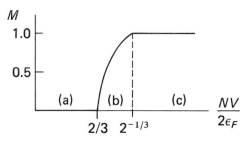

Fig. 17.5 *Relative Magnetization, M, as a Function of the Parameter, $NV/2\epsilon_F$, in System with δ-Function Interaction. (a) Paramagnetic Region, (b) Unsaturated Ferromagnetic, (c) Saturated Ferromagnetic*

It is seen that the transition point where the magnetization first becomes non-zero occurs at the critical density N_c given by:

$$\frac{N_c V}{2\epsilon_F} = \frac{2}{3} \quad \text{or} \quad N_c = \frac{4 \epsilon_F}{3 V}. \tag{17.36}$$

This is the point at which the paramagnet–ferromagnet phase transition takes place. Note that since M changes continuously at N_c, while the derivative of M changes discontinuously, we have a second-order phase transition (see end of §17.1).

It should be observed that $M=0$ is *always* a solution of (17.35), also for $N > N_c$. To show that the finite M solution is the correct one for $N > N_c$, we have to prove that it yields the lowest energy. This is done in MJ, p. 546.

The above calculation is in HF approximation (or 'generalized HF,' since it includes anomalous propagations). To go beyond HF and take correlations into account, we have to include higher order diagrams in the self-energy. This is described briefly at the end of MJ, §7.

17.6 Divergence of the two-particle propagator and scattering amplitude at the transition point

We have seen how the matrix propagator may be used to investigate the properties of the condensed phase. However, in many cases one is only interested in finding the transition points, i.e., the value of the density N_c, and temperature, T_c, where the condensation starts. In the case of second order transitions, the simplest way to do this is to start in the normal phase and calculate the appropriate two-particle propagator G_2 (or scattering amplitude, Γ). Then the transition point is that value of N or T where G_2 (or Γ) becomes infinite.

Let us see how this works in the ferromagnetic case. For simplicity we first take $T=0$. (The finite T case is treated at the end of this section.) The quantity characterizing the magnetic properties in the normal, paramagnetic phase, is the magnetic susceptibility $\chi(\mathbf{k}, \omega)$ which gives the response of the system to a small applied magnetic field $\mathbf{H}(\mathbf{k}, \omega)$ by (cf. (13.12″)):

$$\mathbf{M}(\mathbf{k}, \omega) = \chi(\mathbf{k}, \omega)\,\mathbf{H}(\mathbf{k}, \omega) \tag{17.37}$$

where $\mathbf{M}(\mathbf{k}, \omega)$ is the wave-number and frequency dependent magnetization. We assume the system is isotropic so the susceptibility is the same in all directions.

Since we are interested in static magnetic properties we take $\omega = 0$, and since we consider only uniform magnetization, the wavelength is infinite and the wavenumber $\mathbf{k} = 0$. Now in general, a small applied field $\mathbf{H}$ will cause a small magnetization when the system is in the normal, paramagnetic phase. But if we now start to increase the density N, then we find that as N approaches N_c (the critical density for the ferromagnetic transition) the magnetic susceptibility $\chi(0,0)$ grows larger and larger, and when $N = N_c$, we find $\chi(0,0) = \infty$. This means that when $N = N_c$, no external field at all is required to produce magnetization, i.e., the system magnetizes spontaneously. Hence we can find the transition point, N_c, by calculating the susceptibility and finding that point at which it becomes infinite.

The magnetic susceptibility can be calculated from a closely related function, the 'retarded spin polarization propagator', D_{ij}^R, defined by

$$D_{ij}^R(\mathbf{r}, t) = -i\theta_t \langle \Psi_0 | [S_i(\mathbf{r}, t), S_j(0, 0)]_+ | \Psi_0 \rangle. \tag{17.38}$$

Here $S_i(\mathbf{r})$ is the ith component of the spin density operator:

$$S_i(\mathbf{r}) = (\psi_\uparrow^\dagger(\mathbf{r}), \psi_\downarrow^\dagger(\mathbf{r}))\,\sigma_i \begin{pmatrix} \psi_\uparrow(\mathbf{r}) \\ \psi_\downarrow(\mathbf{r}) \end{pmatrix} \tag{17.39}$$

where σ_i are the Pauli spin matrices and $i = x, y, z, +, -$ $(\sigma_\pm = \sigma_x \pm i\sigma_y)$. $S(\mathbf{r}, t)$ is $S(\mathbf{r})$ in Heisenberg picture (see (9.3)). In an isotropic system, the following relations hold:

$$D_{xx}^R = D_{yy}^R = D_{zz}^R = \tfrac{1}{2}D_{+-}^R = \tfrac{1}{2}D_{-+}^R$$
$$D_{xy}^R = D_{yx}^R = D_{xz}^R = D_{zx}^R = D_{yz}^R = D_{zy}^R = 0. \tag{17.40}$$

The magnetic susceptibility tensor and its Fourier transform are related to D_{ij}^R and its Fourier transform by

$$\chi_{ij}(\mathbf{r}, t) = -\mu^2\,D_{ij}^R(\mathbf{r}, t); \quad \chi_{ij}(\mathbf{q}, \omega) = -\mu^2\,D_{ij}^R(\mathbf{q}, \omega) \tag{17.41}$$

where μ is the magnetic moment of the particles in the system. Thus the problem is to find the retarded spin polarization propagator, D_{ij}^R.

Now D_{ij}^R itself cannot be obtained from Feynman diagrams. However, we can use diagrams to calculate its cousin, the time-ordered spin polarization propagator

$$D_{ij}(\mathbf{r},t) = -i\langle \Psi_0 | T\{S_i(\mathbf{r},t)\, S_j(0,0)\} | \Psi_0 \rangle \qquad (17.42)$$

or its Fourier transform, $D_{ij}(\mathbf{q},\omega)$. Then we can employ the relations (see appendix L.26):

$$\operatorname{Re} D_{ij}(\mathbf{q},\omega) = \operatorname{Re} D_{ij}^R(\mathbf{q},\omega)$$
$$\operatorname{Im} D_{ij}(\mathbf{q},\omega) = \operatorname{sgn}(\omega)\operatorname{Im} D_{ij}^R(\mathbf{q},\omega) \qquad (17.43)$$

to find $D_{ij}^R(\mathbf{q},\omega)$ and thus $\chi_{ij}(\mathbf{q},\omega)$.

Because of (17.40), in an isotropic system we only need to calculate one of the components of D_{ij}^R, so we consider the simplest one, i.e., D_{-+}^R or its time-ordered version D_{-+} given by (use (17.38, 39)):

$$D_{-+}(\mathbf{r},t) = -i\langle \Psi_0 | T\{\psi_\downarrow^\dagger(\mathbf{r},t)\,\psi_\uparrow(\mathbf{r},t)\,\psi_\uparrow^\dagger(0,0)\,\psi_\downarrow(0,0)\} | \Psi_0 \rangle. \qquad (17.44)$$

Note that this is a two-particle propagator similar to the polarization propagator in (13.5). It has the diagram expansion

$$(17.45)$$

Now it turns out (see MJ §8) that as far as getting a transition point is concerned, the approximation for D_{-+} in the normal phase which corresponds to the HF approximation in the condensed phase (§17.5) is just ladder approximation:

$$(17.46)$$

where we have taken the Fourier transform. The particle-hole K-matrix in (17.46) satisfies the integral equation (cf. (10.20)):

$$(17.47)$$

which, since $V=$ constant (cf. exercise 10.4) has the solution

$$K(\mathbf{q}, \omega) = \cfrac{V}{1-(i-V)\int \dfrac{d^3 l}{(2\pi)^3} \int \dfrac{d\beta}{2\pi} iG_0(\mathbf{l}, \beta) iG_0(\mathbf{l}+\mathbf{q}, \beta+\omega)}$$

$$= \cfrac{V}{1-(-iV)\dfrac{(-i\pi_0(\mathbf{q}, \omega))}{(-1)(2)}} = \cfrac{V}{1-\dfrac{V}{2}\pi_0(\mathbf{q}, \omega)}, \qquad (17.48)$$

where we have expressed the integral in the denominator in terms of the pair bubble (9.54) (note factors of (-1) for fermion loop and 2 for spin sum). Translating (17.46) into functions, noting that K is a function only of $\mathbf{q}, \omega$, and expressing integrals over G's in terms of $\pi_0(\mathbf{q}, \omega)$ we easily find:

$$-iD_{-+}(\mathbf{q}, \omega) = \frac{i}{2}\pi_0(\mathbf{q}, \omega) + \left[\frac{i}{2}\pi_0(\mathbf{q}, \omega)\right]^2 [-iK(\mathbf{q}, \omega)],$$

$$D_{-+}(\mathbf{q}, \omega) = \cfrac{-\frac{1}{2}\pi_0(\mathbf{q}, \omega)}{1-\dfrac{V}{2}\pi_0(\mathbf{q}, \omega)}. \qquad (17.49)$$

Now we are interested in the case $\mathbf{q}=0$, $\omega=0$ for which we have from (10.78): $\pi_0(0,0)=3N/2\epsilon_F$. Since this is pure real, so is $D_{-+}(0,0)$, and we have by (17.42) that $D_{-+}(0,0)=D_{-+}^R(0,0)$. That is

$$D_{-+}^R(0,0) = \frac{-3N/4\epsilon_F}{1-V(3N/4\epsilon_F)}. \qquad (17.50)$$

This diverges, so that $\chi_{-+}(0,0)$ $(=-\mu^2 D_{-+}(0,0))$ diverges, when

$$1 = \frac{3NV}{4\epsilon_F} \quad \text{or} \quad N_c = \frac{4\epsilon_F}{3V}, \qquad (17.51)$$

which is exactly the result for the transition point in (17.36).

It is important to observe that the transition point may also be obtained from the retarded K-matrix, $K^R(\mathbf{q}, \omega)$. This is related to the time-ordered $K(\mathbf{q}, \omega)$ in (17.48) in the same way that D^R is related to D in (17.43). Hence we have

$$K^R(0,0) = K(0,0) = \frac{V}{1 - \dfrac{V}{2}\, \pi_0(0,0)} = \frac{V}{1 - \dfrac{3NV}{4\epsilon_F}} \qquad (17.52)$$

so that the retarded K-matrix diverges at just the point where the transition from the paramagnetic to the ferromagnetic phase takes place.

When $N > N_c$, it can be shown that the pole of $K^R(\omega)$ occurs for ω in the upper half-plane, in violation of analyticity, analogous to the superconductivity result (15.43). When N approaches N_c from above, the pole moves down to the point $\omega = 0$. There is no pole when $N < N_c$.

Just as in the superconducting case, §15.7, the argument here is easily generalized to finite temperatures. The K-matrix is, for finite $\mathbf{q}$, ω_m:

$$\mathcal{K}(i\omega_m) = V\left\{ 1 - (-V)\int \frac{d^3\mathbf{p}}{(2\pi)^3}\frac{1}{\beta}\sum_{n=-\infty}^{+\infty}\left(\frac{1}{i\omega_n + i\omega_m - \epsilon_{p+q} + \mu}\right)\left(\frac{1}{i\omega_m - \epsilon_p + \mu}\right) \right\}^{-1}.$$

$$(17.53)$$

Evaluating the frequency sum as in §14.5, and analytically continuing $i\omega_m \to \omega + i\delta$, breaking the integral up into real and imaginary parts, setting $\omega = 0$ and taking the limit $\mathbf{q} \to 0$, we find

$$K^R(0) \equiv \mathcal{K}(0 + i\delta) = V\left\{ 1 + V\int d\epsilon_p\, \rho(\epsilon_p)\frac{df(\epsilon_p)}{d\epsilon_p} \right\}^{-1} \qquad (17.54)$$

where $\rho(\epsilon_p)$ is the density of states for spins in one direction. The chemical potential may be determined from

$$N = 2\int d\epsilon_p\, \rho(\epsilon_p) f(\epsilon_p). \qquad (17.55)$$

To make things easy to evaluate, assume a constant density of states given by $\rho(\epsilon_p) = N/2\epsilon_F$ where ϵ_F is the Fermi energy at $T = 0$. Carrying out the integrals and eliminating μ we find that the temperature T_c for which K^R diverges is given by

$$k_B T_c = \frac{-\epsilon_F}{\ln(1 - 2\epsilon_F/NV)}. \qquad (17.56)$$

These results are easily generalized: in all cases where a second-order phase transition takes place (see MJ, §8), the retarded response function, or its related scattering amplitude, evaluated at $\omega = 0$ becomes infinite when the coupling/density (or temperature) reaches its critical value. The physical interpretation of this is that the system shows a 'response' in the absence of any external field, i.e., it spontaneously goes over into the condensed phase. An example of this was discussed in the case of superconductivity (§15.4) where we showed that the scattering amplitude becomes infinite at the point where Cooper pairs start to form.

Finally, it should be mentioned that an enormous amount of work on phase transitions has been concerned with the behaviour of physical quantities in the neighbourhood of the critical point. For a review of this large subject, the reader is referred to the book of Stanley (1971). A microscopic theory of critical phenomena has been formulated by Wilson (1971a, b, c), and expressed in terms of Feynman diagrams by Wilson (1972) and Tsuneto and Abrahams (1972). The graphical methods used in critical phenomena are reviewed in an article by Brout (1974), and Wilson and Kogut (1974).

Further reading

Mattuck and Johansson (1968).
Brout (1965).
Stanley (1971).
Brout (1974).
Wilson and Kogut (1974).

Exercises

17.1 (a) Show that H in (7.51) is invariant under global gauge transformation (17.1′).
  (b) Use (a) plus the fact that $|\Psi_0\rangle$ is non-degenerate in a normal system to prove that $|\Psi_0\rangle$ is invariant under (17.1′). (Hint: represent the global gauge transformation as an operator T.)
  (c) Show that for a superconductor, H_{red} in (15.16) is invariant under (17.1′) while the BCS $|\Psi_0\rangle$ in (15.29) is not.
17.2 Verify (17.15).
17.3 Verify in detail (17.23) → (17.35).
17.4 Verify (17.40).
17.5 Verify in detail (17.53) → (17.56).

Chapter 18

Feynman Diagrams in the Kondo Problem

18.1 Introduction

The history of mankind has been marked by a series of catastrophes—fires, floods, famines, plagues. Physicists have to add to this list another catastrophe: divergences. The Kondo problem is largely concerned with a so-called 'infra-red divergence' occurring in the theory of dilute magnetic alloys, and the history of the problem is the story of the physicist's fight against this divergence. The problem has attracted a great deal of attention because, like critical phenomena and the X-ray problem, it can be handled only by self-consistent renormalization of all propagators and interactions. In this chapter, we will present a very brief introduction to just the diagrammatic treatment of the problem. For a general review, the reader is referred to Rado and Suhl (1973), Kondo (1969), Heeger (1969), and for details of the approach used here, to Cheung and Mattuck (1970), Mattuck, Hansen, and Cheung (1971), Larsen (1972), Larsen and Mattuck (1974), and Larsen (1974).

A dilute magnetic alloy consists of a few magnetic impurity atoms like Fe, dissolved in a non-magnetic host metal, e.g., Cu. Such systems show an anomalous increase in electrical resistance at low temperatures shown in Fig. 18.1. In 1964 Kondo proposed that this increase was due to exchange scattering between the conduction electrons and the magnetic impurity atoms. We will write his Hamiltonian for the case of a *single* impurity of spin $S=\frac{1}{2}$. (As we noted in connection with the problem of randomly distributed non-

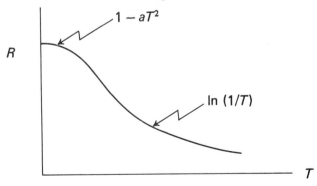

Fig. 18.1 *Resistance, R, due to Magnetic Impurities in Dilute Magnetic Alloys, as a Function of Temperature, T*

magnetic impurities in §3.5, for very low impurity density, the only important terms in the self-energy are those representing repeated scattering from the *same* impurity (see (3.67). Hence we just need to deal with one impurity and multiply the result by a concentration factor.) The Kondo Hamiltonian is

$$H = \sum_{k,\alpha} \epsilon_k c_{k\alpha}^\dagger c_{k\alpha} - J \sum_{\alpha,\alpha',k,k'} (\boldsymbol{\sigma}_{\alpha'\alpha} \cdot \mathbf{S}) c_{k'\alpha'}^\dagger c_{k\alpha}. \tag{18.1}$$

The first term describes a set of conduction s-electrons in a band of width D, with ϵ_k measured relative to the Fermi surface. The second term is the exchange interaction, where $\boldsymbol{\sigma}_{\alpha'\alpha}$ is the Pauli spin matrix for s-electrons, $\mathbf{S}$ is the impurity spin operator and J is the negative (antiferromagnetic) interaction constant.

Kondo's Hamiltonian may be rewritten in the form

$$H' = \sum_{k,\alpha} \epsilon_k c_{k\alpha}^\dagger c_{k\alpha} + \sum_\beta 0 d_\beta^\dagger d_\beta^\dagger - J \sum_{k,k',\alpha,\alpha',\beta,\beta'} (\boldsymbol{\sigma}_{\alpha'\alpha} \cdot \mathbf{S}_{\beta'\beta}) d_{\beta'}^\dagger c_{k'\alpha'}^\dagger c_{k\alpha} d_\beta. \tag{18.2}$$

Here $\mathbf{S}$ has been expressed in second quantized form (see exercise 7.11, with no k-index, since the impurity spin is fixed in configuration space). $\frac{1}{2}\mathbf{S}$ is a Pauli spin matrix, and $d_\beta^\dagger$, d_β create (destroy) a spin in state $\beta (= \pm\frac{1}{2})$ at the impurity site. These d's are called 'pseudofermion operators' (see Abrikosov (1965)). Note that the free pseudofermion energy (second term in (18.2)) is zero because the free spin energy in (18.1) is zero.

The Hamiltonian in form (18.2) allows impurity states with $S=0$ as well as $S=\frac{1}{2}$. Eliminating the effect of these spurious $S=0$ states requires multiplying physical quantities by a correction factor $Q(T)$, as described in detail by Larsen (1972). We will take $Q(T)=$ constant in the following discussion.

Let us represent the s-electrons by solid lines, the impurity spin or 'pseudofermion' (abbreviated 'pf') by dashed lines, and the interaction by the intersection of a solid and a dashed line. Then the exact equation for the s-electron propagator may be written

$$\tag{18.3}$$

$$\tag{18.4}$$

$$(18.5)$$

while that for the *pf* propagator is

$$(18.6)$$

$$(18.7)$$

In these equations, Γ is the *s–pf* scattering amplitude, given by

$$(18.8)$$

The free propagators are

$$G_0(\mathbf{k}, i\omega_n) = \frac{1}{i\omega_n - \epsilon_k}, \quad \mathscr{G}_0(i\omega_m) = \frac{1}{i\omega_m}. \qquad (18.9)$$

Observe that (18.4) is similar to the non-magnetic impurity case for density (N/Ω) small (first four terms on the right-hand side of (3.67')), with Σ_1^s replacing the empty circles. The difference is that, as shown in (18.5), the circles here have very complicated internal structure. This is due to the fact that the exchange term in (18.1) does *not* act as an external potential, since it has an internal degree of freedom, i.e., the impurity spin can flip.

The first-order bubbles in (18.5) and (18.7) are zero on account of the spin sum involved, and will henceforth be omitted.

The resistance can be obtained from the reciprocal lifetime at the Fermi surface, $\tau_{k_F}^{-1}$, and $\tau_{k_F}^{-1}$ is given in turn by the imaginary part of the self-energy, analogous to (3.71). It can be shown that

$$R \propto \tau_{k_F}^{-1} = \text{Im } \Sigma^s(k_F, \epsilon_{k_F} + i\delta), \qquad (18.10)$$

where we have analytically continued the imaginary frequency self energy to just above the real axis: $\Sigma(k, i\omega_n \to \omega + i\delta)$ (i.e., we have the retarded self-energy—see appendix L).

The physical origin of the self-energy can be seen from diagram (18.8(b)). The s-electron comes in, interacts once with the impurity spin, flipping it, and is itself flipped. At a later time, the flipped spin interacts back on the s-electron. The net effect is to alter the s-electron energy and to scatter it out of its original state, giving it a lifetime.

(For those who like analogies, we can imagine the situation in Fig. 18.2. Man M throws hat H onto shelf where it accidentally hits bucket of wet cement C. This flips cement bucket, analogous to spin flip, and cement falls down on man's head, thus increasing his mass (altering his energy) and slowing him down (offering resistance).)

Note that it is this spin flip which makes the Kondo problem a many-body problem, despite the fact that there is no direct interaction between two electrons in the Kondo Hamiltonian. That is, electron 1 comes in, flips the spin, and the flipped spin interacts with electron 2. Thus, two electrons interact indirectly via the impurity spin, as shown in (18.11):

$$(18.11)$$

18.2 Second-order (Born) approximation

Kondo's original calculation was equivalent to the following approximations for (18.3)–(18.8):

$$\text{(A)} \quad \| = 1 \Big/ \Big[\,|^{-1} - \Sigma^s\,\Big], \quad \text{(B)} \quad \Sigma^s \approx \Sigma_1^s ,$$

$$\text{(C)} \quad \Sigma_1^s \approx \Gamma , \quad \text{(D)} \quad \Gamma \approx \times + \bigcirc + \bigcirc \qquad (18.12)$$
$$\qquad\qquad\qquad\qquad\qquad\qquad (a) \qquad (b) \qquad (c)$$

Note that this is a second-order calculation of Γ but is third order for Σ_1^s. Diagram (a) in Γ is just the bare interaction. The s–pf 'pair bubble' diagram (b) is given by

R. D. M.

Fig. 18.2 *Analogue of Electron-Impurity Spin Scattering in Kondo Problem*

$$\equiv J \sum_{\alpha';\beta'} (\boldsymbol{\sigma}_{\alpha\alpha'} \cdot \mathbf{S}_{\beta\beta'})(\boldsymbol{\sigma}_{\alpha'\alpha''} \cdot \mathbf{S}_{\beta'\beta''})$$

$$\times J \sum_{l} \frac{1}{\beta} \sum_{\omega_s} G_0(l,\omega_s) \mathscr{G}_0(-\omega_s+\omega_m),$$

(with diagram labels: $\mathbf{q},\alpha,\omega_n$; $\beta, -\omega_n+\omega_m$; $\mathbf{l},\alpha',\omega_s$; $\beta', -\omega_s+\omega_m$; $\mathbf{p},\alpha'',\omega_i$; $\beta'', -\omega_i+\omega_m$)

(18.13)

The spin sum here is easily shown to be

$$\sum_{\alpha'\beta'} (\boldsymbol{\sigma}_{\alpha\alpha'} \cdot \mathbf{S}_{\beta\beta'})(\boldsymbol{\sigma}_{\alpha'\alpha''} \cdot \mathbf{S}_{\beta'\beta''}) = \tfrac{3}{4}\delta_{\alpha\alpha''}\delta_{\beta\beta''} - \boldsymbol{\sigma}_{\alpha\alpha''} \cdot \mathbf{S}_{\beta\beta''}$$ (18.14)

and the l, ω_s sum is, after performing the frequency sum with the aid of (14.54):

$$K_0(i\omega_m) = J \sum_l \frac{f(\epsilon_l)-f(0)}{i\omega_m-\epsilon_l}, \quad \text{or} \quad K_0(z) = -J \int_{-\infty}^{+\infty} d\epsilon \rho(\epsilon) \frac{[f(\epsilon)-\tfrac{1}{2}]}{\epsilon-z},$$ (18.15)

where we have analytically continued $i\omega_m$ to the whole complex z-plane, and where $\rho(\epsilon)$ is the density of states. We will assume that $\rho(\epsilon)$ is a simple Lorentzian of width $2D$:

$$\rho(\epsilon) = \frac{D/\pi}{\epsilon^2+D^2} = \frac{1}{2\pi i}\left\{\frac{1}{\epsilon-iD} - \frac{1}{\epsilon+iD}\right\}.$$ (18.16)

Placing (18.16) in (18.15) yields

$$K_0(z) = K_0^a(z) + K_0^b(z)$$

$$K_0^a(z) = -\frac{J}{2\pi i} \int_{-\infty}^{+\infty} d\epsilon \frac{[f(\epsilon)-\tfrac{1}{2}]}{(\epsilon-z)(\epsilon-iD)}$$

$$K_0^b(z) = +\frac{J}{2\pi i} \int_{-\infty}^{+\infty} d\epsilon \frac{[f(\epsilon)-\tfrac{1}{2}]}{(\epsilon-z)(\epsilon+iD)}.$$ (18.17)

We will take z to be in the upper half-plane, then set $z=\omega+i\delta$ to get the physically interesting retarded scattering amplitude. Just the case $\omega=0$ will be considered here.

Let us first evaluate $K_0^a(z)$. This has poles at $\epsilon = z$, $\epsilon = +iD$ and $\epsilon = \pi i T(2n+1)$, n = integer (coming from the Fermi function. Note that we set the Boltzmann constant equal to 1). If we close the contour in the lower half-plane, the poles within the contour are at $\epsilon_n = -\pi i T(2n+1)$, $n=0, 1, 2, \dots$ as seen in Fig.18.3.

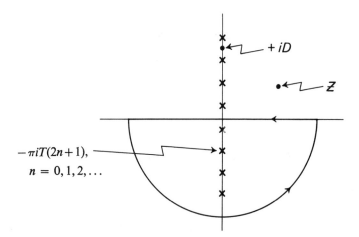

Fig. 18.3 *Contour for s–pf Pair-Bubble Integral*

The integral vanishes on the half-circle, so we have

$$K_0^a(z) = \frac{-J}{2\pi i}(-2\pi i)\sum \left\{ \text{Residues of } \frac{[f(\epsilon)-\tfrac{1}{2}]}{(\epsilon-z)(\epsilon-iD)} \quad \begin{array}{l} \text{at } \epsilon_n = -\pi iT(2n+1), \\ n = 0,1,2,\dots \end{array} \right\}$$

$$= J \sum_{n=0}^{\infty} \lim_{\epsilon\to-\pi iT(2n+1)} \frac{(\epsilon-[-\pi iT(2n+1)])(-T)}{(\epsilon-z)(\epsilon-iD)(\epsilon-[-\pi iT(2n+1)])}. \tag{18.18}$$

This gives us, for $z=0+i\delta$:

$$K_0^a(0+i\delta) = \frac{-J}{(2\pi i)^2 T}\sum_{n=0}^{\infty} \frac{1}{(n+\tfrac{1}{2})}\cdot\frac{1}{(n+\tfrac{1}{2})+D/2\pi T}. \tag{18.19}$$

The second integral, $K_0^b(z)$, may be evaluated similarly, except that we close the contour in the upper half-plane. In this case, we have inside the contour a pole at $\epsilon=z$. For $z=0$, the contribution from this pole vanishes, since $f(z=0)-\tfrac{1}{2}=0$. The rest of the contribution is from the poles at $\epsilon_n = +i\pi T(2n+1)$, $n=0, 1, \dots$ and is exactly the same as (18.19). Hence we just multiply (18.19) by 2 to get the entire K_0 result:

$$K_0(0+i\delta) = K_0^a(0+i\delta) + K_0^b(0+i\delta)$$

$$= \frac{2J}{(2\pi)^2 T}\sum_{n=0}^{\infty} \frac{1}{(n+\tfrac{1}{2})}\cdot\frac{1}{(n+\tfrac{1}{2})+D/2\pi T} \tag{18.20}$$

This can be expressed in terms of the digamma function, given by ($\Gamma(z) =$ gamma function)

$$\psi(z) = \frac{d \ln \Gamma(z)}{dz} = -C + \sum_{n=1}^{\infty} \left\{ \frac{1}{n} - \frac{1}{(z-1)+n} \right\} \quad (18.21)$$

where $C = 0.5772 \ldots$ (Euler's constant). Changing to a sum from 1 to ∞ in (18.20) and expanding in partial fractions gives

$$K(0 + i\delta) = \frac{J}{\pi D} \sum_{n=1}^{\infty} \left\{ \frac{1}{n - \frac{1}{2}} - \frac{1}{(n - \frac{1}{2}) + D/2\pi T} \right\}$$

$$= J\rho[\psi(\tfrac{1}{2} + D/2\pi T) - \psi(\tfrac{1}{2})] \quad (18.22)$$

where $\rho = (\pi D)^{-1} =$ density of states at $\epsilon = 0$. For $D \gg 2\pi T$, the argument of ψ is large and we can use the expansion

$$\psi \text{ (large } z) = \ln z - 1/2z - \cdots \quad (18.23)$$

so

$$K(0 + i\delta) \approx J\rho \ln (D/2\pi T). \quad (18.24)$$

This diverges when $T \to 0$. The origin of the divergence is the $1/\epsilon$ behaviour of the integrand together with the sharp Fermi surface (at $T = 0$) which cuts off the integral at $\epsilon = 0$. It is called an 'infra-red' divergence since it comes from the low-energy range where $\epsilon \to 0$.

This same divergence shows up when we calculate the resistance. Placing (18.12D(a)) plus (18.22) plus a similar contribution from (18.12D(c)) into (18.12C, B, A) and utilizing (18.10), we obtain, after some work:

$$R \propto J^2[1 - J\rho \ln (D/2k_B T)]. \quad (18.25)$$

Thus Kondo's second-order calculation of Γ (or third order for Σ^s) yields the experimentally observed $\ln T$ behaviour. Unfortunately, the $\ln T$ term diverges as $T \to 0$, which indicates that second-order perturbation theory is no longer valid. In the temperature region where result (18.25) matches experiment, it is called the 'Kondo effect'. In the region where it blows up, it is called the 'Kondo problem'.

18.3 Parquet approximation with bare propagators

The standard cure when finite order perturbation theory breaks down is to use partial summation, as we did in the case of the electron gas in RPA. In the present case, Abrikosov (1965) showed that at high T, the dominant diagrams were the 'parquets', i.e., diagrams such as $(18.8(a), (b), (c), (d), (e))$, obtained by 'opening up' more and more interaction vertices and inserting an s–pf pair bubble in them. (Thus $(18.8(f), (g)$ and $(h))$ are not parquets.) Summing over all parquets including terms of leading order in $\ln(D/T)$ yielded:

$$\Gamma_{\alpha\beta\alpha''\beta''}(i\omega_m) = \Gamma(i\omega_m)\,\boldsymbol{\sigma}_{\alpha\alpha''}\cdot\mathbf{S}_{\beta\beta''}, \tag{18.26}$$

where

$$\Gamma(i\omega_m) = J/[1 + 2K_0(i\omega_m)]. \tag{18.27}$$

Using (18.24) we see that $\Gamma(\omega + i\delta)$ diverges when $\omega = 0$, and T is given by

$$1 + 2K_0(0 + i\delta) = 0, \quad \text{or} \quad 1 = 2|J|\rho \ln(D/2k_B T), \quad \text{or} \quad T_K = \frac{D}{2k_B} e^{-1/2|J|\rho}, \tag{18.28}$$

where T_K is called the 'Kondo temperature'. This divergence of Γ at T_K is much worse than Kondo's divergence at $T = 0$. Moreover, there is a corresponding divergence in the resistance at the Kondo temperature. In other words, the cure appears to have killed the patient!

For some time after this result, it was believed that since this divergence at T_K was similar to that occurring at the critical temperature in superconductivity (see §15.7), or in ferromagnetism (see §17.7), it indicated some sort of 'localized phase transition' at the impurity site, i.e., formation of a bound state of the impurity spin and s-electron spins. However, it is now known that the divergence at T_K is spurious, and occurs because we have only included those diagrams which are important at high T, and neglected diagrams which become important at low T.

We can get an idea of which diagrams are significant in the low-temperature region by examining the similar, but much simpler 'X-Ray Problem', which can be solved exactly (see Roulet et al. (1969), Nozières et al. (1969), Nozières and de Dominicis (1969)). In the X-ray problem, an approximation is developed which rigorously takes into account all diagrams contributing to lowest order in the coupling constant. This approximation involves self-consistent renormalization of the propagator lines and vertices. Applied to the Kondo problem, this would mean a self-consistent treatment including diagrams having clothed s-propagators, like $(18.8(f))$, clothed pf-propagators, like $(18.8(g))$, and clothed vertices, like $(18.8(h))$. No one has yet succeeded in carrying out such a full renormalization programme although there have been some attempts (Murata (1971)). However, we can perhaps get a qualitative idea of what to expect from such a programme by examining first the effect of self-consistently renormalizing just the s-electron propagator, and then the effect of renormalizing the pf-propagator and vertices.

18.4 Self-consistently renormalized s-electrons

The self-consistent calculation using just clothed s-electrons (Cheung and Mattuck (1970)) may be summarized in the following three equations:

$$\left\| \approx \right\| + \widehat{\Sigma^s} \tag{18.29}$$

$$\widehat{\Sigma^s} \approx \widehat{\Sigma_1^s} \approx \widehat{\Gamma} \tag{18.30}$$

$$\widehat{\Gamma} \approx \bowtie + \bigcirc + \overset{K}{\bigcirc} + \bigcirc + \cdots + \bigcirc + \cdots. \tag{18.31}$$

The procedure is to calculate Γ in parquet approximation in terms of the unknown clothed s-propagator, G, using (18.31). Then we calculate G in terms of the unknown self-energy, Σ^s, using (18.29). Finally, we calculate Σ^s in terms of Γ, using (18.30). This gives us the self-consistent 'dog-biting-its-own-tail' argument illustrated in Fig. (18.4).

Simultaneous solution of (18.29, 30, 31) shows that the renormalized pair-bubble, K, no longer diverges as $T{\to}0$, but goes to a constant there. Γ no longer diverges at $T=T_K$ but instead has only a weak divergence as $T{\to}0$,

Fig. 18.4 *Self-Consistent Kondo Dog*

indicating the formation of some sort of bound state at $T=0$. The resistance is

$$R \propto \left\{1 - \frac{\ln{(T/T_K)}}{[\ln^2{(T/T_K)} + \pi^2]^{\frac{1}{2}}}\right\} \qquad (18.32)$$

which is finite everywhere. The agreement with experiment is good above T_K but bad as $T \to 0$, where it comes up with infinite slope (cf. the $(1 - aT^2)$ behaviour in Fig. 18.1). (This result was first obtained by Hamann (1967) by using Nagaoka's decoupling of the equations of motion for G^s. This decoupling is equivalent to the partial sum over parquets in (18.31). See appendix M and Theumann (1970)).

The reason why this calculation removes the divergence of Γ at T_K is that there is 'negative feedback' in the self-consistency loop: when Γ tends to become large, so does Σ, which reduces the s-propagator, diminishing K and thus Γ.

(This negative feedback has a striking parallel in the dog-biting-its-own-tail of Fig. 18.4. The harder the dog bites, the more it hurts. The more it hurts, the less hard he bites. This negative feedback explains why dogs do not diverge, i.e., why they do not bite their own tails off!)

18.5 Strong-coupling approximation with self-consistently renormalized pseudo-fermions and vertices

We turn now to the self-consistent calculation using renormalized pf propagators and vertices (Larsen (1974), Larsen and Mattuck (1974)). The expressions for the s- and pf-propagators and self-energies are just (18.3, 4, 6). However, the s-electron self-energy Σ_1^s, pf self-energy Σ^{pf} and the s–pf scattering amplitude, Γ, are not found by the ordinary Feynman diagram expansions (18.5), (18.7), and (18.8). Instead it is convenient to use the *reduced graph expansion* (appendix N) to find the *imaginary* parts of these quantities; the real part then follows from dispersion relations. The reduced graphs for Σ_1^s, Γ, and Σ^{pf} are:

$$\delta \bigcirc \equiv \text{Im} \widehat{\Sigma_1^s} = \quad + \quad + \quad + \cdots \qquad (18.33)$$

$$\eqno(18.34)$$

$$\eqno(18.35)$$

That is, the reduced graph expansion for the imaginary part of each quantity is the sum of all possible reduced graphs having the proper number of s-electrons and pf's in one endpiece and out the other endpiece with all possible intermediate states.

It is assumed that all s-electron–pf interaction vertices occurring within the endpieces of the reduced graphs are clothed, i.e., replaced by full 4-tail Γ's of the type on the left of (18.34). For example, diagram 3 on the right of (18.35) has as a typical contribution the diagram

$$\eqno(18.36)$$

where the circles are the clothed s–pf interaction, and the dashed line denotes intermediate state.

Observe that the higher-order terms in each expansion involve higher and higher-order endpieces, so that the set of equations (18.33–35) is not complete.

In fact we need an infinite number of equations for these higher-order amplitudes. The expansion for the most general many-tail diagram may be written:

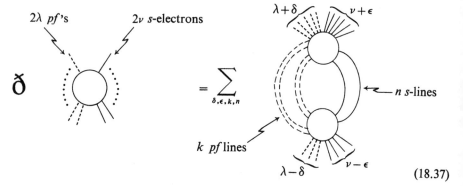

$$(18.37)$$

If we designate the general many-tail diagram as follows:

$$\equiv \Gamma_{m,n}^{k,l} \equiv \Gamma_{(m+n)/2}^{(k+l)/2}$$

$$(18.38)$$

where the latter notation shows $(k+l)/2 =$ total number of pairs of pf lines and $(m+n)/2 =$ total number of s-electron lines, then (18.37) may be written out using rules similar to those in appendix N. This yields the following monster:

$$\delta \sum_{i'} \Gamma_{i',\nu}^{\lambda}(\{y_{2\lambda+2\nu}\}) = \sum_{i,j} A_{ij} \sum_{\delta,\varepsilon,k,n} (-1)^h [e^{Y/T} + (-1)^{p-1}] \int_1 \cdots$$

$$\cdots \int_p \delta(x_1 + \cdots + x_p - Y) \times {}^R \Gamma_{\nu-\varepsilon,n}^{\lambda-\delta,k}(\{y_{\lambda+\nu-\delta-\varepsilon}\}, \{x_p\})$$

$$\times {}^A \Gamma_{n,\nu+\varepsilon}^{k,\lambda+\delta}(\{x_p\}, \{y_{\lambda+\nu+\delta+\varepsilon}\}).$$

$$(18.39)$$

where $p = k + n$ and

$$\int_1 \cdots \int_p = \int_{-\infty}^{+\infty} dx_1 \mathscr{A}(x_1) f(x_1) \cdots \int_{-\infty}^{+\infty} dx_k \mathscr{A}(x_k) f(-x_k)$$

$$\times \int_{-\infty}^{+\infty} dx_{k+1} \rho(x_{k+1}) f(-x_{k+1}) \cdots \int_{-\infty}^{+\infty} dx_{k+n} \rho(x_{k+n}) f(x_{k+n})$$

In this beast, $\mathscr{A}(x)$ is the *pf* spectral function, $\rho(x)$ is the *s*-electron spectral function, and f the Fermi function. Y is a linear combination of external frequencies, $\{y_{\lambda+\nu-\delta-\varepsilon}\}$ and $\{y_{\lambda+\delta+\nu+\varepsilon}\}$ are short for the set of labels on the lines entering at the bottom or the top respectively, $\{x_\rho\}$ are the intermediate state variables, h is the number of hole lines, R means retarded, A means advanced, and the index i, i' or j stands for the particular Feynman diagram contributing to the total amplitude. A_{ij} is equal to $(-1)^f$ times a spin sum factor, where f = number of fermion loops external to the endpiece diagrams.

Fortunately, there turns out to be a relatively non-traumatic path through this chamber of horrors (18.39). Namely, there is a simple class of solutions to (18.39) known as *homogeneous* or *scale-invariant* solutions. A homogeneous function $f(x_1, x_2, \ldots, x_n)$ is defined as follows

$$f(x_1, x_2, \ldots, x_n) = \eta^a f\left(\frac{x_1}{\eta}, \frac{x_2}{\eta}, \ldots, \frac{x_n}{\eta}\right), \qquad (18.40)$$

where η is an arbitrary real number known as the *scaling factor*, and a is a real exponent. A simple example of a homogeneous function is $f(x_1, x_2) = x_1^2 + x_2^2$. Scaling this by η yields $f(x_1/\eta, x_2/\eta) = x_1^2/\eta^2 + x_2^2/\eta^2 = \eta^{-2} f(x_1, x_2)$, so $a = 2$. It is easily verified that if the quantities in (18.39) are homogeneous, and behave as follows when scaled by $|\eta|$ (γ is an as yet unknown constant):

$$\Gamma_\nu^\lambda(\{y_{2\lambda+2\nu}\}) = |\eta|^{1-\lambda\gamma-\nu} \Gamma_\nu^\lambda\left(\left\{\frac{y_{2\lambda+2\nu}}{\eta}\right\}\right) \qquad (18.41)$$

$$\mathscr{A}(x) = |\eta|^{\gamma-1} \mathscr{A}(x/|\eta|) \qquad (18.42)$$

$$\rho(x) = \rho(x/|\eta|) \qquad (18.43)$$

then (18.39) is satisfied. That is, on the left and right of (18.39), we divide x_i, y_i, Y and T by $|\eta|$ wherever they appear. Noting that $f(x_i) \to f(x_i/|\eta|) = f(x_i)$ (since $f(x_i) = [\exp(x_i/T) + 1]^{-1}$,) and $\delta(x_i) \to \delta(x_i/|\eta|) = |\eta|\delta(x_i)$, and substituting the expression in (18.41) for the various scaled quantities yields just (18.39) multiplied on the right by the factor $|\eta|$ to the power

$$\text{Pow(right)} = k+n-1+k(\gamma-1)+1-\gamma\left[\frac{\lambda-\delta+k}{2}\right] - \left[\frac{\nu-\epsilon+n}{2}\right] +$$

$$+1-\gamma\left[\frac{k+\lambda+\delta}{2}\right] - \left[\frac{n+\nu+\epsilon}{2}\right], \qquad (18.44)$$

and on the left $|\eta|$ to the power

$$\text{Pow(left)} = 1-\lambda\gamma-\nu. \qquad (18.45)$$

These two powers are identical, showing that homogeneous functions do indeed satisfy (18.39).

Using results (18.41–43), and under the assumption that the Γ's are functions only of a single variable z (or ω if we continue z to just above or below the real axis), where z is the variable associated with each of the incoming pf lines, and requiring that $\int \mathcal{A}(x)$ and $\int \rho(x)$ be normalized to unity, it can be shown that $\mathcal{A}$ has the form ($\phi = \tan^{-1}(\Delta/\alpha T)$):

$$\mathcal{A}(\omega) = \frac{\gamma(\sin\phi)^\gamma}{4\Delta^\gamma \sin(\phi\gamma)} \{(\alpha T - i\omega)^{\gamma-1} + (\alpha T + i\omega)^{\gamma-1}\} \quad \text{for} \quad |\omega| \leqslant \Delta$$
$$= 0 \quad \text{for} \quad |\omega| > \Delta \tag{18.46}$$

with corresponding pf propagator:

$$\mathcal{G}^R(\omega) = i\Delta^{-\gamma}(\alpha T - i\omega)^{\gamma-1} \frac{(\pi\gamma/2)(\sin\phi)^\gamma}{\sin(\phi\gamma)}, \tag{18.47}$$

and the general many-tail vertex has the form

$$^{R,A}\Gamma_\nu^\lambda \sim \frac{\Delta^{\lambda\gamma}}{\rho^\nu} (\pm i)^{\lambda+\nu-2} (\alpha T \mp i\omega)^{1-\gamma\lambda-\nu}. \tag{18.48}$$

Here the quantity α is a constant. Note that these results are valid only to lowest order in ω/T and ω/Δ. Thus for example, we can see from (18.46) that $\mathcal{A}$ is not scale-invariant for $\omega > \Delta$.

Note that by (18.38, 48), Γ_0^1 is just the pf self-energy, given by:

$$\Sigma^{pf} \equiv \Gamma_0^1 \sim \Delta^\gamma i^{-1}(\alpha T - i\omega)^{1-\gamma}. \tag{18.49}$$

But $\mathcal{G}$ is given by the Dyson equation:

$$\mathcal{G}(\omega) = \frac{1}{\omega - \Sigma^{pf}(\omega)}. \tag{18.50}$$

For this to reduce to the strong-coupling form (18.56), when $\omega \to 0$ we must have

$$\lim_{\omega \to 0} \Sigma^{pf}(\omega) \gg \omega.$$

This requires $\gamma > 0$.

To see the significance of Δ and the exponential γ in (18.46–48), let $T=0$ (then $\phi = \pi/2$) and consider the time transform of $\mathcal{G}(\omega)$. For $\gamma = 1$ we have $\Sigma^{pf} = \Gamma_0^1 \sim -i\Delta$. Placing this in (18.50) yields

$$\mathcal{G}(\omega) = \frac{1}{\omega + i\Delta}. \tag{18.51}$$

Fourier transforming we find

$$\mathscr{G}^R(t) = -i\theta_t e^{-\Delta t}.$$ (18.52)

Similarly, for $\gamma \neq 1$ we find:

$$\mathscr{G}^R(t) \propto \theta_t (\Delta t)^{-\gamma}.$$ (18.53)

Thus, for $\gamma = 1$, the probability amplitude for the *pf* being in, say, a spin up state, decays exponentially, while for $0 < \gamma < 1$ the decay is algebraic, with a long time tail. The associated lifetime of the decay is Δ^{-1}.

If we now expand (18.48) to lowest order in ω/T and place the result on the left of (18.39), we obtain

$$\delta \sum_i \Gamma_\nu^\lambda \sim \frac{\Delta^{\lambda\gamma}}{\rho^\nu} T^{1-\lambda\gamma-\nu} i^{\lambda+\nu+2} \frac{\omega}{\alpha T}, \quad \text{if } \lambda, \nu \text{ both even or both odd}$$

$$\sim \frac{\Delta^{\lambda\gamma}}{\rho^\nu} T^{1-\lambda\gamma-\nu} i^{\lambda+\nu+1}, \quad \text{if } \lambda \text{ odd, } \nu \text{ even or } \lambda \text{ even, } \nu \text{ odd.} \quad (18.54)$$

Placing (18.48) on the right of (18.39) and expanding to lowest order in ω/T we find that every term on the right has exactly form (18.54) so that monster (18.39) is satisfied. Solutions of the form (18.46), (18.48) are known as *strong-coupling* solutions. It should be emphasized that the strong-coupling forms hold only to leading order in ω/Δ and T/Δ, since homogeneity and scale invariance is broken by Δ.

The strong-coupling form is not a complete solution since the constants γ, Δ, α are thus far undetermined. To find them would require actually carrying out the sums in (18.39) in detail with all A_{ij}'s included, which would be a formidable task. However, it is possible to at least determine γ by recourse to an exact numerical result of Wilson (1973) for $S=\frac{1}{2}$ in the following way: First of all, let us get an expression for the magnetic susceptibility χ due to the impurity spin. χ is given by ($H=$ magnetic field):

$$\chi = \lim_{H \to 0} \frac{\langle n_\uparrow \rangle - \langle n_\downarrow \rangle}{H}.$$ (18.55)

Since $\langle n_\uparrow \rangle$ and $\langle n_\downarrow \rangle$ are obtained directly from the single-particle propagator $\mathscr{G}$ (see 11.24)), they will have the same dependence on Δ and T that $\mathscr{G}$ does (see (18.47)), i.e., in leading order at $T \to 0$:

$$\chi \sim \frac{T^{\gamma-1}}{\Delta^\gamma}.$$ (18.56)

Now a numerical result of the $T=0$ calculation of Wilson shows that χ is finite at $T=0$. Hence in (18.56) we must have $\gamma=1$. From (18.52), this means a *pf* spin state decaying exponentially with lifetime Δ^{-1} and propagator (18.51). We will, in what follows, take the corresponding spectral density to be the Lorentzian

$$\mathscr{A}(\omega) = \frac{\Delta}{\omega^2+\Delta^2} \qquad (18.57)$$

rather than the square form which (18.46) reduces to when $\gamma=1$.

To get a crude idea of the other constants, i.e., Δ, α, we approximate (18.34, 35) by the first diagram on the right. These diagrams are the most important at high temperature. At low T, as pointed out after (18.54), in the ω, $T\rightarrow0$ limit, all diagrams have the same strong-coupling form. Hence the first diagram on the right of (18.34, 35) will also go over to the strong-coupling form when $T\rightarrow0$. The coefficients will of course be wrong, but since the $T\rightarrow0$ and the high T forms will be correct, we may expect to get a useful interpolation formula.

In functional form, using just the first diagram on the right, and taking the case $\omega=0$, (18.34, 35) are

$$\partial\Sigma^{pf}(0) = \tfrac{3}{2}|\Gamma(0)|^2\delta B_L(0) \qquad (18.58)$$

$$\partial\Gamma(0) = -2|\Gamma(0)|^2\delta K(0) \qquad (18.59)$$

where $|\Gamma(0)|^2=\Gamma^R(0)\,\Gamma^A(0)$ comes from the two endpieces, and $\delta B_L(0)$, $\delta K(0)$ are the factors coming from the intermediate states. Using the dispersion relation

$$\Gamma(z)-\Gamma(\infty) = -\frac{1}{\pi}\int dx\,\frac{\partial\Gamma(x)}{x-z}$$

(18.59) becomes

$$\Gamma^R(0) = \frac{1}{1/J+2K^R(0)}, \qquad (18.60)$$

where $K(0)$ is the pair-bubble with clothed *pf* propagator, evaluated using the Lorentzian density of states (18.57):

$$K^R(0) = \rho\left[\ln\frac{D}{2\pi T} - \psi\left(\frac{1}{2}+\frac{\Delta}{2\pi T}\right)\right] \approx \rho\left[\ln\frac{D}{\Delta} - \frac{\pi^2 T^2}{6\Delta^2}\right]. \qquad (18.61)$$

This is (apart from a factor of J) just the generalization of the bare pair-bubble (18.22).

We now put (18.60, 61) in (18.58) above and note by (18.51) that $\delta \Sigma^{pf} = \Sigma^{pf} = i\Delta$. This gives us the following equation for Δ:

$$\Delta = \tfrac{3}{8}\Delta \frac{\left\{\dfrac{\pi T}{\Delta} + \psi\left(\dfrac{1}{2} + \dfrac{\Delta}{2\pi T}\right) - \psi\left(1 + \dfrac{\Delta}{2\pi T}\right)\right\}}{\left\{\ln \dfrac{2\pi T}{T_K} + \psi\left(\dfrac{1}{2} + \dfrac{\Delta}{2\pi T}\right)\right\}^2} \tag{18.62}$$

which has the solution

$$\Delta(T) \simeq T_K + \frac{\pi\sqrt{3}}{4} T, \quad \text{for } T \ll T_K \tag{18.63}$$

$$\Delta(T) \simeq \frac{3\pi}{8} \frac{T}{\ln^2(T/T_K)}, \quad \text{for } T \gg T_K \tag{18.64}$$

where T_K is the Kondo temperature.

If we place this into (18.61, 60), we find that as $T \to 0$, $\Gamma(0)$ has the strong coupling form (cf. (18.48)):

$$\Gamma \sim \frac{\Delta}{\rho \alpha T}, \quad \alpha = \frac{\pi\sqrt{3}}{4}. \tag{18.65}$$

If (18.63) is put into the susceptibility (18.56) we find

$$\chi \sim \frac{1}{T_K + \dfrac{\pi\sqrt{3}}{4} T} \tag{18.66}$$

i.e., a Curie-Weiss like behaviour.

The resistance may be calculated from $\mathrm{Im}\,\Sigma_1^s$ in (18.33), using the first diagram on the right for a crude approximation. The result is

$$R \simeq 1 - \frac{\pi^2}{4} \frac{T^2}{T_K^2} \quad \text{for } T \ll T_K \tag{18.67}$$

$$R \simeq \frac{3\pi^2}{16 \ln^2(T/T_K)} \quad \text{for } T \gg T_K \tag{18.68}$$

These results are in good agreement with experiment. The whole resistance curve as a function of temperature is compared with the experiments of Loram, Whall, and Ford (1971) in Fig. 18.5.

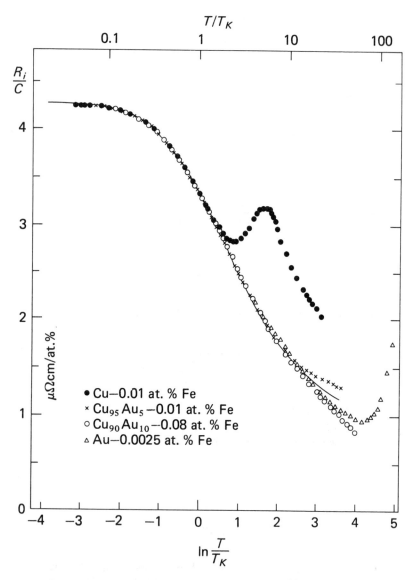

Fig. 18.5 *Comparison of Strong Coupling Theory with Experiment.*
Points: Experimental Results. Solid Line: Theory

Further reading

Rado and Suhl (1973).
Kondo (1969).
Heeger (1969).

Exercises

18.1 Verify in detail (18.13)–(18.25).

Chapter 19

The Renormalization Group

19.1 Introduction

During the years 1964–74, a new tool known as the 'renormalization group' was introduced into the attack on the many-body problem. The method was originally developed by Gell-Mann and Low (1954) to improve on perturbation theory in particle physics, and later applied to infra-red divergences by Bogoliubov and Shirkov (1959), and to quantum electrodynamics and high-energy physics by Wilson (1970). In many-body theory, Bonch-Bruevich (1962) employed it on the electron gas, and it has been applied to the Kondo problem by Abrikosov and Migdal (1970), Fowler and Zawadowski (1971), Fowler (1972), Zawadowski (1973), and to one-dimensional metals by Menyhárd and Sólyom (1973). (Another type of renormalization group has been used with great success by Wilson and others (Wilson (1971a, b, c), (1972)) to tackle the problem of critical phenomena, but this type will not be discussed here.)

In this chapter, we will give a brief introduction to the renormalization group, following Bonch-Bruevich (1962) and Wilson (1970). To keep the beginner from getting lost in what at first sight appears to be an impenetrable jungle of variables and parameters, we will restrict ourselves to discussing just one simple, exactly soluble case, i.e., the high-density electron gas.

The renormalization group is a set of transformations of the quantities in the Dyson equation. These transformations are simple multiplications by a factor, and they have the property that they leave the form of the Dyson equation unchanged. The transformed quantities obey differential equations called 'Lie equations', which turn out in many cases to be a lot simpler to solve than the Dyson equation itself. This constitutes one advantage of the method. Another advantage is that it may lead to a new series expansion which is an improvement over the original expansion.

We will first review the theory of the effective interaction in a high-density electron gas, then define the renormalization group in this case. Finally, we derive the Lie equation for the group and show how to solve it.

19.2 Review of effective interaction in the high-density electron gas

In the high-density limit, the effective interaction is given exactly by the Dyson equation

$$\mathbf{k}, \omega \qquad\qquad\qquad\qquad\qquad\qquad\qquad\qquad (19.1)$$

which may be translated into (for simplicity we consider only the $\omega = 0$ case, so the ω-dependence will be omitted from all that follows)

$$V_{\text{eff}}(\mathbf{k}, \omega = 0) = V(\mathbf{k}, 0) + V_{\text{eff}}(\mathbf{k}, 0) \pi_0(\mathbf{k}, 0) V(\mathbf{k}, 0). \qquad (19.2)$$

Using (10.40) and (10.82) this may be written out in detail, letting $k \equiv |\vec{k}|$:

$$V_{\text{eff}}(k) = \frac{4\pi e^2}{k^2} + V_{\text{eff}}(k) \frac{mk_F}{2\pi^2 \hbar^2} \left[-1 + \frac{1}{k'}(1 - \tfrac{1}{4}k'^2) \ln \left| \frac{1 - \tfrac{1}{2}k'}{1 + \tfrac{1}{2}k'} \right| \right] \frac{4\pi e^2}{k^2}, \quad (19.3)$$

where $k' = k/k_F$. It will be convenient in what follows to re-write this equation in terms of the following quantities:

(a) The 'dimensionless charge': $g^2 = \dfrac{e^2 m}{\hbar^2 k_F}$ $\qquad\qquad\qquad$ (19.4)

(b) The 'bare interaction propagator': $D_0 = 1/k^2$ $\qquad\qquad$ (19.5)

(c) The 'effective interaction propagator': $D = V_{\text{eff}}(k)/4\pi e^2$ $\qquad$ (19.6)

(d) The pair-bubble function:

$$f(k^2/W) = -\frac{2W}{\pi k^2} \left\{ -1 + \left(\frac{W}{k^2} \right)^{\frac{1}{2}} \left(1 - \frac{1}{4}\frac{k^2}{W} \right) \ln \left| \frac{1 - \tfrac{1}{2}(k^2/W)^{\frac{1}{2}}}{1 + \tfrac{1}{2}(k^2/W)^{\frac{1}{2}}} \right| \right\} \quad (19.7)$$

where $W = k_F^2$. Note that $f(\infty) = 0$, $f(0) = \infty$, and f is always positive. Then the Dyson equation (19.3) becomes:

$$D = D_0 - Dg^2[k^2 f(k^2/W)] D_0. \qquad (19.8)$$

It will be convenient to re-write this using the quantity

$$d(k^2/W, g^2) \equiv k^2 D(k^2, W, g^2), \quad d_0 \equiv k^2 D_0 = 1, \qquad (19.9)$$

in terms of which the Dyson equation becomes

$$d = d_0 - dg^2 f(k^2/W) d_0, \quad \text{or} \quad d = 1 - dg^2 f(k^2/W). \qquad (19.10)$$

Note that $d = d(k^2/W, g^2)$. Equation (19.10) may be immediately solved to yield

$$d(k^2/W, g^2) = \frac{1}{1 + g^2 f(k^2/W)}. \qquad (19.11)$$

This d is just the 'screening' or 'shielding' factor described in (10.35) and (10.37). Since $f(0) = \infty$ and $f(\infty) = 0$, we have that $d = 0$ for $k \to 0$ (corresponding to large distances, where we have total screening of the electron charge) and

$$d = 1 \quad \text{for} \quad k \to \infty, \tag{19.12}$$

corresponding to short distances where the charge is unscreened.

Note that all of these results are exact in the case of the high density electron gas which we are considering here.

In the succeeding sections of this chapter, we will pretend that we don't know the solution (19.11), and show how to solve (19.10) using the renormalization group.

19.3 Renormalization group for interaction propagators in the high-density electron gas

The renormalization group in the present case is defined as the set of transformations

$$D \to zD(\equiv \tilde{D}), \quad D_0 \to zD_0(\equiv \tilde{D}_0), \quad g^2 \to z^{-1}g^2(\equiv \tilde{g}^2) \tag{19.13}$$

where the multiplier z is a real number. Since z can vary continuously, the renormalization group is a continuous or 'Lie' group. Note that the quantities D, D_0, g^2 are the true 'physical' quantities, while $zD, zD_0, z^{-1}g^2$ are unphysical.

The Dyson equation (19.8) is invariant under the renormalization group transformation:

$$\tilde{D} = \tilde{D}_0 - \tilde{D}\tilde{g}^2[k^2 f(k^2/W)] \tilde{D}_0, \tag{19.14}$$

i.e., the unphysical or 'renormalized' quantities obey the same Dyson equation as the physical ones. The renormalization group transformation may also be stated in terms of the quantity $d(k)$ defined in (19.10)

$$d \to zd(\equiv \tilde{d}), \quad d_0 \to zd_0(\equiv \tilde{d}_0), \quad g^2 \to z^{-1}g^2(\equiv \tilde{g}^2) \tag{19.15}$$

which leaves the Dyson equation in form (19.10) invariant.

It is useful to define a quantity called the 'invariant charge', given by

$$g_{\text{inv}}^2 = g^2 d = \tilde{g}^2 \tilde{d}. \tag{19.16}$$

This is also invariant under the renormalization group transformation.

Next we express the multiplier z in terms of a parameter, λ, as follows:

$$z(\lambda/W, g^2) = \frac{1}{d(\lambda/W, g^2)}. \tag{19.17}$$

Note that this $z(\lambda/W, g^2)$ is an unknown function at the outset, since $d(k^2/W, g^2)$ is unknown (here, as mentioned above, we are just pretending that it is unknown). Hence the transformed quantities become:

$$\tilde{d}_\lambda \equiv \tilde{d}(k^2/\lambda, W/\lambda, \tilde{g}_\lambda^2) = z(\lambda/W, g^2)\, d(k^2/W, g^2)$$

$$= \frac{d(k^2/W, g^2)}{d(\lambda/W, g^2)} = \frac{d(k^2/\lambda \cdot \lambda/W, g^2)}{d(\lambda/W, g^2)}, \qquad (19.18)$$

$$\tilde{g}_\lambda^2 = \frac{g^2}{z(\lambda/W, g^2)} = g^2\, d(\lambda/W, g^2). \qquad (19.19)$$

Observe that $\tilde{d}_\lambda$ is a function of k^2/λ and W/λ, and that it is expressed as a function of $\tilde{g}_\lambda^2$ rather than g^2. This means that we should imagine that (19.19) is solved for g^2 as a function of $\tilde{g}_\lambda^2$, W/λ, then place the result in (19.18). This is what is done in practice (see after (19.34)). Note that $\tilde{d}$ obeys the so-called 'normalization condition', i.e., that $\tilde{d} = 1$ when $k^2 = \lambda$:

$$\tilde{d}(1, W/\lambda, \tilde{g}_\lambda^2) = 1. \qquad (19.20)$$

The identity transformation of the renormalization group is obtained when $z = 1$. Since, by (19.12), $d(\lambda/W, g^2) \to 1$ when $\lambda \to \infty$, we have

$$z = 1 \quad \text{for} \quad \lambda \to \infty, \qquad (19.21)$$

showing that $\lambda = \infty$ is the value of the parameter for the identity transformation.

19.4 Transforming from one transformed quantity to another: the functional equation of the renormalization group

Now let us see how to go from one transformed quantity $\tilde{d}_\lambda$ to another, $\tilde{d}_{\lambda'}$. We have by (19.18)

$$\tilde{d}(k^2/\lambda', W/\lambda', \tilde{g}_{\lambda'}^2) = \frac{d(k^2/W, g^2)}{d(\lambda'/W, g^2)}, \qquad (19.22a)$$

$$\tilde{d}(k^2/\lambda, W/\lambda, \tilde{g}_\lambda^2) = \frac{d(k^2/W, g^2)}{d(\lambda/W, g^2)}. \qquad (19.22b)$$

Dividing (19.22a) by (19.22b) yields

$$\tilde{d}(k^2/\lambda', W/\lambda', \tilde{g}_{\lambda'}^2) = \frac{d(\lambda/W, g^2)}{d(\lambda'/W, g^2)}\, \tilde{d}(k^2/\lambda, W/\lambda, \tilde{g}_\lambda^2)$$

$$= \tilde{d}(\lambda/\lambda', W/\lambda', \tilde{g}_{\lambda'}^2)\, \tilde{d}(k^2/\lambda, W/\lambda, \tilde{g}_\lambda^2) \qquad (19.23)$$

and

$$\tilde{g}_{\lambda'}^2 = \frac{1}{\tilde{d}(\lambda/\lambda', W/\lambda', \tilde{g}_{\lambda'}^2)}\, \tilde{g}_\lambda^2. \qquad (19.24)$$

It is convenient here to relabel the various quantities in (19.23) and (19.24) as follows:

$$x \equiv k^2/\lambda', \quad y \equiv W/\lambda', \quad \tilde{g}_{\lambda'}^2 \equiv J^2, \quad \lambda/\lambda' \equiv t. \tag{19.25}$$

Then (19.24) becomes:

$$\tilde{g}_\lambda^2 = J^2 \tilde{d}(t, y, J^2) \tag{19.26}$$

and (19.23) is

$$\tilde{d}(x, y, J^2) = \tilde{d}(t, y, J^2) \tilde{d}(x/t, y/t, J^2 \tilde{d}(t, y, J^2)). \tag{19.27}$$

Equation (19.27) is the functional equation of the renormalization group in this particular case. It has the same form as (11.16) in Bonch-Bruevich (1962) and as (13) in Fowler and Zawadowski (1971), so actually it is quite general. However, it should be carefully noted that these authors use the single symbol d or J to denote *both* the physical quantity and the transformed, multiplicatively renormalized quantity. To avoid the profound confusion this can lead to, we have followed Wilson (1970) and used two different symbols, d and $\tilde{d}$ (or g^2 and $\tilde{g}^2$). (Wilson uses d_c for the physical and d for the transformed quantities.) In particular, it should be noted that

$$\tilde{d}(k^2/\lambda, W/\lambda, \tilde{g}_\lambda^2) \neq d(k^2/\lambda, W/\lambda, \tilde{g}_\lambda^2) \tag{19.27'}$$

i.e., the transformed d is not simply the physical d with the variables k, W, g^2 replaced by k^2/λ, W/λ, $\tilde{g}_\lambda^2$.

19.5 Lie equation for the renormalization group

We now cast the functional equation (19.27) into differential form. Differentiating both sides with respect to x (let $s \equiv x/t$):

$$\frac{\partial \tilde{d}(x, y, J^2)}{\partial x} = \tilde{d}(t, y, J^2) \left[\frac{\partial \tilde{d}(s, y/t, J^2 \tilde{d}(t, y, J^2))}{\partial s} \right] \frac{1}{t}$$

and setting $t = x$ on both sides (so $s = 1$) to get an equation in the single variable, x, yields the *Lie equation*:

$$\frac{\partial \tilde{d}(x, y, J^2)}{\partial x} = \frac{\tilde{d}(x, y, J^2)}{x} \left[\frac{\partial \tilde{d}(s, y/x, J^2 \tilde{d}(x, y, J^2))}{\partial s} \right]_{s=1}. \tag{19.28}$$

The bracketed quantity on the right is called the '*infinitesimal generator*'. Equation (19.28) may be written in the alternative form:

$$\frac{\partial \ln \tilde{d}(x, y, J^2)}{\partial x} = \frac{1}{x} \left[\frac{\partial \ln \tilde{d}(s, y/x, J^2 \tilde{d}(x, y, J^2))}{\partial s} \right]_{s=1}, \tag{19.28'}$$

where we have used the fact that $\tilde{d}(s = 1, \ldots) = 1$.

Note that (19.28) is an equation for the transformed, unphysical function, $\tilde{d}$ as a function of x. Once we have found this function, the physical d can be found from (19.21) together with (19.19):

$$\tilde{d}_{\lambda \to \infty} = d(k^2/W, g^2); \quad \tilde{g}^2_{\lambda \to \infty} = g^2. \tag{19.29}$$

That is, the transformed, unphysical $\tilde{d}_\lambda$, $\tilde{g}^2_\lambda$ go into the physical d, g^2 when $\lambda \to \infty$.

Another way of writing the Lie equation is in terms of the invariant charge, (19.16). Multiplying both sides of (19.28) by $\tilde{J}^2$ and noting that by (19.16) and (19.25) we have

$$g^2_{\text{inv}} = \tilde{g}^2_\lambda \tilde{d}_\lambda = \tilde{g}^2_{\lambda'} \tilde{d}_{\lambda'} \equiv \tilde{g}^2_{\lambda'} \tilde{d}(k^2/\lambda', W/\lambda', \tilde{g}^2_{\lambda'}) = \tilde{J}^2 \tilde{d}(x, y, \tilde{J}^2),$$

which yields:

$$\frac{\partial g^2_{\text{inv}}}{\partial x} = \frac{g^2_{\text{inv}}}{x} \left[\frac{\partial \tilde{d}(s, y/x, g^2_{\text{inv}})}{\partial s} \right]_{s=1}, \tag{19.30}$$

which is an equation for the invariant charge.

19.6 Solution of the Lie equation

The usual procedure in solving the Lie equation is to make some approximation for $\tilde{d}$ in the 'generator' on the right-hand side of the equation, then solve the equation subject to the boundary conditions (19.20, 21) to obtain a better approximation for $\tilde{d}$. In the present case, we will use a crude first-order perturbation approximation, valid only for $k \gg k_F$, for d in the generator. When the Lie equation is solved, we find that the result is valid for *all k* values, and is in fact exact (due to the simplicity of our example).

Let us approximate the effective interaction by the first-order expression

$$\text{〰〰〰〰} \approx \text{〜〜〜} + \text{〜〜〰(}\bigcirc\text{)〜〰〜}, \tag{19.31}$$

which corresponds to setting $d \approx 1$ on the right side of the Dyson equation (19.10) yielding:

$$d(k^2/W, g^2) \approx 1 - g^2 f(k^2/W). \tag{19.32}$$

This is valid when $g^2 f \ll 1$, which means that k^2/W cannot be too close to 0, since $f(0) = \infty$. The corresponding approximate $\tilde{d}_\lambda$, $\tilde{g}^2_\lambda$ may be found by substituting (19.32) in (19.18, 19):

$$\tilde{d}_\lambda \equiv \tilde{d}(k^2/\lambda, W/\lambda, \tilde{g}^2_\lambda) \approx \frac{1 - g^2 f(k^2/W)}{1 - g^2 f(\lambda/W)} \tag{19.33}$$

$$\tilde{g}_\lambda^2 \approx g^2[1-g^2 f(\lambda/W)]. \tag{19.34}$$

To get $\tilde{d}$ expressed in terms of $\tilde{g}_\lambda^2$ rather than g^2, we solve (19.34) for g^2 in terms of $\tilde{g}_\lambda^2$, and substitute the result in (19.33). Retaining only the lowest-order terms in $\tilde{g}_\lambda^2$ in the result, we find for our approximate $\tilde{d}$:

$$\tilde{d}(k^2/\lambda, W/\lambda, \tilde{g}_\lambda^2) \approx 1 - \tilde{g}_\lambda^2[f(k^2/W) - f(\lambda/W)], \tag{19.35}$$

or, using (19.25)

$$\tilde{d}(x, y, \tilde{J}^2) \approx 1 - \tilde{J}^2[f(x/y) - f(1/y)]. \tag{19.36}$$

We now calculate the generator on the right of the Lie equation (19.28). Substituting $x \to s$, $y \to y/x$, etc., in (19.36) yields

$$\tilde{d}(s, y/x, \tilde{J}^2 \tilde{d}(x, y, \tilde{J}^2) \approx 1 - \tilde{J}^2 \tilde{d}(x, y, \tilde{J}^2) \left[f\left(\frac{sx}{y}\right) - f\left(\frac{x}{y}\right) \right]. \tag{19.37}$$

Note that this is an expansion in terms of the invariant charge $\tilde{g}_{inv}^2 = \tilde{J}^2 \tilde{d}$ (see (19.16)). Taking the derivative:

$$\left[\frac{\partial \tilde{d}}{\partial s} \right]_{s=1} = -\tilde{J}^2 \tilde{d}(x, y, \tilde{J}^2) \cdot \frac{x}{y} \cdot \frac{df(x/y)}{d(x/y)}. \tag{19.38}$$

Substituting this into (19.28) gives

$$\frac{\partial \tilde{d}(x, y, \tilde{J}^2)}{\partial x} = -[\tilde{d}(x, y, \tilde{J}^2)]^2 \tilde{J}^2 \frac{\partial f(x/y)}{\partial x}, \tag{19.39}$$

which is easily solved to yield

$$\tilde{d}(x, y, \tilde{J}^2) = \frac{1}{B + \tilde{J}^2 f(x/y)}. \tag{19.40}$$

Using the normalization condition (19.20):

$$\tilde{d}(1, y, \tilde{J}^2) = 1 = \frac{1}{B + \tilde{J}^2 f(1/y)}, \quad \text{or} \quad B = 1 - \tilde{J}^2 f(1/y). \tag{19.41}$$

Hence

$$\tilde{d}(x, y, \tilde{J}^2) = \frac{1}{1 + \tilde{J}^2[f(x/y) - f(1/y)]}, \tag{19.42}$$

or, using (19.25)

$$\tilde{d}(k^2/\lambda, W/\lambda, \tilde{g}_\lambda^2) = \frac{1}{1 + \tilde{g}_\lambda^2[f(k^2/W) - f(\lambda/W)]}. \tag{19.42'}$$

To get the physical d from the solution for the transformed $\tilde{d}$ in (19.43), we simply use (19.29), which yields (note that $f(\lambda/W) \to 0$ when $\lambda \to \infty$):

$$\tilde{g}^2_{\lambda \to \infty} = g^2, \quad \tilde{d}(\lambda \to \infty) = \tilde{d}(k^2/W, g^2) = \frac{1}{1 + g^2 f(k^2/W)}, \quad (19.43)$$

which is the exact result (19.11)!

In order to see why it is that we can get (in this super-simple case) an exact result starting from the crude first-order approximation (19.32) let us see what happens if we start with the second-order approximation:

$$d(k^2/W, g^2) \approx 1 - g^2 f(k^2/W) + [g^2 f(k^2/W)]^2. \quad (19.44)$$

Going through the same procedure as in the first-order case, yields, instead of (19.37), the expression

$$\tilde{d}(s, y/x, \tilde{J}^2 \tilde{d}(x, y, \tilde{J}^2)) \approx 1 - \tilde{J}^2 \tilde{d}(x, y, \tilde{J}^2)[f(sx/y) - f(x/y)] + $$
$$+ \{\tilde{J}^2 \tilde{d}(x, y, \tilde{J}^2)[f(sx/y) - f(x/y)]\}^2. \quad (19.45)$$

Taking the derivative, we find

$$\left[\frac{\partial \tilde{d}}{\partial s} \right]_{s=1} = -\tilde{J}^2 \tilde{d}(x, y, \tilde{J}^2) \cdot \frac{x}{y} \cdot \frac{df(x/y)}{d(x/y)} + $$
$$+ 2[\tilde{J}^2 d(x, y, \tilde{J}^2)]^2 [f(sx/y) - f(x/y)]_{s=1} \cdot \frac{x}{y} \cdot \frac{df(x/y)}{d(x/y)}. \quad (19.46)$$

The second-order term is evidently zero, so we are left with just result (19.38). In a similar way, all higher-order terms yield zero when we calculate the derivative, so the first-order result (19.38) is exact.

Further reading

Bonch-Bruevich (1962).
Wilson (1970).
Fowler and Zawadowski (1971).
Zawadowski (1973).
Bogoliubov and Shirkov (1959).

Exercises

19.1 Verify (19.45).

Appendices

Appendix $\mathscr{A}$

Finding Fictitious Particles with the Canonical Transformation

$\mathscr{A}.1$ Canonical transformation method of solving the many-body problem

In this appendix we will pin down some of our qualitative statements in chapter 0 with the aid of the canonical transformation technique. The idea is to transform the equations for the system to a new set of coordinates such that the interaction term becomes small. In other words, we transform from the coordinates of *strongly* interacting *real* particles, to coordinates of *weakly* interacting *fictitious* particles, i.e., quasi particles and collective excitations. Although the transformation method is not as systematic as the propagator technique described in chapter 0, it nevertheless gives excellent insight into the nature of many-body systems, so it will be presented briefly here.

Let us first express quantitatively our remarks about a system of N non-interacting particles. Suppose that these particles have masses $m_1, m_2, \ldots, m_N$, and that they are placed in a time-independent external force field $\mathbf{F}(\mathbf{r})$, with associated potential $V(\mathbf{r})$. Our problem is to find out how they behave. In the classical case, the system is described by the N independent equations of motion

$$\mathbf{F}(\mathbf{r}_i) = m_i \frac{d^2 \mathbf{r}_i}{dt^2}, \quad i = 1, 2, \ldots, N, \qquad (\mathscr{A}.1)$$

with solutions of the form $\mathbf{r}_i = \mathbf{r}_i(t)$. In the quantum mechanical case, this is replaced by the Schrödinger equation

$$H\Psi = E\Psi, \quad \text{with} \quad H = \sum_{i=1}^{N} H_i, \quad H_i = \frac{p_i^2}{2m_i} + V(\mathbf{r}_i), \qquad (\mathscr{A}.2)$$

where H, Ψ, E are the total system Hamiltonian, wave function, and energy, respectively, and H_i, $\mathbf{p}_i$, $V(\mathbf{r}_i)$ are the single particle Hamiltonian, momentum, and potential energy. If Ψ is written in the product form (neglect symmetry requirements for simplicity):

$$\Psi(\mathbf{r}_1 \ldots \mathbf{r}_N) = \prod_{i=1}^{N} \phi_{k_i}(\mathbf{r}_i) \qquad (\mathscr{A}.3)$$

and substituted into ($\mathscr{A}$.2), it is found that ($\mathscr{A}$.2) separates into the N single-particle Schrödinger equations

$$H_i \phi_{k_i}(\mathbf{r}_i) = E_{k_i} \phi_{k_i}(\mathbf{r}_i), \quad i = 1, \ldots, N \qquad (\mathscr{A}.4)$$

and the total energy is just the sum of the single-particle energies:

$$E = \sum_i E_{k_i}. \qquad (\mathscr{A}.5)$$

Thus, in both the classical and quantum mechanical non-interacting systems we find N independent equations, which means that we have just N one-body problems.

In the real many-body problem, we have seen that the particles interact with one another. This means that in the classical case it is necessary to solve the N coupled equations

$$\mathbf{F}(\mathbf{r}_i) + \sum_{j=1}^{N} \mathbf{F}(\mathbf{r}_i, \mathbf{r}_j) = m_i \frac{d^2 \mathbf{r}_i}{dt^2}, \quad i = 1, 2, \ldots, N, \qquad (\mathscr{A}.6)$$

where $\mathbf{F}(\mathbf{r}_i, \mathbf{r}_j)$ is the interaction force between two particles. In the quantum case, we find a non-separable Schrödinger equation:

$$\left\{ \sum_{i=1}^{N} \left[\frac{p_i^2}{2m_i} + V(\mathbf{r}_i) \right] + \tfrac{1}{2} \sum_{\substack{i,j=1 \\ (i \neq j)}}^{N} V(\mathbf{r}_i, \mathbf{r}_j, \mathbf{p}_i, \mathbf{p}_j) \right\} \Psi(\mathbf{r}_1 \ldots \mathbf{r}_N) = E\Psi(\mathbf{r}_1 \ldots \mathbf{r}_N)$$
$$(\mathscr{A}.7)$$

where $V(\mathbf{r}_i, \mathbf{r}_j, \mathbf{p}_i, \mathbf{p}_j)$ is the two-particle interaction potential, considered to be momentum-dependent, for the sake of generality.

It will be convenient in what follows to make a distinction between 'weak' and 'strong' interactions. We will call the interaction term in ($\mathscr{A}$.7) 'weak' or 'small' if it causes just a small perturbation to the solutions in the non-interacting case, i.e., if it can be handled by ordinary finite-order perturbation theory. If it cannot be handled this way, the interaction will be referred to as 'strong' or 'large'. Most of the interactions we have to deal with are of the strong type. For example, consider the Coulomb interaction between two electrons in a metal. This has the form

$$V(\mathbf{r}_i, \mathbf{r}_j) = \frac{e^2}{|\mathbf{r}_i - \mathbf{r}_j|}. \qquad (\mathscr{A}.8)$$

If the ground state energy of the system is calculated using this as the perturbation, we obtain the result

$$E_0 = E_0^{(0)} + E_0^{(1)} + \infty + \infty + \infty + \ldots,$$

i.e., infinity to all orders of perturbation theory above the first! (It should be noted that there are some interactions which are small in magnitude but which cannot be treated as small perturbations, such as the small attraction between electrons which gives rise to superconductivity. According to the above scheme, these will also be classified as 'strong'.)

Since visualizing the motion of a strongly interacting N particle system is roughly like trying to watch a basketball game played with N balls, the early attacks on the many-body problem tried to reduce it to a one-body problem by getting rid of the interaction. This could be done either by pretending it didn't exist, or more legally, by transforming it away. In the classical case, such a transformation implies that we change to a new set of co-ordinates in which the equations of motion ($\mathscr{A}$.6) become approximately decoupled. For instance, in simple situations it might be possible to define new co-ordinates R_k given by

$$\mathbf{r}_i = \mathbf{r}_i(\mathbf{R}_1, \mathbf{R}_2, \ldots, \mathbf{R}_k, \ldots, \mathbf{R}_N) \qquad (\mathscr{A}.9)$$

which, when introduced into ($\mathscr{A}$.6) change it into:

$$f(\mathbf{R}_k) = M_k \frac{d^2 \mathbf{R}_k}{dt^2} + \underbrace{\sum_l h(\mathbf{R}_k, \mathbf{R}_l)}_{small}. \qquad (\mathscr{A}.10)$$

The M_k are referred to as 'effective masses'. Equation ($\mathscr{A}$.10) constitutes a set of N nearly independent 'one-body' problems which can be more or less accurately solved, the remaining small interaction $h(\mathbf{R}_k, \mathbf{R}_l)$ being treated by perturbation methods. The 'bodies' here are of course not real but rather *fictitious* bodies.

The idea of transforming a system of interacting real particles into a system of approximately non-interacting fictitious bodies can be easily illustrated by means of two interacting particles in a gravitational field. Let the masses of the two particles be m_1, m_2, positions $\mathbf{r}_1$, $\mathbf{r}_2$, gravitational force $\mathbf{F}_1 = m_1\mathbf{g}$, $\mathbf{F}_2 = m_2\mathbf{g}$, and interaction force $\mathbf{F}(\mathbf{r}_1 - \mathbf{r}_2)$. These obey the coupled equations of motion

$$\begin{aligned} \mathbf{F}_1 + \mathbf{F}(\mathbf{r}_1 - \mathbf{r}_2) &= m_1\ddot{\mathbf{r}}_1 \\ \mathbf{F}_2 - \mathbf{F}(\mathbf{r}_1 - \mathbf{r}_2) &= m_2\ddot{\mathbf{r}}_2. \end{aligned} \qquad (\mathscr{A}.11)$$

The problem may be reduced to two independent one-body problems by making the transformation to new co-ordinates

$$\begin{aligned} (m_1 + m_2)\mathbf{R} &= m_1\mathbf{r}_1 + m_2\mathbf{r}_2 \quad \text{(centre of mass co-ordinate)} \\ \mathbf{r} &= \mathbf{r}_1 - \mathbf{r}_2 \qquad \text{(relative co-ordinate)}. \end{aligned} \qquad (\mathscr{A}.12)$$

Adding the two equations yields

$$\mathbf{F}_{\text{total}} = (m_1 + m_2)\ddot{\mathbf{R}}, \qquad (\mathscr{A}.13)$$

while multiplying by m_2 and m_1 and subtracting gives

$$\mathbf{F}(\mathbf{r}) = \frac{m_1 m_2}{m_1 + m_2}\ddot{\mathbf{r}}. \qquad (\mathscr{A}.14)$$

Thus, we have two new independent equations: one for a fictitious particle-like body of effective mass $m_1 + m_2$ with position at mass centre $\mathbf{R}$ and the other for a fictitious particle-like body of effective mass $m_1 m_2/(m_1 + m_2)$ and position at the relative co-ordinate $\mathbf{r}$.

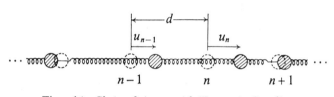

Fig. $\mathscr{A}$.1   *Chain of Atoms with Harmonic Coupling*

The new non-interacting fictitious bodies are not always 'particle-like' in their appearance, however. Consider for instance the well-known case of a one-dimensional chain of N atoms of mass m coupled by nearest-neighbour harmonic forces shown in Fig. $\mathscr{A}$.1. The classical equation of motion for such a system is

$$-2k_c u_n + k_c(u_{n+1} + u_{n-1}) = m\frac{d^2 u_n}{dt^2} \qquad (\mathscr{A}.15)$$

where u_n is the displacement of the nth atom from equilibrium and k_c is the spring constant. The transformation $(\mathscr{A}.9)$ here is just the simple Fourier transform:

$$u_n = \sum_k U_k e^{-iknd}. \qquad (\mathscr{A}.16)$$

Substituting this in $(\mathscr{A}.15)$ produces decoupled equations of form $(\mathscr{A}.10)$:

$$2k_c(\cos kd - 1)U_k = m\frac{d^2 U_k}{dt^2} + 0, \qquad (\mathscr{A}.17)$$

which have the solution

$$U_k = A_k e^{i\omega_k t}; \quad \omega_k = \sqrt{\left\{\frac{2k_c}{m}(1 - \cos kd)\right\}}. \qquad (\mathscr{A}.18)$$

If all the A_k's are zero except one, then ($\mathscr{A}$.16) becomes

$$u_n = A_k e^{-i(knd - \omega_k t)} \qquad (\mathscr{A}.19)$$

which is just the equation of a sinusoidal longitudinal vibration or 'sound wave' on the chain, of frequency ω_k and wavelength $\lambda = 2\pi/k$. Thus, the fictitious 'single bodies' here are wave-like instead of particle-like.

The quantum case may often be handled in a parallel fashion by transforming the Hamiltonian, H, canonically, so that its interaction term becomes small. Introducing the new co-ordinates $\mathbf{R}_k$ and momenta $\mathbf{P}_k$ by

$$\begin{aligned} \mathbf{r}_i &= \mathbf{r}_i(\mathbf{R}_1, \ldots, \mathbf{R}_k, \ldots, \mathbf{R}_N; \mathbf{P}_1, \ldots, \mathbf{P}_k, \ldots, \mathbf{P}_N) \\ \mathbf{p}_i &= \mathbf{p}_i(\mathbf{R}_1, \ldots, \mathbf{R}_N; \mathbf{P}_1, \ldots, \mathbf{P}_N), \end{aligned} \qquad (\mathscr{A}.20)$$

(Note: The definition of a canonical transformation is that the new co-ordinates, $\mathbf{R}_i$, $\mathbf{P}_i$, must obey the same canonical commutation relations which are obeyed by $\mathbf{r}_i$, $\mathbf{p}_i$, i.e.

$$[\mathbf{r}_i, \mathbf{p}_j] = i\hbar\delta_{ij}, \quad [\mathbf{r}_i, \mathbf{r}_j] = 0, \quad [\mathbf{p}_i, \mathbf{p}_j] = 0.) \qquad (\mathscr{A}.20')$$

yields the transformation

$$\begin{aligned} H &= \sum_i H(\mathbf{p}_i, \mathbf{r}_i) + \tfrac{1}{2}\sum_{i,j} V(\mathbf{r}_i, \mathbf{r}_j; \mathbf{p}_i, \mathbf{p}_j) \\ &\to H' = \sum_q H'(\mathbf{P}_q, \mathbf{R}_q) + \tfrac{1}{2}\sum_{q,m} \underbrace{V'(\mathbf{R}_q, \mathbf{R}_m; \mathbf{P}_q, \mathbf{P}_m)}_{small}. \end{aligned} \qquad (\mathscr{A}.21)$$

If this can be done, then the problem reduces to the set of N approximately independent one-body problems:

$$H'(\mathbf{P}_q, \mathbf{R}_q)\phi'_q = E'_q \phi'_q \qquad (\mathscr{A}.22)$$

with V' as small perturbation. These are just the equations for the independent fictitious bodies in the quantum case. The energy levels of the quantum system are the sum of the fictitious body energies:

$$E = \sum_q E'_q. \qquad (\mathscr{A}.23)$$

A simple example of transformation ($\mathscr{A}$.21) is provided by the system of two interacting particles treated quantum mechanically. The system obeys the equation (assume no gravitational field)

$$H\Psi(\mathbf{r}_1, \mathbf{r}_2) = E\Psi(\mathbf{r}_1, \mathbf{r}_2) \qquad (\mathscr{A}.24)$$

where

$$H = \frac{p_1^2}{2m_1} + \frac{p_2^2}{2m_2} + V(\mathbf{r}_1 - \mathbf{r}_2); \quad \mathbf{p} = \frac{\hbar}{i}\boldsymbol{\nabla}. \qquad (\mathscr{A}.25)$$

Making the same transformation ($\mathscr{A}$.12), to centre of mass and relative co-ordinates, we find

$$H = H(\mathbf{R}) + H(\mathbf{r})$$

where

$$H(\mathbf{R}) = \frac{p_R^2}{2(m_1+m_2)}, \quad H(\mathbf{r}) = \frac{p_r^2}{2\left(\dfrac{m_1 m_2}{m_1+m_2}\right)} + V(\mathbf{r}) \qquad (\mathscr{A}.26)$$

which yields the two Schrödinger equations

$$\begin{aligned} H(\mathbf{R})\,\phi_k(\mathbf{R}) &= E_k\,\phi_k(\mathbf{R}) \\ H(\mathbf{r})\,\gamma_l(\mathbf{r}) &= E_l\gamma_l\,(\mathbf{r}) \end{aligned} \qquad (\mathscr{A}.27)$$

showing two fictitious particle-like objects just like those in the corresponding classical case.

We can also demonstrate the transformation ($\mathscr{A}$.21) for the chain of atoms in Fig. $\mathscr{A}$.1 (for details, see Ziman (1962), p. 6 ff., or H. Højgaard Jensen (1964), p. 1). The Hamiltonian for the chain is (neglecting end effects)

$$\begin{aligned} H &= \sum_{n=1}^{N}\left(\frac{p_n^2}{2m}+k_c u_n^2\right)-\tfrac{1}{2}k_c\sum_{n=1}^{N}(u_n u_{n+1}+u_n u_{n-1}) \\ &= \sum_{n=1}^{N}\left(\frac{p_n^2}{2m}+k_c u_n^2\right)-k_c\sum_{n=1}^{N}u_n u_{n+1}. \end{aligned} \qquad (\mathscr{A}.28)$$

The first term describes a set of independent harmonic oscillators, one on each atom site, while the second is the interaction between each atom and its nearest neighbour. The transformation ($\mathscr{A}$.20) in this case turns out to be

$$u_n = \frac{1}{\sqrt{N}}\sum_q\left(\mathscr{U}_q\cos qnd-\frac{1}{m\omega_q}\mathscr{P}_q\sin qnd\right) \qquad (\mathscr{A}.29)$$

$$p_n = \frac{1}{\sqrt{N}}\sum_q(m\omega_q\mathscr{U}_q\sin qnd+\mathscr{P}_q\cos qnd) \qquad (\mathscr{A}.30)$$

where

$$\omega_q = \sqrt{\left\{\frac{2k_c}{m}(1-\cos qd)\right\}}. \qquad (\mathscr{A}.31)$$

The $\mathscr{U}_q$, $\mathscr{P}_q$ obey the canonical commutation rules

$$[\mathscr{U}_q,\mathscr{P}_{q'}] = i\hbar\delta_{qq'}, \quad [\mathscr{U}_q,\mathscr{U}_{q'}] = 0, \quad [\mathscr{P}_q,\mathscr{P}_{q'}] = 0, \qquad (\mathscr{A}.32)$$

which may be proved by substituting ($\mathscr{A}$.29, 30) into ($\mathscr{A}$.20′), with $\mathbf{r}_i \equiv u_i$. Placing ($\mathscr{A}$.29, 30) into ($\mathscr{A}$.28) and using ($\mathscr{A}$.32) we get, after some tedious calculation, the result:

$$H' = \sum_q\left[\frac{\mathscr{P}_q^2}{2m}+\frac{m\omega_q^2}{2}\mathscr{U}_q^2\right], \qquad (\mathscr{A}.33)$$

which is just the sum of a set of ordinary harmonic oscillator Hamiltonians of various frequencies. (Note: the usual transformation which is made is not ($\mathscr{A}$.29, 30) but rather

$$u_n = \frac{1}{\sqrt{N}} \sum_q U_q e^{-iqnd}, \quad p_n = \frac{1}{\sqrt{N}} \sum_q P_q e^{+iqnd}.$$

However, this leads to a Hamiltonian which is not completely decoupled since it mixes modes of wavenumber q and $-q$. Transformation ($\mathscr{A}$.29, 30) produces complete decoupling.) The Schrödinger equation and energies of the qth oscillator are the well-known

$$H_q' \phi_{n_q}' = \left(\frac{\mathscr{P}_q^2}{2m} + \frac{m\omega_q^2}{2} \mathscr{U}_q^2 \right) \phi_{n_q}' = E_{n_q}' \phi_{n_q}'$$

$$E_{n_q}' = \hbar\omega_q[n_q + \tfrac{1}{2}], \quad n_q = 0, 1, 2, 3, \ldots, \qquad (\mathscr{A}.34)$$

so that the total system energy and wave functions are

$$E = \sum_q \hbar\omega_q(n_q + \tfrac{1}{2})$$

$$\Psi = \phi_{n_{q_1}}' \phi_{n_{q_2}}' \cdots \phi_{n_{q_t}}' \cdots. \qquad (\mathscr{A}.35)$$

Thus the independent 'fictitious bodies' here are just quantized longitudinal sound waves.

It will be useful for us to recall the operator algebra method of deriving results ($\mathscr{A}$.34) (see, for example, Park (1964) p. 110 ff.). This is done by performing an additional transformation on H_q' in ($\mathscr{A}$.34) which involves the 'ladder' operators, b_q, $b_q^\dagger$ defined by:

$$b_q = \sqrt{\frac{1}{2m\hbar\omega_q}} \mathscr{P}_q - i\sqrt{\frac{m\omega_q}{2\hbar}} \mathscr{U}_q$$

$$b_q^\dagger = \sqrt{\frac{1}{2m\hbar\omega_q}} \mathscr{P}_q + i\sqrt{\frac{m\omega_q}{2\hbar}} \mathscr{U}_q. \qquad (\mathscr{A}.36)$$

From the commutation rules ($\mathscr{A}$.32), it is easily verified that the b_q's obey the rules:

$$(a)\ [b_q, b_{q'}^\dagger] = \delta_{qq'}, \quad (b)\ [b_q, b_{q'}] = 0, \quad (c)\ [b_q^\dagger, b_{q'}^\dagger] = 0. \qquad (\mathscr{A}.37)$$

Substituting the inverse of transformation ($\mathscr{A}$.36) into ($\mathscr{A}$.34) and using ($\mathscr{A}$.37a), yields the new Hamiltonian and Schrödinger equation

$$H_q'' = \hbar\omega_q(b_q^\dagger b_q + \tfrac{1}{2}), \quad H_q'' \phi_{n_q}'' = E_{n_q}' \phi_{n_q}''. \qquad (\mathscr{A}.38)$$

Using ($\mathscr{A}$.37, 38), it may be shown (see above reference) that if we let ψ_{0q} be the lowest-energy eigenfunction of H_q'', then all the other eigenfunctions are given by

$$\phi''_{n_q} \equiv \psi_{n_q} = \frac{1}{\sqrt{n_q!}} (b_q^\dagger)^{n_q} \psi_{0q}, \quad n_q = 0, 1, 2, \ldots. \qquad (\mathscr{A}.39a)$$

The corresponding energies are

$$E'_{n_q} = \hbar\omega_q(n_q + \tfrac{1}{2}), \qquad (\mathscr{A}.39b)$$

thus establishing ($\mathscr{A}.34$). It also follows that the b-operators have the following important properties when operating on the energy eigenfunctions:

$$
\begin{aligned}
(a)\ & b_q^\dagger \psi_{n_q} = \sqrt{n_q + 1}\, \psi_{n_q+1}, \\
(b)\ & b_q \psi_{n_q} = \sqrt{n_q}\, \psi_{n_q-1}, \\
(c)\ & b_q^\dagger b_q \psi_{n_q} = n_q \psi_{n_q}.
\end{aligned}
\qquad (\mathscr{A}.40)
$$

Observe that the additional transformation, ($\mathscr{A}.36$), changes H' in ($\mathscr{A}.33$) to

$$H'' = \sum_q \hbar\omega_q(b_q^\dagger b_q + \tfrac{1}{2}) = \underbrace{\tfrac{1}{2}\sum_q \hbar\omega_q}_{E_0} + \sum_q \hbar\omega_q b_q^\dagger b_q, \qquad (\mathscr{A}.41)$$

with associated Schrödinger equation

$$H'' \Psi_{n_{q_1}, n_{q_2}, \ldots, n_{q_i}, \ldots} = E \Psi_{n_{q_1}, n_{q_2}, \ldots, n_{q_i}, \ldots},$$

where

$$\Psi_{n_{q_1}, n_{q_2}, \ldots, n_{q_i}, \ldots} = \psi_{n_{q_1}} \psi_{n_{q_2}} \ldots \psi_{n_{q_i}} \ldots. \qquad (\mathscr{A}.42)$$

$\mathscr{A}.2$ Elementary excitations

The results in the last section show how the problem of a strongly coupled classical or quantum many-body system may often be solved by transforming to a set of approximately (or exactly, in the simple cases considered) independent fictitious bodies. There is a more modern (but equivalent) way of viewing these fictitious bodies which has the advantage of giving a unified picture of many-body systems. It is based on the concept of *elementary excitations*, and is the viewpoint which is taken throughout this book.

In order to understand what an elementary excitation is, and to see how it is related to the fictitious bodies, let us look at a specific case, the *phonon*. Recall first that light waves may be regarded either as quantized radiation oscillators or as consisting of particles ('quanta') called photons each of which has energy $\hbar\omega$. This suggests that it may be possible to look at sound waves in the same way. If we examine result ($\mathscr{A}.34$), we see that instead of regarding the sound wave of wavenumber q as one fictitious body (the harmonic oscillator) having quantized energy $E'_q = \hbar\omega_q(n_q + \tfrac{1}{2})$, we could alternatively regard it as a set of n_q quanta each having energy $\hbar\omega_q$, together with a ground state of energy $\frac{1}{2}\hbar\omega_q$. These quanta of the sound wave are called *phonons* and, like photons, they behave very much as particles (i.e., particles in the quantum-mechanical sense—they are not necessarily localized, although they can be—see Højgaard Jensen (1964)). It should be pointed out here that it is common to call a phonon a 'quantized sound wave', but that according to the above, this is not correct. For a given n_q there is only one quantized sound wave of wavenumber q (this is just the fictitious body of energy E_q'), but there are many, i.e., n_q, phonons of wavenumber q. Hence it is more proper to call the phonon a quantum or particle of sound.

The energy $\hbar\omega_q$ is evidently just the least unit of excitation energy above the zero-point energy $\frac{1}{2}\hbar\omega_q$. Since the phonon carries this least unit, it is referred to as an '*elementary excitation*'. The 'compound excitations' are then just excitations involving many phonons.

The phonon way of viewing the sound wave provides a new interpretation of the wave function and ladder operators, $b_q^\dagger$, b_q. Evidently, $\Psi_{n_{q_1}, n_{q_2}, \ldots, n_q, \ldots}$ in $(\mathscr{A}.42)$ describes a system with n_{q_1} phonons of wavenumber q_1, n_{q_2} of wavenumber q_2, etc. Therefore, by $(\mathscr{A}.40)$, $b_q^\dagger$ is an operator which creates a phonon of wavenumber q, while b_q destroys such a phonon, and $b_q^\dagger b_q$ is the 'number operator' for phonons of wavenumber q.

An important feature of phonons, in contrast to the quantized sound waves, is that the total system energy is equal to the sum of the energies of all the quantized sound waves, by $(\mathscr{A}.35)$, but it is not equal to the sum of the energies for all the phonons. That is, $(\mathscr{A}.23)$ does not hold true for phonons. This is obvious from $(\mathscr{A}.41)$ where we see that to get the total energy, we must add the ground state energy to that of the phonons. Thus $(\mathscr{A}.41)$ gives us a picture of any excited state of the system as being composed of a ground state plus a collection of independent phonons above the ground state.

This result for the phonon case turns out to be extremely general. It appears now that, in most many-body systems, it is possible to perform a transformation from the system of strongly interacting particles to a set of approximately independent elementary excitations above the ground state. Thus, we may write, using $(\mathscr{A}.21)$ and by analogy with $(\mathscr{A}.41)$, the transformation in the form:

$$H = \sum_i H(\mathbf{p}_i, \mathbf{r}_i) + \tfrac{1}{2}\sum_{ij} V(\mathbf{r}_i, \mathbf{r}_j; \mathbf{p}_i, \mathbf{p}_j)$$

$$\rightarrow H' = E_0 + \sum_q \epsilon_q' A_q^\dagger A_q + \underbrace{f(\ldots A_k \ldots A_k^\dagger \ldots)}_{\text{small}} \qquad (\mathscr{A}.43)$$

where E_0 is the ground state energy of the interacting system, ϵ_q' is the energy of the elementary excitation, $A_q^\dagger$, A_q, $A_q^\dagger A_q$ are the creation, destruction and number operators for the elementary excitations. (The ϵ_q' is often called the 'dispersion law' or 'excitation spectrum'.)

The small term in ($\mathscr{A}$.43) describes the interactions between elementary excitations. These give rise to a broadening, $\Delta\epsilon_q'$, of the energy levels, ϵ_q'. By the uncertainty principle, we have $\Delta E\Delta t \sim \hbar$, which means that the elementary excitations have a *lifetime*, $\tau_q \sim \hbar(\Delta\epsilon_q')^{-1}$. This lifetime must of course be reasonably long (i.e., level width $\hbar\tau_q^{-1}$ must be $\ll \epsilon_q'$) for the elementary excitations to really be considered well-defined and independent of one another.

Elementary excitations are often (see chapter 0) divided into two general types: '*quasi particles*' and '*collective excitations*', although this practice is not universal. (For example, some writers, like Ter Haar (1960) call all elementary excitations quasi particles.) Various examples of elementary excitations were discussed in chapter 0, such as quasi electrons, quasi nucleons, plasmons, phonons, and nuclear quanta.

It is important to notice that there may be more than one kind of elementary excitation in a given system. For example, in the electron gas in a metal, there are both quasi electrons and plasmons, while in a nucleus we have quasi nucleons and vibrational and rotational collective excitations existing simultaneously. Thus, in general, the right-hand side of ($\mathscr{A}$.43) may be written

$$H' = E_0 + \underbrace{\sum_q \epsilon_q' A_q^\dagger A_q}_{\text{quasi particles}} + \underbrace{\sum_k \epsilon_k'' B_k^\dagger B_k}_{\text{collective excitations}} + \text{small terms.} \qquad (\mathscr{A}.44)$$

Exercises

$\mathscr{A}$.1 Prove the commutation rule ($\mathscr{A}$.37a).

$\mathscr{A}$.2 Verify ($\mathscr{A}$.38).

$\mathscr{A}$.3 Calculate: $b_{q_2}^\dagger \psi_{3q_2}$.

$\mathscr{A}$.4 Prove ($\mathscr{A}$.40c) from ($\mathscr{A}$.40a, b).

$\mathscr{A}$.5 Calculate the following:

  (a) $b_{q_2}^\dagger \Psi_{0_{q_1}, 3_{q_2}, 0_{q_3}, 0, 0, \ldots}$

  (b) $b_{q_2} \Psi_{0_{q_1}, 0_{q_2}, 0, 0 \ldots}$

  (c) $b_{q_1}^\dagger b_{q_2} b_{q_1}^\dagger \Psi_{1_{q_1}, 5_{q_2}, 0_{q_3}, 0, 0 \ldots}$

$\mathscr{A}$.6 How many phonons are in the atomic chain described by the wave function: $\Psi_{0_{q_1}, 0_{q_2}, 2_{q_3}, 0_{q_4}, 3_{q_5}, 0, 0 \ldots}$

$\mathscr{A}$.7 What is the total energy of the system in Ex. $\mathscr{A}$.6?

Appendix A

Dirac Formalism

The three most popular formulations of quantum theory are Schrödinger's wave mechanics, Heisenberg's matrix mechanics, and Dirac's abstract vector space method. They are all equivalent, but Dirac's formulation has the advantage of being more compact and general than the first two. We shall show that the equations of the Dirac formalism bear the same relation to those of Schrödinger and Heisenberg, as vector equations like $\mathbf{A}+\mathbf{B}=\mathbf{C}$ bear to $A_x+B_x=C_x$, $A_y+B_y=C_y$, $A_z+B_z=C_z$ (i.e., the same vector equation written out in terms of its components in some orthogonal system of unit basis vectors). We shall work with a one-particle system; the generalization to the many-particle system is straightforward.

In the Schrödinger scheme the state of the system at time t is described by the *wave function*, and its complex conjugate

$$\psi(\mathbf{r},t), \quad \psi^*(\mathbf{r},t) \tag{A.1}$$

while dynamical variables are given by *differential operators*

$$\mathbf{r}, \mathbf{\nabla}_r, \quad \alpha = \alpha(\mathbf{r}, -i\mathbf{\nabla}_r); \text{ Example: Energy operator:}$$

$$H = -\frac{1}{2m}\nabla_r^2 + V(\mathbf{r}), \quad (\hbar = 1). \tag{A.2}$$

The wave function obeys the time-dependent Schrödinger equation

$$-iH\psi = \frac{\partial\psi}{\partial t}. \tag{A.3}$$

The observable values of any operator α are the eigenvalues a_i of

$$\alpha\phi_i(\mathbf{r}) = a_i\phi_i(\mathbf{r}), \tag{A.4}$$

where the ϕ_i are the corresponding eigenfunctions. They satisfy the ortho-normality relations:

$$\int \phi_i^*(\mathbf{r})\phi_j(\mathbf{r})d^3\mathbf{r} = \delta_{ij}. \tag{A.5}$$

The probability of observing the eigenvalue a_i is $|A_i(t)|^2$, where $A_i(t)$ is the coefficient of $\phi_i(\mathbf{r})$ in the following expansion of the wave function:

$$\psi(\mathbf{r},t) = \sum_i A_i(t)\phi_i(\mathbf{r}),$$

i.e.,

$$A_i(t) = \int \phi_i^*(\mathbf{r})\psi(\mathbf{r},t)d^3\mathbf{r}. \tag{A.6}$$

In the Heisenberg matrix method, ψ, ψ^* are replaced by the *column* and *row matrices*

$$\begin{pmatrix}\psi_1\\\psi_2\\\vdots\\\psi_k\\\vdots\end{pmatrix}, \quad (\psi_1^*,\psi_2^*,\ldots,\psi_k^*,\ldots). \tag{A.7}$$

These may be obtained from ψ, ψ^* by choosing any complete orthogonal set of functions, $\eta_k(\mathbf{r})$ (which are usually eigenfunctions of some Schrödinger operator, β); then we have

$$\psi_k = \int \eta_k^*(\mathbf{r}) \psi(\mathbf{r}, t) d^3\mathbf{r}. \tag{A.8}$$

Similarly, the dynamical variables are *square matrices*

$$\begin{pmatrix} \alpha_{11} & \alpha_{12} & \cdots & \cdots \\ \alpha_{21} & \alpha_{22} & \cdots & \cdots \\ \vdots & \vdots & & \\ \cdots & \cdots & \alpha_{kl} & \cdots \\ \vdots & \vdots & & \\ \cdots & \cdots & \cdots & \cdots \end{pmatrix} \tag{A.9}$$

where

$$\alpha_{kl} = \int \eta_k^*(\mathbf{r}) \, \alpha(\mathbf{r}, \nabla_r) \, \eta_l(\mathbf{r}) \, d^3\mathbf{r}. \tag{A.10}$$

The column matrices of (A.7) obey

$$-i \begin{pmatrix} H_{11} & H_{12} & \cdots \\ H_{21} & H_{22} & \cdots \\ \vdots & \vdots & \end{pmatrix} \begin{pmatrix} \psi_1 \\ \psi_2 \\ \vdots \end{pmatrix} = \frac{\partial}{\partial t} \begin{pmatrix} \psi_1 \\ \psi_2 \\ \vdots \end{pmatrix} \tag{A.11}$$

analogous to (A.3). The eigenvalue equation (A.4) becomes:

$$\begin{pmatrix} \alpha_{11} & \alpha_{12} & \cdots \\ \alpha_{21} & \alpha_{22} & \cdots \\ \vdots & \vdots & \end{pmatrix} \begin{pmatrix} \phi_1^i \\ \phi_2^i \\ \vdots \end{pmatrix} = a_i \begin{pmatrix} \phi_1^i \\ \phi_2^i \\ \vdots \end{pmatrix} \tag{A.12}$$

and the orthonormality relation (A.5) has the form

$$(\phi_1^{i*}, \phi_2^{i*}, \ldots) \begin{pmatrix} \phi_1^j \\ \phi_2^j \\ \vdots \end{pmatrix} = \delta_{ij}. \tag{A.13}$$

The probability of observing a_i is given again by $|A_i(t)|^2$ in (A.6) with row and column matrices substituted for ϕ_i and ψ, and the integral removed.

Since the Dirac abstract vector space description is essentially a generalization of vectors and tensors (dyadics) in ordinary space, let us review these first. Consider an ordinary two-dimensional plane. As is well known, we may construct vectors and tensors in this plane which may be combined in the following ways:

(a) Addition: $\mathbf{A} + \mathbf{B} = \mathbf{C}$ (a vector)

(b) Scalar product: $\mathbf{A} \cdot \mathbf{B} = |\mathbf{A}| \, |\mathbf{B}| \cos \theta_{AB} = c$

$\qquad\qquad\qquad\qquad = $ 'component of $\mathbf{A}$ on $\mathbf{B}$'

$\qquad\qquad\qquad\qquad = $ a number or 'scalar'.

(1) normalized vector, $\mathbf{A}$: $\mathbf{A} \cdot \mathbf{A} = |\mathbf{A}|^2 = 1$

(2) orthogonal vectors, $\mathbf{A}$, $\mathbf{B}$: $\mathbf{A} \cdot \mathbf{B} = 0$

(3) orthonormal set, $\mathbf{A}_i$: $\mathbf{A}_i \cdot \mathbf{A}_j = \delta_{ij}$

(c) Multiplication by tensor (dyadic) $\mathbf{D}$:

$$\mathbf{B} = \mathbf{D}\mathbf{A} \quad \text{(dyadic stretches and rotates vector).} \tag{A.14}$$

For purposes of calculation, it is convenient to choose an arbitrary pair of orthogonal and normal ('orthonormal') vectors as 'basis' vectors—call them $\mathbf{u}_1$ and $\mathbf{u}_2$—and express all other vectors and dyadics in terms of their components or 'representatives' in the chosen basis. When this is done, the vectors become column matrices and the dyadics square matrices, thus:

(a)
$$\mathbf{A} \equiv \begin{pmatrix} \mathbf{A} \cdot \mathbf{u}_1 \\ \mathbf{A} \cdot \mathbf{u}_2 \end{pmatrix} \equiv \begin{pmatrix} A_1 \\ A_2 \end{pmatrix},$$

(b)
$$\mathbf{D} \equiv \begin{pmatrix} \mathbf{u}_1 \cdot \mathbf{D}\mathbf{u}_1 & \mathbf{u}_1 \cdot \mathbf{D}\mathbf{u}_2 \\ \mathbf{u}_2 \cdot \mathbf{D}\mathbf{u}_1 & \mathbf{u}_2 \cdot \mathbf{D}\mathbf{u}_2 \end{pmatrix} \equiv \begin{pmatrix} D_{11} & D_{12} \\ D_{21} & D_{22} \end{pmatrix},$$

$$\tag{A.15}$$

and may be manipulated by the ordinary rules for matrix algebra. Substituting these in (A.14) produces

(a)
$$\begin{pmatrix} A_1 \\ A_2 \end{pmatrix} + \begin{pmatrix} B_1 \\ B_2 \end{pmatrix} = \begin{pmatrix} A_1 + B_1 \\ A_2 + B_2 \end{pmatrix} = \begin{pmatrix} C_1 \\ C_2 \end{pmatrix}$$

(b)
$$(A_1, A_2)\begin{pmatrix} B_1 \\ B_2 \end{pmatrix} = A_1 B_1 + A_2 B_2 = C \tag{A.16}$$

(c)
$$\begin{pmatrix} B_1 \\ B_2 \end{pmatrix} = \begin{pmatrix} D_{11} & D_{12} \\ D_{21} & D_{22} \end{pmatrix}\begin{pmatrix} A_1 \\ A_2 \end{pmatrix} = \begin{pmatrix} D_{11} A_1 + D_{12} A_2 \\ D_{21} A_1 + D_{22} A_2 \end{pmatrix}.$$

Of course, the actual numbers appearing in the matrices depend on which $\mathbf{u}_1, \mathbf{u}_2$ have been chosen as basis.

The primary differences between the Dirac abstract vector space and the ordinary one above are first, the Dirac space has an infinite number of dimensions and second, Dirac's vectors (called '*ket*' vectors, $|A\rangle$) are complex so they each have a complex conjugate '*bra*' vector, $\langle A| = \overline{|A\rangle}$. (More precisely, the bra is the 'conjugate imaginary' of the ket. See Dirac (1947), p. 20.) Parallel to (A.14) these Dirac vectors may be combined as follows:

(a) Addition: $|A\rangle + |B\rangle = |C\rangle$

(b) Scalar product of $|A\rangle$ and $|B\rangle$: $\langle A|B\rangle = c$ (a number)

(1) normalized: $\langle A|A\rangle = 1$

(2) orthogonal: $\langle A|B\rangle = 0$

(3) orthonormal set, A_i: $\langle A_i|A_j\rangle = \delta_{ij}$

(c) Multiplication by linear operator α:

$$|B\rangle = \alpha|A\rangle. \tag{A.17}$$

In addition, Dirac algebra has three new concepts not found in ordinary vector algebra:

(d) Operator product of $|A\rangle$ and $|B\rangle$:

$$\beta = |A\rangle\langle B| = \text{a linear operator},$$

because $\qquad \beta|R\rangle = |A\rangle\langle B|R\rangle = \text{vector}.$

(e) Multiplication by unit operator:

If $|\eta_i\rangle$ is an orthonormal set, then the operator

$$\sum_i |\eta_i\rangle\langle\eta_i| \quad \left(\text{or} \int |\eta_i\rangle\langle\eta_i|\, d\eta_i\right)$$

is a unit operator, i.e., it $= 1$. This is because

$$\sum_i |\eta_i\rangle\langle\eta_i|\eta_j\rangle = \sum_i |\eta_i\rangle\delta_{ij} = |\eta_j\rangle$$

so that:

$$\sum_i |\eta_i\rangle\langle\eta_i| = 1.$$

Thus we have:

$$\langle A|B\rangle = \sum \langle A|\eta_i\rangle\langle\eta_i|B\rangle.$$

(f) Complex conjugation (c.c.): α may be written:

$$\alpha = \text{Re}\,\alpha + i\,\text{Im}\,\alpha, \quad \text{so that} \quad \bar{\alpha} = \text{Re}\,\alpha - i\,\text{Im}\,\alpha.$$

Further, $\overline{|A\rangle} = \langle A|$. It may then be shown that to get the c.c. of any product of bras, kets and operators, take c.c. of each factor and reverse the order. Thus, for example:

$$\overline{\langle B|A\rangle} = \langle A|B\rangle, \quad \overline{\langle P|\alpha} = \bar{\alpha}|P\rangle,$$

$$\overline{\langle P|\alpha|B\rangle} = \langle B|\bar{\alpha}|P\rangle, \quad \overline{\alpha\beta} = \bar{\beta}\bar{\alpha}, \quad \text{etc.} \tag{A.18}$$

Using this new algebra, Dirac builds up quantum mechanics precisely parallel to the Schrödinger and Heisenberg schemes. The states of a system are described by *abstract vectors* and their complex conjugates

$$|\psi(t)\rangle, \quad \langle\psi(t)|, \tag{A.19}$$

and dynamical variables by *abstract linear operators*

$$\mathbf{r}, \mathbf{p}, \alpha(\mathbf{r}, \mathbf{p}); \quad \text{Example:} \quad H = \frac{p^2}{2m} + V(\mathbf{r}). \tag{A.20}$$

The state vector obeys:

$$-iH|\psi\rangle = \frac{\partial|\psi\rangle}{\partial t}. \tag{A.21}$$

The observable values of α are the eigenvalues a_i of

$$\alpha|\phi_i\rangle = a_i|\phi_i\rangle \tag{A.22}$$

where the eigenvectors ϕ_i satisfy the orthonormality relation

$$\langle\phi_i|\phi_j\rangle = \delta_{ij}. \tag{A.23}$$

For example, the position operator, $\mathbf{r}$, has the eigenvalue equation

$$\mathbf{r}|\mathbf{r}_i\rangle = \mathbf{r}_i|\mathbf{r}_i\rangle$$

with

$$\langle\mathbf{r}_i|\mathbf{r}_j\rangle = \delta(\mathbf{r}_i - \mathbf{r}_j), \tag{A.24}$$

(note that $\mathbf{r}_i$ is a continuous eigenvalue). The probability of observing eigenvalue a_i is $|A_i(t)|^2$ where

$$|\psi(t)\rangle = \sum_i A_i(t)|\phi_i\rangle, \quad A_i(t) = \langle\phi_i|\psi(t)\rangle. \tag{A.25}$$

The relation of Dirac's abstract vector formulation to the Schrödinger and Heisenberg formulations is precisely the same as the relation of the ordinary vector algebra (A.14) to vector algebra in terms of a basis (A.16). Thus, the Schrödinger wave mechanics is just the Dirac abstract vector formulation expressed in position basis. This means that the basis vectors (which correspond to the $\mathbf{u}_1, \mathbf{u}_2$ in (A.15)) are the eigenvectors $|\mathbf{r}_i\rangle$ of the position operator $\mathbf{r}$ as given in (A.24), and that all other vectors and operators are expressed in terms of their components along the $|\mathbf{r}_i\rangle$. This is often called the 'position representation'. Then it can be shown that the Schrödinger wave function is just the components of the Dirac state vector $|\psi(t)\rangle$ in the position basis:

$$\psi(\mathbf{r}, t) = \langle\mathbf{r}|\psi(t)\rangle, \quad \psi^*(\mathbf{r}, t) = \langle\psi(t)|\mathbf{r}\rangle \tag{A.26}$$

analogous to (A.15a).

The Schrödinger operators are given by (see Merzbacher (1970), p. 326 ff., or Dirac (1947), p. 90)

$$\langle\mathbf{r}'|\alpha(\mathbf{r}, \mathbf{p})|\mathbf{r}\rangle = \alpha(\mathbf{r}, -i\nabla_r)\,\delta(\mathbf{r}' - \mathbf{r}) \tag{A.27}$$

analogous to (A.15b). For example, the Dirac eigenvalue equation (A.22) can be transcribed into the Schrödinger one (A.4) by multiplying on the left by $\langle\mathbf{r}|$ and inserting the unit operator

$$1 = \int |\mathbf{r}_k\rangle\langle\mathbf{r}_k|\,d^3\mathbf{r}_k, \tag{A.28}$$

as follows:

$$\int d^3\mathbf{r}_k \langle \mathbf{r}|\alpha|\mathbf{r}_k\rangle \langle \mathbf{r}_k|\phi_i\rangle = a_i\langle \mathbf{r}|\phi_i\rangle$$

or

$$\int d^3\mathbf{r}_k \,\alpha(\mathbf{r}_k, -i\nabla_{r_k})\,\delta(\mathbf{r}_k-\mathbf{r})\,\langle \mathbf{r}_k|\phi_i\rangle = a_i\langle \mathbf{r}|\phi_i\rangle$$

so

$$\alpha(\mathbf{r}, -i\nabla_r)\,\phi_i(\mathbf{r}) = a_i\phi_i(\mathbf{r}). \tag{A.29}$$

Similarly, to get the Heisenberg matrix method we just express the Dirac vectors and operators in an arbitrary basis—call it $|\eta_i\rangle$. Then (using (A.28)) the component ψ_k in (A.7) is

$$\psi_k = \langle \eta_k|\psi\rangle = \int d^3\mathbf{r}\langle \eta_k|\mathbf{r}\rangle \langle \mathbf{r}|\psi\rangle$$

$$= \int d^3\mathbf{r}\eta_k^*(\mathbf{r})\,\psi(\mathbf{r}) \tag{A.30}$$

which is just (A.8), while α_{kl} in (A.9) is

$$\alpha_{kl} = \langle \eta_k|\alpha(\mathbf{r},\mathbf{p})|\eta_l\rangle = \int d^3\mathbf{r}\,d^3\mathbf{r}'\langle \eta_k|\mathbf{r}'\rangle \langle \mathbf{r}'|\alpha(\mathbf{r},\mathbf{p})|\mathbf{r}\rangle \langle \mathbf{r}|\eta_l\rangle$$

$$= \int d^3\mathbf{r}\,d^3\mathbf{r}'\,\eta_k^*(\mathbf{r}')\,\alpha(\mathbf{r}, -i\nabla_r)\,\delta(\mathbf{r}'-\mathbf{r})\,\eta_l(\mathbf{r})$$

$$= \int \eta_k^*(\mathbf{r})\,\alpha(\mathbf{r}, -i\nabla_r)\,\eta_l(\mathbf{r})\,d^3\mathbf{r}, \tag{A.31}$$

which is just (A.10). Equation (A.31) shows that $\langle \eta_k|\alpha|\eta_l\rangle$ may be used as a shorthand way of writing the ordinary Heisenberg matrix elements.

Appendix B

The Time Development Operator, $U(t)$

This appendix is the first step of the labyrinthine argument (appendices B → G) necessary to derive the diagrammatic expansion of the vacuum amplitude and the propagator from the time-dependent Schrödinger equation. In order to minimize the risk of the reader getting appendixitis on the way through, we start with a diagram of the labyrinth. The letters in each box refer to the various appendices. (Note that appendix C is not included, although certain results in it will be used.)

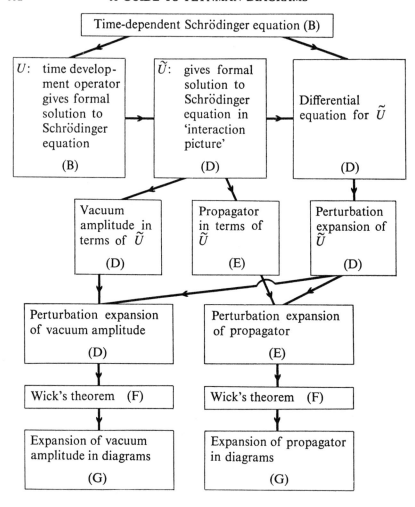

Let us go on now to the time development operator, $U(t-t_0)$. Suppose that $\Psi(t_0)$, $\Psi(t)$ are the wave functions at initial time t_0 and later time t. Then the time development operator is defined as the operator which produces $\Psi(t)$ when it operates on $\Psi(t_0)$:

$$\Psi(t) = U(t-t_0)\,\Psi(t_0)$$

$$(\text{or } |\Psi(t)\rangle = U(t-t_0)|\Psi(t_0)\rangle \text{ in Dirac notation}). \qquad (B.1)$$

We can find a formal expression for $U(t-t_0)$ by using the fact that Ψ obeys the time-dependent Schrödinger equation

$$-iH\Psi(t) = \frac{\partial}{\partial t}\Psi(t)$$

$$\text{(or } -iH|\Psi(t)\rangle = \frac{\partial}{\partial t}|\Psi(t)\rangle \text{ in Dirac notation)} \qquad \text{(B.2)}$$

where it is assumed that H is time independent. It follows that

$$U(t-t_0) = e^{-iH(t-t_0)} \equiv 1 - iH(t-t_0) + \frac{i^2}{2!}H^2(t-t_0)^2 + \cdots, \qquad \text{(B.3)}$$

where the expansion gives the meaning of the exponential operator. The correctness of (B.3) may be established by substituting it in (B.1) and differentiating:

$$\frac{\partial}{\partial t}|\Psi(t)\rangle = \frac{\partial}{\partial t}U(t-t_0)|\Psi(t_0)\rangle$$

$$= \frac{\partial}{\partial t}e^{-iH(t-t_0)}|\Psi(t_0)\rangle$$

$$= -iH|\Psi(t)\rangle \qquad \text{(B.4)}$$

in agreement with (B.2).

As an example of how to work with $U(t-t_0)$, suppose at t_0, which we take equal to zero, the system is in state Ψ_n, which is an eigenstate of H so that

$$H|\Psi_n\rangle = E_n|\Psi_n\rangle. \qquad \text{(B.5)}$$

We ask for the value of

$$|\Psi(t)\rangle = e^{-iHt}|\Psi_n\rangle. \qquad \text{(B.6)}$$

This is just

$$|\Psi(t)\rangle = \left[1 - iHt + \frac{i^2}{2!}H^2t^2 + \cdots\right]|\Psi_n\rangle$$

$$= \left[1 - iE_nt + \frac{i^2}{2!}E_n^2t^2 + \cdots\right]|\Psi_n\rangle$$

$$= e^{-iE_nt}|\Psi_n\rangle = |\Psi_n\rangle e^{-iE_nt}. \qquad \text{(B.7)}$$

We shall also have occasion to evaluate

$$\langle\Psi(t)| = \langle\Psi_n|e^{iHt}. \qquad \text{(B.8)}$$

This is done by using the complex conjugate (see $(A.18f)$):

$$\langle\Psi(t)| = \overline{|\Psi(t)\rangle} = \overline{e^{-iHt}|\Psi_n\rangle} = \overline{e^{-iE_nt}|\Psi_n\rangle} = e^{+iE_nt}\langle\Psi_n|. \qquad \text{(B.9)}$$

Appendix C

Finding the Ground State Energy from the Vacuum Amplitude

(Readers going systematically through the appendices according to the master plan in appendix B may skip this appendix on first reading!)

We wish to prove that

$$E_0 = W_0 + i\left[\frac{d}{dt}\ln R(t)\right]_{t\to\infty(1-i\eta)}, \tag{C.1}$$

where E_0, W_0 are the ground state energies of the interacting and non-interacting systems respectively and η is a positive infinitesimal such that $\eta \times \infty = \infty$. We begin with the vacuum amplitude as defined in (5.3). This may be written out

$$R(t) = e^{+iW_0 t}\langle\Phi_0|U(t)|\Phi_0\rangle = e^{iW_0 t}\langle\Phi_0|e^{-iHt}|\Phi_0\rangle, \tag{C.2}$$

where (B.3) has been used. Let us expand this in terms of the exact eigenstates $|\Psi_n\rangle$ of the Hamiltonian H. Introducing the unit operator

$$\sum_n |\Psi_n\rangle\langle\Psi_n|$$

(see (A.18e)) gives

$$R(t) = e^{iW_0 t}\sum_n \langle\Phi_0|e^{-iHt}|\Psi_n\rangle\langle\Psi_n|\Phi_0\rangle$$

$$= e^{iW_0 t}\sum_n \langle\Phi_0|\Psi_n\rangle\langle\Psi_n|\Phi_0\rangle e^{-iE_n t}. \tag{C.3}$$

Taking logarithms of both sides, differentiating, and putting in the limit:

$$\left[\frac{d}{dt}\ln R(t)\right]_{\substack{T(1-i\eta)\\T\to\infty}} = iW_0 + \left[\frac{\sum_n |\langle\Phi_0|\Psi_n\rangle|^2 e^{-iE_n t}(-iE_n)}{\sum_n |\langle\Phi_0|\Psi_n\rangle|^2 e^{-iE_n t}}\right]_{\substack{T(1-i\eta)\\T\to\infty}}$$

$$= iW_0 + \left[\frac{\sum_n |\langle\Phi_0|\Psi_n\rangle|^2(-iE_n) e^{-iE_n T} e^{-E_n \eta T}}{\sum_n |\langle\Phi_0|\Psi_n\rangle|^2 e^{-iE_n T} e^{-E_n \eta T}}\right]_{T\to\infty}. \tag{C.4}$$

Now as $T \to \infty$, the quantity $\eta T \to \infty$ by assumption, and all the exponentials decay to zero. But the exponential with lowest value of E_k, e.g., E_0, will decay the slowest, so that only the $n=0$ term will be important, giving us

$$
\left[\frac{d}{dt} \ln R(t)\right]_{\substack{T(1-i\eta) \\ T \to \infty}} = iW_0 + \frac{|\langle \Phi_0 | \Psi_0 \rangle|^2 (-iE_0) e^{-iE_0 \, \infty} e^{-E_0 \, \eta \times \infty}}{|\langle \Phi_0 | \Psi_0 \rangle|^2 e^{-iE_0 \infty} e^{-E_0 \, \eta \times \infty}}
$$

$$
= iW_0 - iE_0, \tag{C.5}
$$

from which theorem (C.1) follows immediately.

It should be noted that for the limiting process in (C.4), (C.5) to be valid we must have

$$
\langle \Phi_0 | \Psi_0 \rangle \neq 0. \tag{C.6}
$$

In the case of a system in which the interacting ground state $|\Psi_0\rangle$ has a different symmetry from that of the non-interacting ground state, $|\Phi_0\rangle$, we find that the two states are orthogonal, so $\langle \Phi_0 | \Psi_0 \rangle = 0$ and (C.1) does not hold. This is the situation in which the original uniform non-interacting system undergoes a *change of phase* as a result of the interaction. This is what happens, for example, when we turn on an attractive force between originally non-interacting electrons in an electron gas; the gas then undergoes a phase change to the superconducting state and (C.1) cannot be used to get the new ground state energy. (See chapter 17, Mattuck and Johansson (1967).)

It should be remarked that the method used in this appendix is due to Thouless (1961) and we have chosen it because of its great mathematical simplicity. The more 'physical' way of getting the relation between E_0 and $R(t)$ uses the 'adiabatic theorem' (see Schweber (1961), p. 316 ff.). This involves (1) slowly (adiabatically) turning on the interaction potential in the non-interacting system in its ground state, so as to avoid transitions to excited states, (2) the assumption that the interacting ground state evolves continuously from the non-interacting one when (1) is carried out (adiabatic hypothesis). In the Thouless method, the imaginary time limit corresponds to (1) since it eliminates all excited states, while the condition $\langle \Phi_0 | \Psi_0 \rangle \neq 0$ corresponds to (2), since it means that the interacting and non-interacting ground states must overlap each other.

Appendix D

The $\tilde{U}(t)$ Operator and its Expansion

D.1 At zero temperature

The operator $\tilde{U}$ is a first cousin to the time development operator U (appendix B). It is in fact itself often called the time development operator because, as will be seen, it gives the time development of the wave function when the wave function is expressed in 'interaction picture'. Its utility lies in the fact that the vacuum amplitude, R, and the Green's function, G, may be expressed directly in terms of it; this means that when we find the perturbation expansion of $\tilde{U}$, we automatically have the expansion for both R and G.

The $\tilde{U}$ operator is defined by

$$\tilde{U}(t, t_0) = e^{iH_0 t} U(t - t_0) e^{-iH_0 t_0}$$

$$= e^{iH_0 t} e^{-iH(t - t_0)} e^{-iH_0 t_0}. \tag{D.1}$$

It should be noted that the exponents in this expression cannot be combined since H, H_0 do not commute! That is, using (B.3):

$$e^A e^B = 1 + (A + B) + \frac{1}{2!}(A^2 + 2AB + B^2) + \cdots$$

while

$$e^{A+B} = 1 + (A + B) + \frac{1}{2!}(A^2 + AB + BA + B^2) + \cdots \tag{D.2}$$

which are not equal if $AB \neq BA$.

The expansion for $\tilde{U}$ is obtained by using the Schrödinger equation to get a differential equation for $\tilde{U}$, changing this to an integral equation, and iterating. The first step is to transform the Schrödinger equation

$$-i(H_0 + H_1)|\Psi(t)\rangle = \frac{\partial}{\partial t}|\Psi(t)\rangle \tag{D.3}$$

and its formal solution (B.1)

$$|\Psi(t)\rangle = U(t - t_0)|\Psi(t_0)\rangle \tag{D.4}$$

into 'interaction picture'. This is defined by (see Schrieffer (1964a), p. 104)

$$\hat{\mathcal{O}}(t) = e^{+iH_0 t} \mathcal{O} e^{-iH_0 t} \tag{D.5}$$

for an arbitrary operator $\mathcal{O}$ and

$$|\hat{\Psi}(t)\rangle = e^{+iH_0 t}|\Psi(t)\rangle \tag{D.6}$$

for an arbitrary state vector $|\Psi\rangle$. The purpose of this transformation is to get rid of the explicit appearance of H_0 in (D.3). Using it, we find that (D.3) becomes

$$\frac{\partial}{\partial t}|\hat{\Psi}(t)\rangle = -i\hat{H}_1(t)|\hat{\Psi}(t)\rangle \tag{D.7}$$

which is easily checked by employing (D.5, 6) and differentiating. Equation (D.4) becomes

$$|\hat{\Psi}(t)\rangle = \tilde{U}(t, t_0)|\hat{\Psi}(t_0)\rangle. \tag{D.8}$$

(Note by (D.1, 5) that $\tilde{U} \neq \hat{U}$!) Substituting (D.8) into (D.7) and cancelling $|\Psi(t_0)\rangle$ gives the basic equation for $\tilde{U}$

$$\frac{d}{dt} \tilde{U}(t, t_0) = -i\hat{H}_1(t) \tilde{U}(t, t_0). \tag{D.9}$$

A series expansion for $\tilde{U}(t, t_0)$ may be obtained by integrating (D.9) and iterating:

$$\tilde{U}(t, t_0) = \tilde{U}(t_0, t_0) - i \int_{t_0}^{t} \hat{H}_1(t_1) \tilde{U}(t_1, t_0) \, dt_1$$

$$= \tilde{U}(t_0, t_0) - i \tilde{U}(t_0, t_0) \int_{t_0}^{t} \hat{H}_1(t_1) \, dt_1 +$$

$$+ i^2 \int_{t_0}^{t} dt_1 \int_{t_0}^{t_1} dt_2 \, \hat{H}_1(t_1) \hat{H}_1(t_2) \tilde{U}(t_2, t_0)$$

$$= 1 - i \int_{t_0}^{t} \hat{H}_1(t_1) \, dt_1 + i^2 \int_{t_0}^{t} dt_1 \int_{t_0}^{t_1} dt_2 \, \hat{H}_1(t_1) \hat{H}_1(t_2) + \cdots \tag{D.10}$$

where $U(t_0, t_0) = 1$ by (D.1). The order of factors is important, since $H(t_1)$, $H(t_2)$ do not commute. Dyson has shown that the nth-order term in (D.10) may be rewritten so each t_i goes from t_0 to t (see Schweber (1961), p. 332, for proof):

$$\int_{t_0} dt_1 \ldots \int_{t_0}^{t_{n-1}} dt_n \, \hat{H}_1(t_1) \ldots \hat{H}_1(t_n) = \frac{1}{n!} \int_{t_0}^{t} dt_1 \ldots \int_{t_0}^{t} dt_n T_D \times$$

$$\times [\hat{H}_1(t_1) \ldots \hat{H}_1(t_n)], \tag{D.11}$$

where T_D stands for 'Dyson time-ordered product', which means operators in brackets re-arranged so times decrease from left to right. Thus, in second order:

$$T_D[\hat{H}_1(t_1) \hat{H}_1(t_2)] = \theta_{t_1-t_2} \hat{H}_1(t_1) \hat{H}_1(t_2) + \theta_{t_2-t_1} \hat{H}_1(t_2) \hat{H}_1(t_1) \tag{D.12}$$

with θ as in (5.25). Here, since we assume H has an even number of creation and destruction operators, T_D is the same as the ordinary Wick time-ordered product (9.4), so from now on we drop the D-subscript. Substituting (D.11) in (D.10) yields the perturbation expansion of the $\tilde{U}$ operator

$$\tilde{U}(t, t_0) = \sum_{n=0}^{\infty} \frac{(-i)^n}{n!} \int_{t_0}^{t} dt_1 \ldots \int_{t_0}^{t} dt_n T\{\hat{H}_1(t_1) \ldots \hat{H}_1(t_n)\}. \tag{D.13}$$

D.2 At finite temperature

The finite temperature $\tilde{U}$-operator is defined and expanded in a way which runs precisely parallel to the zero temperature case. It is used to find the finite temperature imaginary time propagator and vacuum amplitude (chapter 14).

As shown in appendix B, in the $T=0$ case the $\tilde{U}$ operator and diagram expansions may all be traced back to the time-dependent Schrödinger equation. In the $T \neq 0$ case, everything goes back to the Bloch equation, (14.21):

$$\frac{\partial \rho}{\partial \beta} = -(H-\mu N)\rho. \tag{D.14}$$

This may be obtained from the Schrödinger equation (D.3) by making the replacements:

$$it \to \beta, \quad H \to H-\mu N, \quad \Psi \to \rho. \tag{D.15}$$

Hence we define the $T \neq 0$ $\tilde{U}$-operator by using these replacements in (D.1), yielding:

$$\tilde{U}(\beta) = e^{\beta(H_0-\mu N)} e^{-\beta(H-\mu N)} \tag{D.16}$$

where $it_0 \to \beta_0$ has been set $=0$. (Note that since $\beta_0 = 1/kT_0$, it can never be negative.) Similarly the $T \neq 0$ 'interaction picture' is given by

$$\hat{\mathcal{O}}(\beta) = e^{\beta(H_0-\mu N)} \mathcal{O} e^{-\beta(H_0-\mu N)} \tag{D.17}$$

so that, for example,

$$\hat{H}_1(\tau) = e^{\tau(H_0-\mu N)} H_1 e^{-\tau(H_0-\mu N)}. \tag{D.18}$$

Finally, making replacements (D.15) in (D.13) yields the expansion of the $\tilde{U}$-operator for $T \neq 0$:

$$\tilde{U}(\beta) = \sum_{n=0}^{\infty} \frac{(-1)^n}{n!} \int_0^{\beta} d\tau_1 \ldots \int_0^{\beta} d\tau_n T\{\hat{H}_1(\tau_1)\ldots\hat{H}_1(\tau_n)\}. \tag{D.19}$$

Exercises

D.1 Derive (D.7), (D.8), and (D.9).

D.2 Derive (D.11) for the case $n=2$. (Note that $\hat{H}_1(t_1)\hat{H}_1(t_2) \neq \hat{H}_1(t_2)\hat{H}_1(t_1)$! Why?) (Answer: Fetter and Walecka (1971), p. 57; Raimes (1972), p. 100.)

Appendix E

Expansion of the Single-particle Propagator and Vacuum Amplitude

E.1 Propagator expansion at $T=0$

The single-particle propagator can be expanded in a perturbation series with the aid of the operator $\tilde{U}$ defined in (D.1). We will confine ourselves to $G^+(k_2,k_1,t_2-t_1)$ since G^- follows by exactly the same type of argument. It will first be shown, using the method of Thouless (1972), p. 89, that G^+ can be expressed in terms of $\tilde{U}$ by:

$$G^+(k_2,k_1,t_2-t_1) = \lim_{\substack{T_1\to-\infty(1-i\eta)\\T_2\to+\infty(1-i\eta)}} Q(T_2,T_1,k_2,k_1,t_2-t_1) \qquad (E.1)$$

where

$$Q = -i\frac{\langle\Phi_0|\,T\{\tilde{U}(T_2,T_1)\,\hat{c}_{k_2}(t_2)\,\hat{c}^\dagger_{k_1}(t_1)\}\,|\Phi_0\rangle}{\langle\Phi_0|\,\tilde{U}(T_2,T_1)\,|\Phi_0\rangle}. \qquad (E.2)$$

In (E.2), since we are dealing with G^+, we must have $t_2 > t_1$, and we assume $T_2 > t_2$, $T_1 < t_1$, so that:

$$T_2 > t_2 > t_1 > T_1. \qquad (E.3), (E.4)$$

The $\hat{c}_k$'s are the ordinary c_k's written in interaction picture as defined by (D.5), and we have $\eta \times \infty = \infty$.

Recall that G^+ is defined by (9.6)

$$G^+(k_2,k_1,t_2-t_1) = -i\theta_{t_2-t_1}\langle\Psi_0|\,c_{k_2}(t_2)\,c^\dagger_{k_1}(t_1)\,|\Psi_0\rangle, \qquad (E.5)$$

so that to prove (E.1) we need to do some juggling to get from the interaction picture $\hat{c}_k$'s in (E.2) to the Heisenberg c_k's, (9.3), in (E.5), and from the non-interacting $|\Phi_0\rangle$ to the interacting $|\Psi_0\rangle$. Begin by breaking up $\tilde{U}$ into

$$\tilde{U}(T_2,T_1) = \tilde{U}(T_2,t_2)\,\tilde{U}(t_2,t_1)\,\tilde{U}(t_1,T_1) \qquad (E.6)$$

which follows immediately from (D.1). Insert this into (E.2) and rearrange the terms in proper time order, using the given order in (E.4), so that the T-symbol may be dropped:

$$Q = \frac{-i\langle\Phi_0|\,\tilde{U}(T_2,t_2)\,\hat{c}_{k_2}(t_2)\,\tilde{U}(t_2,t_1)\,\hat{c}^\dagger_{k_1}(t_1)\,\tilde{U}(t_1,T_1)\,|\Phi_0\rangle}{\langle\Phi_0|\,\tilde{U}(T_2,T_1)\,|\Phi_0\rangle}. \qquad (E.7)$$

Writing out the $\hat{c}(t)$'s and using (D.1) again, this may be expressed in terms of U instead of $\tilde{U}$:

$$Q = \frac{-i\,e^{+iW_0(T_2-T_1)}\langle\Phi_0|\,U(T_2-t_2)\,c_{k_2}\,U(t_2-t_1)\,c^\dagger_{k_1}\,U(t_1-T_1)\,|\Phi_0\rangle}{e^{+iW_0(T_2-T_1)}\langle\Phi_0|\,U(T_2-T_1)\,|\Phi_0\rangle}. \qquad (E.8)$$

Applying the definition of U in (B.3) gives

$$Q = \frac{-i\langle\Phi_0| e^{-iHT_2} c_{k_2}(t_2) c_{k_1}^\dagger(t_1) e^{+iHT_1}|\Phi_0\rangle}{\langle\Phi_0| e^{-iHT_2} e^{+iHT_1}|\Phi_0\rangle}, \tag{E.9}$$

where the c_k's are now in Heisenberg picture. Inserting the unit operator over the exact eigenstates of H, $\sum_n |\Psi_n\rangle\langle\Psi_n|$, yields

$$Q = \frac{-i\sum_{n,m} \langle\Phi_0|\Psi_n\rangle\langle\Psi_m|\Phi_0\rangle e^{-i(E_n T_2 - E_m T_1)}\langle\Psi_n| c_{k_2}(t_2) c_{k_1}^\dagger(t_1)|\Psi_m\rangle}{\sum_n \langle\Phi_0|\Psi_n\rangle\langle\Psi_n|\Phi_0\rangle e^{-i(E_n T_2 - E_n T_1)}}. \tag{E.10}$$

The limit is taken as in (C.4, 5). Only the $m=0$, $n=0$ terms survive, there is some cancellation and theorem (E.1) is proved.

The perturbation series for $G^+(k_2, k_1, t_2 - t_1)$ is then obtained simply by substituting the expansion of $\tilde{U}$, (D.13) into (E.1, 2):

$$G(k_2, k_1, t_2 - t_1) = \frac{\left\{ -i\sum_{n=0}^\infty \frac{(-i)^n}{n!} \int_{-\infty(1-i\eta)}^{+\infty(1-i\eta)} dt_1' \ldots \int dt_n' \times \atop \times \langle\Phi_0| T\{\hat{H}_1(t_1') \ldots \hat{H}_1(t_n')\, \hat{c}_{k_2}(t_2)\, \hat{c}_{k_1}^\dagger(t_1)\}|\Phi_0\rangle \right\}}{\left\{ \sum_{n=0}^\infty \frac{(-i)^n}{n!} \int_{-\infty(1-i\eta)}^{+\infty(1-i\eta)} dt_1' \ldots \int dt_n' \times \atop \times \langle\Phi_0| T\{\hat{H}_1(t_1') \ldots \hat{H}_1(t_n')\}|\Phi_0\rangle \right\}}. \tag{E.11}$$

The $+$ has been dropped since the result holds also true for G^-.

It must be carefully noted that (C.6) must hold, for the limit of (E.10) to yield (E.1), i.e., we must have

$$\langle\Phi_0|\Psi_0\rangle \neq 0 \tag{E.12}$$

for the perturbation expansion of G to be valid (see §15.4).

E.2 The vacuum amplitude expansion at $T=0$

The vacuum amplitude is easily expressed in terms of $\tilde{U}$ as follows:

$$\begin{aligned} R(t-t_0) &= e^{iW_0(t-t_0)}\langle\Phi_0| U(t-t_0)|\Phi_0\rangle \\ &= e^{iW_0(t-t_0)}\langle\Phi_0| e^{-iH_0 t} e^{+iH_0 t} U(t-t_0) e^{-iH_0 t_0} e^{+iH_0 t_0}|\Phi_0\rangle \\ &= \langle\Phi_0| \tilde{U}(t,t_0)|\Phi_0\rangle \end{aligned} \tag{E.13}$$

where (B.7), (B.9) have been used. Thus, taking matrix elements on both sides of (D.13):

$$R(t-t_0) = \sum_{n=0}^{\infty} \frac{(-i)^n}{n!} \int_{t_0}^{t} dt_1 \ldots \int_{t_0}^{t} dt_n \langle \Phi_0 | T\{\hat{H}_1(t_1) \ldots \hat{H}_1(t_n)|\Phi_0\rangle \quad (E.14)$$

which is the desired perturbation expansion of the vacuum amplitude.

E.3 Expansion at finite temperatures

As mentioned in §14.3, the imaginary time propagator

$$\mathcal{G}(k_2, k_1, \tau_2 - \tau_1) = -\langle T\{c_{k_2}(\tau_2) c_{k_1}^\dagger(\tau_1)\}\rangle \quad (E.15)$$

may be expanded the same way the $T=0$ propagator was. Parallel to (E.1), (E.2), we obtain the theorem

$$\begin{aligned}\mathcal{G}(k_2, k_1, \tau_2 - \tau_1) \\ (\text{for } \tau_2 > \tau_1)\end{aligned} = \frac{-\langle T\{\tilde{U}(\beta)\hat{c}_{k_2}(\tau_2)\hat{c}_{k_1}^\dagger(\tau_1)\}\rangle_0}{\langle \tilde{U}(\beta)\rangle_0} \quad (E.16)$$

where $\langle\ \rangle_0$ denotes average over ensemble of non-interacting systems at temperature T, and $\tilde{U}(\beta)$ is defined in (D.16). Using (D.19) then yields

$$\mathcal{G}(k_2, k_1, \tau_2 - \tau_1) = \frac{-\left\{ \sum_{n=0}^{\infty} \frac{(-1)^n}{n!} \int_0^\beta d\tau_1' \ldots \int_0^\beta d\tau_n' \times \\ \times \langle T\{\hat{H}_1(\tau_1')\ldots\hat{H}_1(\tau_n')\hat{c}_{k_2}(\tau_2)\hat{c}_{k_1}^\dagger(\tau_1)\rangle_0 \right\}}{\sum_{n=0}^{\infty} \frac{(-1)^n}{n!} \int_0^\beta d\tau_1' \ldots \int_0^\beta d\tau_n' \langle T\{\hat{H}_1(\tau_1')\ldots\hat{H}_1(\tau_n')\rangle_0}. \quad (E.17)$$

The useful form of the finite temperature vacuum amplitude (14.40) is

$$\mathcal{R}(\beta) = \langle \tilde{U}(\beta)\rangle_0. \quad (E.18)$$

This is evidently just the denominator of (E.16, 17) so that we have the perturbation expansion:

$$\mathcal{R}(\beta) = \sum_{n=0}^{\infty} \frac{(-1)^n}{n!} \int_0^\beta d\tau_1' \ldots \int_0^\beta d\tau_n' \langle T\{\hat{H}_1(\tau_1')\ldots\hat{H}_1(\tau_n')\}\rangle_0. \quad (E.19)$$

Exercises

E.1 Derive (E.7), (E.8), and (E.9).

E.2 In (E.10), let $T_1 = -a(1-i\eta)$ and $T_2 = +a(1-i\eta)$ and carry out the limit of Q as $a \to \infty$. Show that the result is just $G^+(k_2, k_1, t_2 - t_1)$, provided $\langle \Phi_0|\Psi_0\rangle \neq 0$.

Appendix F

Evaluating Matrix Elements by Wick's Theorem

In order to do any good deeds with the perturbation expansion for $R(t-t_0)$ in (E.14) or that for $G(k_2, k_1, t_2 - t_1)$ in (E.11), it is necessary to evaluate the matrix elements occurring in these expressions. We will first do this by the tedious ordinary method, then show how to save time by using one of Wick's tricks called 'Wick's theorem'.

For simplicity, let us assume that the Hamiltonian is for N non-interacting fermions in an external potential:

$$H = H_0 + H_1$$

where

$$H_0 = \sum_k \epsilon_k c_k^\dagger c_k, \quad H_1 = \sum_{k,l} V_{kl} c_k^\dagger c_l. \tag{F.1}$$

In interaction picture, H_1 is

$$\begin{aligned}
\hat{H}_1(t) &= e^{+iH_0 t} \sum_{k,l} V_{kl} c_k^\dagger c_l e^{-iH_0 t} \\
&= \sum_{k,l} V_{kl} e^{+iH_0 t} c_k^\dagger e^{-iH_0 t} e^{+iH_0 t} c_l e^{-iH_0 t} \\
&= \sum_{k,l} V_{kl} \hat{c}_k^\dagger(t) \hat{c}_l(t) \\
&= \sum_{k,l} V_{kl} [\hat{a}_k^\dagger(t) \hat{a}_l(t) + \hat{a}_k^\dagger(t) \hat{b}_l^\dagger(t) + \hat{b}_k(t) \hat{a}_l(t) + \hat{b}_k(t) \hat{b}_l^\dagger(t)].
\end{aligned} \tag{F.2}$$

The $\hat{c}_k^\dagger(t)$, $\hat{c}_l(t)$ may be simplified in the following way. We have

$$\begin{aligned}
\hat{c}_l(t) &= e^{+iH_0 t} c_l e^{-iH_0 t} \\
&= c_l + i[H_0, c_l] t + \frac{i^2}{2!} [H_0, [H_0, c_l]] t^2 + \cdots,
\end{aligned} \tag{F.3}$$

where the exponentials have been expanded by (B.3). The commutator is

$$[H_0, c_l] = \sum_k \epsilon_k (c_k^\dagger c_k c_l - c_l c_k^\dagger c_k). \tag{F.4}$$

But by (7.32)

$$c_k^\dagger c_k c_l = -c_k^\dagger c_l c_k = [-\delta_{k,l} + c_l c_k^\dagger] c_k, \tag{F.5}$$

so that

$$[H_0, c_l] = -\sum_k \epsilon_k \delta_{k,l} c_k = -\epsilon_l c_l. \tag{F.6}$$

Substituting this for the commutators in (F.3) yields

$$\hat{c}_l(t) = c_l - i\epsilon_l t c_l + \frac{i^2}{2!} \epsilon_l^2 t^2 c_l + \cdots$$

or

$$\hat{c}_l(t) = c_l e^{-i\epsilon_l t}. \tag{F.7}$$

Similarly, the $\hat{a}(t)$ and $\hat{b}(t)$ operators have the simple forms:

$$\hat{a}_k^\dagger(t) = a_k^\dagger e^{+i\epsilon_k t}, \quad \hat{b}_k^\dagger(t) = b_k^\dagger e^{-i\epsilon_k t},$$

$$\hat{a}_k(t) = a_k e^{-i\epsilon_k t}, \quad \hat{b}_k(t) = b_k e^{+i\epsilon_k t}. \tag{F.8}$$

Let us now use these results to evaluate some of the matrix elements appearing in the numerator of (E.11). The zeroth order term is easy enough. Assume $t_2 > t_1$. Then, for $k_2, k_1 > k_F$,

$$
\begin{aligned}
M^{(0)} &= \langle \Phi_0 | T\{\hat{a}_{k_2}(t_2)\,\hat{a}_{k_1}^\dagger(t_1)\} | \Phi_0 \rangle \\
&= \langle \Phi_0 | a_{k_2} a_{k_1}^\dagger | \Phi_0 \rangle e^{-i(\epsilon_{k_2} t_2 - \epsilon_{k_1} t_1)} \\
&= \delta_{k_1, k_2} e^{-i\epsilon_{k_1}(t_2 - t_1)}
\end{aligned} \tag{F.9}
$$

(see (7.35)). This is just the free propagator, $iG_0^+(k_2, k_1, t_2 - t_1)$.

The first-order element in the numerator of (E.11), assuming $t_2 > t_1' > t_1$ (write t for t_1' from now on) and that $k_1, k_2 > k_F$, is:

$$
\begin{aligned}
M^{(1)} &= \langle \Phi_0 | T\{\hat{H}_1(t)\,\hat{a}_{k_2}(t_2)\,\hat{a}_{k_1}^\dagger(t_1)\} | \Phi_0 \rangle \\
&= \sum_{k,l} V_{kl} [\langle 0 | T\{\hat{a}_k^\dagger(t)\,\hat{a}_l(t)\,\hat{a}_{k_2}(t_2)\,\hat{a}_{k_1}^\dagger(t_1)\} | 0 \rangle + \\
&\quad + \cdots + \langle 0 | T\{\hat{b}_k(t)\,\hat{b}_l^\dagger(t)\,\hat{a}_{k_2}(t_2)\,\hat{a}_{k_1}^\dagger(t_1)\} | 0 \rangle],
\end{aligned} \tag{F.9'}
$$

or, using (F.2, 8):

$$
M^{(1)} = \sum_{k,l} V_{kl}
\begin{bmatrix}
\langle 0 | a_{k_2} a_k^\dagger a_l a_{k_1}^\dagger | 0 \rangle \\
+ \langle 0 | a_{k_2} a_k^\dagger b_l^\dagger a_{k_1}^\dagger | 0 \rangle \\
+ \langle 0 | a_{k_2} b_k a_l a_{k_1}^\dagger | 0 \rangle \\
+ \langle 0 | a_{k_2} b_k b_l^\dagger a_{k_1}^\dagger | 0 \rangle
\end{bmatrix}
e^{-i(\epsilon_{k_2} t_2 - \epsilon_k t + \epsilon_l t - \epsilon_{k_1} t_1)} \tag{F.10}
$$

where $|0\rangle$ is the Fermi vacuum. The elements are evaluated by using the commutation rules (7.73) to systematically bring all destruction operators to the right where they operate on the vacuum and produce zero by (4.20). The first element is

$$
\begin{aligned}
\langle 0 | a_{k_2} a_k^\dagger a_l a_{k_1}^\dagger | 0 \rangle &= \langle 0 | a_{k_2} a_k^\dagger [\delta_{l, k_1} - a_{k_1}^\dagger a_l] | 0 \rangle \\
&= \langle 0 | a_{k_2} a_k^\dagger | 0 \rangle \delta_{l, k_1} - \langle 0 | a_{k_2} a_k^\dagger a_{k_1}^\dagger a_l | 0 \rangle \tag{F.11} \\
&= \delta_{k_2, k} \delta_{l, k_1} - 0.
\end{aligned}
$$

The second and third terms in (F.10) are equal to zero since they do not destroy the same number of particles and holes they produce. The fourth is found to be:

$$\langle 0 | \, a_{k_2} b_k b^\dagger_l a^\dagger_{k_1} | 0 \rangle = \delta_{k_1, k_2} \delta_{k, l}. \tag{F.12}$$

Hence we obtain the final result

$$M^{(1)} = V_{k_2 k_1} e^{-i\epsilon_{k_2}(t_2-t)} e^{-i\epsilon_{k_1}(t-t_1)} + \sum_k V_{kk} e^{-i\epsilon_{k_1}(t_2-t_1)} \delta_{k_1, k_2}. \tag{F.13}$$

Now, if we had infinite patience, infinite energy, and zero imagination, we could plough blindly ahead through the second-, third-, and higher-order elements, evaluating them by brute force using the commutation relations as above. However, there is a much easier way, using Wick's theorem.

Wick's theorem is essentially based on the idea of bringing all destruction operators to the right where they operate on the Fermi vacuum $|0\rangle$ and give zero. Thus the central concept employed is that of the *normal product* which has all destruction operators on the right. It is defined by

$$N[ABC^\dagger DF^\dagger G^\dagger \ldots] = (-1)^P [C^\dagger F^\dagger G^\dagger \ldots ABD \ldots] \tag{F.14}$$

where P is the number of interchanges of neighbouring operators required to get from the given order on the left to the 'normal order' on the right. For example,

$$N[\hat{a}_k(t_3) \, \hat{b}^\dagger_l(t_2) \, \hat{a}^\dagger_m(t_1)] = + \hat{b}^\dagger_l(t_2) \, \hat{a}^\dagger_m(t_1) \, \hat{a}_k(t_3)$$

$$= - \hat{a}^\dagger_m(t_1) \, \hat{b}^\dagger_l(t_2) \, \hat{a}_k(t_3). \tag{F.15}$$

Note that normal order has nothing to do with time order, and that order among the different creation (or destruction) operators is immaterial, provided we affix the proper sign.

The importance of the normal product lies in the fact that

$$\langle 0 | \, N[ABC^\dagger D \ldots] | 0 \rangle = 0. \tag{F.16}$$

If there is at least one destruction operator, this is obvious from (4.20). If there are only creation operators, the element vanishes since

$$\langle 0 | \, a^\dagger_k = \overline{a_k | 0 \rangle} = 0, \quad \text{etc.}$$

It is also necessary to introduce the contracted product or '*contraction*' of two operators, which is the difference between the time-ordered and normal product:

$$\overline{AB} = T[AB] - N[AB]. \tag{F.17}$$

Thus, for example,

$$\overline{\hat{a}_k(t_2)\,\hat{a}_l^\dagger}(t_1) = \hat{a}_k(t_2)\,\hat{a}_l^\dagger(t_1) - (-1)\,\hat{a}_l^\dagger(t_1)\,\hat{a}_k(t_2)$$
$$(\text{for } t_2 > t_1)$$

$$= (a_k a_l^\dagger + a_l^\dagger a_k)\,e^{-i(\epsilon_k t_2 - \epsilon_l t_1)}$$

$$= \delta_{kl}\,e^{-i\epsilon_k(t_2-t_1)}. \tag{F.18}$$

In a similar fashion, using (9.4), it is found that

$$\overline{\hat{a}_k(t_2)\,\hat{a}_l^\dagger}(t_1) = 0, \quad \text{for } t_2 \leqslant t_1$$

$$\overline{\hat{b}_k(t_2)\,\hat{b}_l^\dagger}(t_1) = \delta_{k,\,l}\,e^{+i\epsilon_k(t_2-t_1)}, \quad \text{for } t_2 \geqslant t_1$$

$$\overline{\hat{b}_k(t_2)\,\hat{b}_l^\dagger}(t_1) = 0 \quad \text{for } t_2 < t_1$$

$$\overline{\hat{a}_k(t_2)\,\hat{a}_l}(t_1) = \overline{\hat{b}_k(t_2)\,\hat{b}_l}(t_1) = \overline{\hat{a}_k(t_2)\,\hat{b}_l^\dagger}(t_1) = \text{etc.} = 0. \tag{F.19}$$

Note that

$$\overline{\hat{a}_{k_2}(t_2)\,\hat{a}_{k_1}^\dagger}(t_1) = iG_0^+(k_2, k_1, t_2 - t_1)$$

$$\overline{\hat{b}_{k_1}(t_1)\,\hat{b}_{k_2}^\dagger}(t_2) = -iG_0^-(k_2, k_1, t_2 - t_1). \tag{F.19'}$$

Evidently, the contraction is a number, not an operator. Because of this, taking matrix elements of both sides of (F.17) and using (F.16), yields

$$\overline{AB} = \langle 0|\,T\{AB\}\,|0\rangle \tag{F.20}$$

which is a short way of writing (F.19). Comparing this with (9.1) shows that the contraction is just $i \times$ the unperturbed propagator.

Finally, one also needs normal products with contractions like, for example,

$$N[\overline{A}B\overline{C}D] = -N[\overline{A}\overline{C}BD] = -\overline{AC}N[BD]. \tag{F.21}$$

Wick's theorem then states that a time-ordered product of operators may be decomposed into a sum over pure normal products, normal products with one or more contractions, and completely contracted normal products as follows:

$$T[UVW\ldots XYZ] = N[UVW\ldots XYZ]$$

$$+ N[\overline{UV}W\ldots XYZ] + N[\overline{U}V\overline{W}\ldots XYZ] + \cdots$$

$$+ N[\overline{UV}\overline{W}\ldots XYZ] + N[\overline{U}\overline{VW}\ldots XYZ] + \cdots$$

$$+ \cdots$$

$$+ N[\overline{U}\overline{V}\overline{W}\ldots X\overline{Y}\overline{Z}] + N[\overline{UVW}\ldots X\overline{Y}Z] + \cdots \tag{F.22}$$

where the last line has only completely contracted products. We see that (F.17) is a special case of this (see Fetter (1971), p. 91 for general proof). If we now take the two sides of (F.22) between Fermi vacuum states, the non-fully contracted normal products vanish by (F.16), leaving:

$$\langle 0| T\{UVW\ldots XYZ\} |0\rangle = \overline{UV}\;\overline{W\ldots X}\;\overline{YZ} + \overline{U}\overline{V}W\ldots \overline{X}\overline{YZ} + \cdots$$

$$= \text{sum over all possible fully contracted}$$
$$\text{products.} \qquad\qquad (F.23)$$

Equation (F.23) eliminates a lot of the drudgery involved in the evaluation of matrix elements. As an illustration of how it works, let us use it to calculate the elements in (F.9), (F.9′) which were calculated before by the old method. Employing (F.23), (F.18), (F.19) we have for (F.9) (remember that we are assuming $k_1, k_2 > k_F$!)

$$\langle 0| T\{\hat{a}_{k_2}(t_2)\,\hat{a}^\dagger_{k_1}(t_1)\} |0\rangle = \overline{\hat{a}_{k_2}(t_2)\,\hat{a}^\dagger_{k_1}(t_1)} = \delta_{k_1,\,k_2} e^{-i\epsilon_{k_1}(t_2-t_1)}. \qquad (F.24)$$

For (F.9′) we find

$$\langle 0| T\{\hat{a}^\dagger_k(t)\hat{a}_l(t)\hat{a}_{k_2}(t_2)\,\hat{a}^\dagger_{k_1}(t_1)\} |0\rangle = +\hat{a}_l(t)\hat{a}^\dagger_k(t)\hat{a}^\dagger_{k_1}(t_1)\hat{a}_{k_2}(t_2)$$

$$(t_2 > t > t_1) \qquad\qquad +\hat{a}^\dagger_k(t)\hat{a}^\dagger_{k_1}(t_1)\hat{a}_l(t)\hat{a}_{k_2}(t)$$

$$+\hat{a}_{k_2}(t_2)\hat{a}^\dagger_k(t)\hat{a}_l(t)\hat{a}^\dagger_{k_1}(t_1)$$

$$= 0+0+(-1)^2\,\delta_{k_2,\,k}\,e^{-i\epsilon_{k_2}(t_2-t)} \times$$

$$\times\,\delta_{l,\,k_1}\,e^{-i\epsilon_{k_1}(t-t_1)}. \qquad (F.25)$$

Similarly, the second and third terms give zero while the fourth yields

$$\langle 0| T\{\hat{b}_k(t)\hat{b}^\dagger_l(t)\hat{a}_{k_2}(t_2)\,\hat{a}^\dagger_{k_1}(t_1)\} |0\rangle = \overline{\hat{b}_k(t)\hat{b}^\dagger_l(t)}\;\overline{\hat{a}_{k_2}(t_2)\,\hat{a}^\dagger_{k_1}(t_1)} + \cdots$$

$$= \delta_{k,\,l}\delta_{k_1,\,k_2}\,e^{-i\epsilon_{k_1}(t_2-t_1)} + 0 + 0. \qquad (F.26)$$

Substituting (F.25, 26) into (F.9′) produces just the result we got before, i.e., (F.13). (Actually, this case is a bit too simple—the real power of the method is first revealed in the higher orders.)

We conclude this section with a note regarding why the time-ordering operator for equal times is defined as in (9.4). If we substitute expression (F.2) for $\hat{H}_1(t)$ in matrix element (F.9′), we see that the equal times operators like $\hat{c}^\dagger_k(t)\hat{c}_l(t)$, $\hat{a}^\dagger_k(t)a_l(t)$, $\hat{a}^\dagger_k(t)\hat{b}^\dagger_l(t)$, etc., must always occur in just the order in which they appear in the Hamiltonian, i.e., $c^\dagger$ (or $a^\dagger$, b) to the left of c (or a, $b^\dagger$), as stated in (9.4). It follows that the propagator for equal times must have $c^\dagger$ to the left of c, i.e., it is a hole propagator.

Exercises

F.1 Verify (F.12).

F.2 (a) Consider a many-body system with two-body interaction given by H_1 in (7.48), and no external perturbing potential. Write out the matrix element in the first-order term in the numerator of G in (E.11), expressed in terms of $\hat{c}_k^\dagger(t)$ and $\hat{c}_k(t)$.

 (b) Write in particle–hole notation the term in (a) corresponding to $k, m > k_F$, $l, n < k_F$, $k_2, k_1 > k_F$. (Answer: (G.6').)

 (c) Write out (b) in the case $t_2 > t_1' > t_1$.

 (d) Evaluate (c) with the help of (F.8) and the commutation rules. Show that (c) equals

$$\tfrac{1}{2} \sum_{klmn} V_{klmn}\, \delta_{mk_1}\, \delta_{ln}\, \delta_{k_2 k}\, e^{-i\varepsilon_{k_2}(t_2 - t)}\, e^{-i\varepsilon_{k_1}(t - t_1)}$$

F.3 Verify (F.19).

F.4 Evaluate Ex. F.2(c) with contractions and show that the result is the same as obtained in F.2(d).

Appendix G

Derivation of the Graphical Expansion for Propagator and Vacuum Amplitude

In appendices B, D, E, F we have seen how to expand the Green's function and vacuum amplitude in a perturbation series and how to evaluate the matrix elements in the expansion by Wick's theorem. The remaining step is to show how the perturbation series may be expressed in graphical form.

We will only sketch the process of translating the perturbation series (E.11) into diagrams for the simplest case, i.e., N non-interacting fermions in an external potential. The diagram expansion for this case was derived in the first part of the book by the monkey argument, and appeared in (4.34). Our object now is to take the matrix elements for this case as in (F.9), (F.9'), show how to draw them diagrammatically, and thus re-derive (4.34).

The method becomes transparent if we just remember that by (F.20), a contraction is $i \times$ propagator. Hence: each contraction in (F.23) will be represented in a diagram by means of a directed line. Furthermore, for the present external interaction, each interaction V_{kl} will be represented by a dot. Consider now as a concrete example the $M^{(1)}$ term in (F.25), which has the non-vanishing contraction on line 3. Including the $\sum_{kl} V_{kl}$ factor from (F.9') we can write/draw this:

$$\sum_{kl} \hat{a}_{k_2}(t_2)\, \hat{a}_k^\dagger(t)\, (-iV_{kl})\, \hat{a}_l(t)\, \hat{a}_{k_1}^\dagger(t_1) \equiv \quad (G.0)$$

which reveals directly how diagrams come from contractions.

In (G.1) we show the graphs for $M^{(0)}$, as obtained from (F.9, 24), and for $M^{(1)}$, as obtained from (F.9', 25, 26):

$$\text{(G.1)}$$

$$M^{(0)} \qquad\qquad M^{(1)}$$

The first two graphs are just what appear in the diagram series for G^+ in (4.34). However, the third diagram is not found in (4.34). It is of the 'unlinked' variety, and in a minute we shall show how to get rid of it. Equation (G.1) makes the meaning of the graphs clear: each graph is in 1–1 correspondence with a non-vanishing matrix element in the perturbation expansion of G in (E.11).

It should be noted that there is a factor of $(-i)^n/n!$ in nth order appearing in (E.11). We can get the $(-i)^n$ factor from the diagrams by simply associating a factor $(-iV_{kl})$ with each dot, instead of V_{kl}.

The $n!$ is cancelled in the following way: Consider the second order matrix element in the numerator of (E.11) as an example:

$$M^{(2)} = \langle 0|T\{\hat{H}_1(t_1')\,\hat{H}_2(t_2')\,\hat{a}_{k_2}(t_2)\,\hat{a}_{k_1}^\dagger(t_1)\}|0\rangle, \text{s.}$$

This is a sum over all possible time orders. Select any one of these—for example $t_2 > t_2' > t_1' > t_1$. It will be sufficient to illustrate the point if we restrict ourselves to just linked diagrams. The two sets of contractions giving rise to linked diagrams here are (omit ⌢ over a, b for clarity):

$$\equiv \sum_{klmn} V_{mn} V_{kl} \langle 0|a_{k_2}(t_2)\,a_m^\dagger(t_2')\,a_n(t_2')\,a_k^\dagger(t_1')\,a_l(t_1')\,a_{k_1}^\dagger(t_1)|0\rangle \qquad \text{(G.1a)}$$

$$\equiv \sum_{klmn} V_{mn} V_{kl} \langle 0|a_{k_2}(t_2)\,b_m(t_2')\,a_n(t_2')\,a_k^\dagger(t_1')\,b_l^\dagger(t_1')\,a_{k_1}^\dagger(t_1)|0\rangle. \qquad \text{(G.1b)}$$

Now consider another particular time order, i.e. $t_2 > t'_1 > t'_2 > t_1$. The linked diagrams here are

$$\equiv \sum_{klmn} V_{kl}\, V_{mn} \langle 0| \overline{a_{k_2}(t_2) a_k^\dagger(t'_1)}\, \overline{a_l(t'_1) a_m^\dagger(t'_2)}\, \overline{a_n(t'_2) a_{k_1}^\dagger(t_1)} |0\rangle \qquad \text{(G.1a$'$)}$$

$$\equiv \sum_{klmn} V_{kl}\, V_{mn} \langle 0| a_{k_2}(t_2)\, b_k(t'_1)\, a_l(t'_1)\, a_m^\dagger(t'_2)\, b_n^\dagger(t'_2)\, a_{k_1}^\dagger(t_1) |0\rangle. \qquad \text{(G.1b$'$)}$$

Comparing (G.1a) with (G.1a$'$), we see that they are identical, except that the time labels are permuted. Hence their values are equal, so we may retain just one of them and multiply its value by 2, thus cancelling the factor 2! The same holds true for (G.1b) and G.1b$'$). In general, in the nth order matrix element, we write out all different time orders, contract each time order in all possible ways, and draw the corresponding diagrams, both linked and unlinked. We will always find that for any type of diagram, there are $n!$ diagrams which are identical except that the times are permuted, so they give equal contributions. Hence we retain just one of them and cancel the $n!$.

Note that the diagrams we end up with after all this are time-ordered. They may then be added together as we did in §9.5 to form the non-time-ordered Feynman diagrams. (The usual method of eliminating the $n!$ involves permuting indices of the interaction Hamiltonian, rather than time label permutations. This produces Feynman diagrams directly. See Fetter and Walecka (1971), p. 96 ff.).

Continuing the process of evaluating the higher-order matrix elements in the numerator of (E.11) by means of (F.23), we obtain

$$i \times \text{Numerator of } G = \quad \cdots \qquad \text{(G.2)}$$

It is easy to see that, analogous to the case of the vacuum amplitude in (5.23, 24), the unlinked diagrams here may be factored into the product of the separate links, so that (G.2) may be written

$$
i \times \begin{array}{c} \text{Numerator} \\ \text{of } G \end{array} = \Big| \times \Big[1 + \bigcirc + \Big(\Big) + \begin{array}{c}\bigcirc\\\bigcirc\end{array} + \cdots \Big] +
$$

$$
\Big| \times \Big[1 + \bigcirc + \Big(\Big) + \begin{array}{c}\bigcirc\\\bigcirc\end{array} + \cdots \Big] + \cdots . \quad (G.3)
$$

The terms in brackets are just the diagrams in the vacuum amplitude expansion for this case, (5.20).

Now look at the denominator of (E.11). By (E.14), this is just the vacuum amplitude expansion, so the diagrams are exactly the same and we may write

$$
\begin{array}{c} \text{Denominator} \\ \text{of } G \end{array} = 1 + \bigcirc + \Big(\Big) + \begin{array}{c}\bigcirc\\\bigcirc\end{array} + \cdots . \quad (G.4)
$$

Thus the denominator precisely cancels the bracketed terms in the numerator, and all unlinked diagrams thereby vanish, yielding

$$
iG = \frac{i \times \text{Num. } G}{\text{Denom. } G} = \Big| + \Big| + \Big| + \bigtriangleup + \Big| + \cdots \quad (G.5)
$$

corroborating the result found in (4.34). This result is often called the 'linked cluster theorem for the propagator'.

In the case of an interaction Hamiltonian for a system of mutually interacting fermions, such as the H_1 in (7.48), the evaluation of the expansion of G in (E.11) by Wick's theorem (F.23) follows identical lines and we obtain the result:

$$
iG = \frac{i\,\text{Num. } G}{\text{Denom. } G} = \frac{\Big| \times \Big[1 + \bigcirc\!\!\!\sim\!\!\!\bigcirc + \bigcirc + \bigotimes + \cdots \Big] + \Big|\!\sim\!\bigcirc \times \Big[1 + \bigcirc\!\!\!\sim\!\!\!\bigcirc + \bigcirc + \bigotimes + \cdots \Big] + \cdots}{\Big[1 + \bigcirc\!\!\!\sim\!\!\!\bigcirc + \bigcirc + \bigotimes + \cdots \Big]}
$$

$$
= \Big| + \Big|\!\sim\!\bigcirc + \sim\!\!\!\!\sim + \Big|\!\!\sim\!\!\!\bigcirc + \Big|\!\!\sim\!\Big) + \bigotimes + \cdots \quad (G.6)
$$

confirming the result in (4.63).

In (G.6) we note that, for example, the bubble diagram (4.54a) comes from the following first-order term in the numerator of G:

$$M^{(1)}_{\substack{\text{inter-}\\\text{acting}}} = -i \int dt \sum_{\substack{k,m > k_F\\l,n < k_F}} \left(\frac{-i}{2}\right) V_{klmn}$$

$$\langle 0 | T\{\hat{b}_l(t)\hat{a}^\dagger_k(t)\hat{a}_m(t)\hat{b}^\dagger_n(t)\hat{a}_{k_2}(t_2)\hat{a}^\dagger_{k_1}(t_1)\} | 0 \rangle. \qquad \text{(G.6')}$$

The fully contracted product which corresponds to (4.54a) is the one connecting $\hat{b}_l$ to $\hat{b}^\dagger_n$, $\hat{a}^\dagger_k$ to $\hat{a}_{k_2}$ and $\hat{a}_m$ to $\hat{a}^\dagger_{k_1}$.

Note on fermion loops

If we try to calculate the third diagram in (G.1) by using dictionary Table 4.2, it is found that

$$
\begin{aligned}
\vphantom{x} &= iG^+_0(k_1, t_2 - t_1) \times \sum_k (-i) V_{kk} iG^-_0(k, t - t)\\[6pt]
&= e^{-i\epsilon_{k_1}(t_2 - t_1)}(-i) \sum_k V_{kk}(-e^{-i\epsilon_k(t-t)})\\[6pt]
&= +i \sum_k V_{kk} \times e^{-i\epsilon_{k_1}(t_2 - t_1)}. \qquad \text{(G.7)}
\end{aligned}
$$

Let us compare this with the result obtained by the present method. Consider the second term in (F.13) which this graph comes from, and put in a $(-i)$-factor for the V_{kk} to take care of the $(-i)^1$ in first order appearing in (E.11). This yields

$$-iM^{(1)} = -i \sum_k V_{kk} \times e^{-i\epsilon_{k_1}(t_2 - t_1)} \qquad \text{(G.8)}$$

which differs by a (-1) from (G.7). This is a typical situation, and it can be shown that the way to get the correct sign in general is to associate a factor of (-1) with each fermion loop. Thus, the diagram in (G.7) has one fermion loop, $\bigcirc$, so gets an extra factor of (-1). The fundamental reason for this rule may be obtained from a careful analysis of the number of interchanges of operators required when (F.23) is applied—each such interchange produces a factor of (-1). (See Fetter and Walecka (1971), p. 98; Raimes (1972), p. 139.)

Note on diagrams which violate the Pauli exclusion principle

It was pointed out that when $k_1 = k_2$, diagram (4) in (4.34) violates the Pauli exclusion principle. If we examine the expansion (G.2) for the numerator of G, we see that the diagram in question is actually cancelled by another diagram, an unlinked one, thus:

$$(G.9)$$

as is easily seen by translating the diagrams into functions. Note especially the (-1) coming from the fermion loop. Despite this cancellation, it is necessary to keep the exclusion-principle-violating diagrams in order to prove the 'linked cluster theorem for the propagator', $(G.5)$ (cf. comment after (5.13) regarding diagrams which violate conservation of particle number).

These statements are general. In the linked cluster expansion of the propagator $(G.5)$, or the second line of $(G.6)$, there will always be an infinite number of diagrams which violate the Pauli exclusion principle. Such diagrams cancel in the unlinked expansion, but must be kept if we wish to use the much simpler linked expansion.

Exercises

G.1 Show using Ex. F.4 that the matrix element in Ex. F.2(c) corresponds to the diagram:

G.2 For H_1 in (7.76) and $k_1 = k_2 > k_F$ and $t_2 > t_1' > t_1$, write out all 16 first-order matrix elements in the numerator of the propagator expansion, $(E.11)$, using particle–hole notation.

G.3 Show with the help of $(F.8)$ that 11 of the elements in Ex. G.2 are equal to zero.

G.4 Evaluate the non-zero elements in Ex. G.2 with the aid of contractions, and draw the corresponding Feynman diagrams.

Appendix H

The Spectral Density Function

1. Single-particle propagator

For brevity, we shall derive only the expression for $A^+(\mathbf{k}, \omega)$; that for A^- is found in exactly the same way. Call Ψ_n^N, E_n^N the exact eigenstates and energies of the Hamiltonian H of the interacting N-particle system. The propagator

$G^+(\mathbf{k},t)$ may be expressed as a sum over these exact states by inserting the unit operator

$$\sum_{n,N'} |\Psi_n^{N'}\rangle\langle\Psi_n^{N'}|$$

(see appendix A) into (9.7), letting $t_1=0$, $t_2=t$, and noting that in the sum over N', all terms are zero except those for which $N'=N+1$:

$$G^+(\mathbf{k},t) = -i\theta_t \sum_n \langle\Psi_0^N| e^{iHt} c_k |\Psi_n^{N+1}\rangle\langle\Psi_n^{N+1}| e^{-iHt} c_k^\dagger |\Psi_0^N\rangle$$

$$= -i\theta_t \sum_n |\langle\Psi_n^{N+1}|c_k^\dagger|\Psi_0^N\rangle|^2 e^{-i(E_n^{N+1}-E_0^N)t}$$

$$= -i\theta_t \sum_n |(c_k^\dagger)_{n0}|^2 e^{-i(E_n^{N+1}-E_0^N)t}. \tag{H.1}$$

Taking the Fourier transform of the above yields

$$G^+(\mathbf{k},\omega) = \sum_n |(c_k^\dagger)_{n0}|^2 \frac{1}{\omega-(E_n^{N+1}-E_0^N)+i\delta}. \tag{H.2}$$

This shows that the poles of G^+ occur at the energies of the interacting $N+1$-particle system minus the ground state energy of the interacting N-particle system, thus proving (3.14) for the case of a system with no external potential. A similar proof holds in the general case.

The exponentials in the above may be expressed in terms of the chemical potential defined in (9.23), thus:

$$E_n^{N+1}-E_0^N = \underbrace{E_n^{N+1}-E_0^{N+1}}_{\omega_{n0}^{N+1}} + \underbrace{E_0^{N+1}-E_0^N}_{\mu^{N+1}}. \tag{H.3}$$

For large N (like for electron gas or nuclear matter, but not for atoms or finite nuclei!) we have

$$\mu^{N+1} \approx \mu^N \equiv \mu$$
$$\omega_{n0}^{N+1} \approx \omega_{n0}^N \equiv \omega_{n0}, \tag{H.4}$$

giving

$$G^+(\mathbf{k},t) = -i\theta_t \sum_n |(c_k^\dagger)_{n0}|^2 e^{-i(\omega_{n0}+\mu)t} \tag{H.5}$$

$$G^+(\mathbf{k},\omega) = \sum_n |(c_k)_{n0}|^2 \frac{1}{\omega-(\omega_{n0}+\mu)+i\delta}. \tag{H.6}$$

In a system with large volume, the energy levels are so closely spaced that we can go from a sum to an integral by introducing the *spectral density function* (see below)

$$A^+(\mathbf{k},\omega)\,d\omega = \sum_{\omega<\omega_{n0}<\omega+d\omega} |(c_k^\dagger)_{n0}|^2 \tag{H.7}$$

or, equivalently,

$$A^+(\mathbf{k}, \omega) = \sum_n |(c_k^\dagger)_{n0}|^2 \delta(\omega - \omega_{n0}). \qquad (\text{H.8})$$

This function is defined only for $\omega \geqslant 0$ because by (H.3, 4), $\omega_{n0} \geqslant 0$. It gives the probability that the state $|\Psi_0^N\rangle$ with an added particle in state $\mathbf{k}$ is an exact eigenstate of the $N+1$-particle system with energy between ω and $\omega + d\omega$. Substituting (H.7) in (H.5, 6) gives

$$G^+(\mathbf{k}, t) = -i\theta_t \int\limits_0^\infty A^+(\mathbf{k}, \omega) e^{-i(\omega + \mu)t} \, d\omega \qquad (\text{H.9})$$

$$G^+(\mathbf{k}, \omega) = \int\limits_0^\infty d\omega' \frac{A^+(\mathbf{k}, \omega')}{\omega - \omega' - \mu + i\delta} \qquad (\text{H.10})$$

as in (9.22, 24).

It is important to observe that a profound change takes place when we go from the sum (H.6) to the integral (H.10). The sum (H.6) has an infinite number of *real* poles, whereas the integral (H.10) has a small number of *complex* poles. For example, as mentioned after (9.27), if we use a Lorentzian for $A^+(\mathbf{k}, \omega)$ in (H.10), we obtain the quasi particle propagator (8.37), which has a complex pole at $\omega = \epsilon_k' - i\tau_k$. This was also mentioned after (3.70).

The physical meaning of the appearance of complex poles when we go from a sum to an integral may be seen by looking at the corresponding expressions for G in the time domain, i.e., (H.5) and (H.9), Consider the sum (H.5) first. To analyse its behaviour, we note that there are two characteristic energies involved: First, if there is no interaction, then $(c_k^\dagger)_{n0} = \delta_{kn}$, i.e., $(c_k^\dagger)_{n0}$ is finite only for a single energy level. But with interaction, in typical cases $(c_k^\dagger)_{n0}$ is spread out over a band of energy levels from say n' to n'', having width $\Delta E = \omega_{n'',0} - \omega_{n',0}$. Secondly, there is the characteristic spacing between adjacent energy levels, $\Delta \epsilon \sim \omega_{n+1,0} - \omega_{n,0}$.

Now, at $t = 0$, all the terms in (H.5) are in phase and $G^+(t)$ is maximum. As t increases, the terms in (H.5) start to get out of phase with each other, and $G^+(t)$ decays in a characteristic time given by $\tau \sim \hbar/\Delta E$. However, if we wait a length of time $T \sim \hbar/\Delta\epsilon$, then the exponentials will start to get in phase with each other again, and $G^+(t)$ builds up again to its value at $t = 0$. (This is just the 'beat' phenomenon observed when we add two signals $\cos(2\pi\nu_1 t)$ and $\cos(2\pi\nu_2 t)$: the beat frequency is $\nu_2 - \nu_1$ and the corresponding period for build-up of the beat is $T = 1/(\nu_2 - \nu_1)$.) Thus, the Green's function shows periodic behaviour.

The above holds for a finite system, with corresponding finite distance between energy levels. But if we go to the infinite volume limit, then $(\omega_{n+1,0} - \omega_{n,0}) \to 0$, and the build-up time $T \to \infty$. That is, $G^+(t)$ becomes aperiodic, decaying to zero in a time of order τ, and never building up again. This is just the quasi particle behaviour. Thus the discontinuous change from real poles to complex poles in the infinite volume limit, is associated with the discontinuous change of the propagator from a periodic to an aperiodic function.

In practice, it is not necessary to have volume→∞ since for a typical large system, we find that T is so large compared with the times involved in the experiment that build-up will not be observed. However, in small systems, like atoms and light nuclei, the above considerations are not valid: the energy levels are widely spaced, the propagator poles are real, and the quasi particle picture does not hold.

2. Two-particle propagator

Consider first the particle–particle propagator. It will be sufficient to treat just the case (see (13.4)) with $t_4 = t_2 = t$ and $t_3 = t_1 = 0$, and $t > 0$. Using $c_{k_i}(t_i)$, $c_{k_i}^\dagger(t_i)$ instead of the ψ's in (13.4), this becomes

$$G_2^+(t) = i\theta_t \langle \Psi_0^N | c_{k_4}(t) c_{k_2}(t) c_{k_3}^\dagger(0) c_{k_1}^\dagger(0) | \Psi_0^N \rangle. \qquad (H.11)$$
$$\binom{\text{particle} -}{\text{particle}}$$

Inserting the unit operator gives

$$\begin{aligned} G_2^+(t) &= i\theta_t \sum_{\substack{n, N'}} \langle \Psi_0^N | e^{+iHt} c_{k_4} c_{k_2} e^{-iHt} | \Psi_n^{N'} \rangle \langle \Psi_n^{N'} | c_{k_3}^\dagger c_{k_1}^\dagger | \Psi_0^N \rangle \\ &= i\theta_t \sum_n \langle \Psi_0^N | c_{k_4} c_{k_2} | \Psi_n^{N+2} \rangle \langle \Psi_n^{N+2} | c_{k_3}^\dagger c_{k_1}^\dagger | \Psi_0^N \rangle e^{-i(E_n^{N+2} - E_0^N)t} \\ &\equiv i\theta_t \sum_n D_n e^{-i(E_n^{N+2} - E_0^N)t}. \qquad (H.12) \end{aligned}$$

Taking the Fourier transform just as in (H.1) we find:

$$\underset{(p-p)}{G_2^+(\omega)} = -\sum_n D_n \frac{1}{\omega - (E_n^{N+2} - E_0^N) + i\delta}, \qquad (H.13)$$

showing that the poles of the particle–particle propagator occur at the excited state energies of the interacting $N+2$-particle system minus the ground energy of the interacting N-particle system.

Similarly, we can analyse the polarization propagator, which is a special case of the particle–hole propagator, with $t_3 = t_4 = t$, $t_1 = t_2 = 0$ and $t > 0$ given by (see (13.5)):

$$F^+(t) = -i \langle \Psi_0 | c_{k_3}^\dagger(t) c_{k_4}(t) c_{k_1}^\dagger(0) c_{k_2}(0) | \Psi_0 \rangle. \qquad (H.14)$$

Inserting unit operators, and Fourier transforming, we find

$$F^+(\omega) = \sum_n B_n \frac{1}{\omega - (E_n^N - E_0^N) + i\delta} \tag{H.15}$$

showing that the poles of $F^+(\omega)$ yield the excitation energies of an N-particle system.

Exercises

H.1 Show that the spectral representations of $G^-(k, t)$ and $G^-(k, \omega)$ are given by the second term in (9.22), (9.24) respectively.

H.2 Verify the sum rule (9.25). [Hint: Convert (9.25) from an integral to a sum, then use commutation rule for c_k, $c_k^\dagger$.]

H.3 Assume that we have a system of N interacting fermions in a box of side L, and that these are approximately describable as non-interacting quasi particles of mass m^*. (a) What is ω_{n0} in (H.5) in this case? (b) What is the spacing between adjacent energy levels, and the corresponding 'build-up' time for the single particle propagator? (Note that (H.5) here is just a Fourier series, and the build-up time is just the period for the series.) (c) What is the time required for a particle at the Fermi surface to traverse the box? (d) Compare the traverse time and build-up time for an electron in a 1 cm³ metal, and for a nucleon in a heavy nucleus.

Appendix I

How the iδ Factor is Used

The simplest example of the use of the $i\delta$ factor in propagator calculations is to make the inverse Fourier transformation from $G_0^+(k, \omega)$ in (3.13) back to $G_0^+(k, t)$ (set $t = t_2 - t_1$) in (3.12'). The Fourier transform is given by

$$G_0^+(k, t) = \int_{-\infty}^{+\infty} \frac{d\omega}{2\pi} e^{-i\omega t} G_0^+(k, \omega)$$

$$= \int_{-\infty}^{+\infty} \frac{d\omega}{2\pi} \frac{e^{-i\omega t}}{\omega - \epsilon_k + i\delta}. \tag{I.1}$$

The integrand has a pole at $\omega = \epsilon_k - i\delta$. The integral may be evaluated by integrating along the following contour surrounding the pole

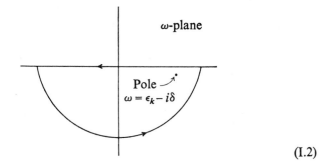

$$\text{(I.2)}$$

By the theorem of residues:

$$\oint = \int_{+\infty}^{-\infty} + \int_{\smile} = 2\pi i \sum_n \text{residue at pole } n$$

$$= 2\pi i \lim_{\omega \to \epsilon_k - i\delta} \frac{(\omega - \epsilon_k + i\delta)e^{-i\omega t}}{2\pi(\omega - \epsilon_k + i\delta)}$$

$$= i e^{-i(\epsilon_k - i\delta)t}. \qquad \text{(I.3)}$$

The integral around $\smile$ vanishes because $t > 0$ (by definition of G^+ in (3.1)), and therefore the 'convergence factor' $e^{-i\omega t}$, goes to zero in (I.1). Hence we obtain

$$-\int_{-\infty}^{+\infty} \frac{d\omega}{2\pi} \frac{e^{-i\omega t}}{\omega - \epsilon_k + i\delta} = i e^{-i(\epsilon_k - i\delta)t}. \qquad \text{(I.4)}$$

For $t < 0$, we complete the contour in the upper half-plane, where there are no poles, so the integral vanishes. Hence we obtain

$$G_0^+(k,t) = -i\theta_t e^{-i(\epsilon_k - i\delta)t} \qquad \text{(I.5)}$$

confirming (3.12'). This method may also be used to take the transform of $G_0(k,\omega)$ in (8.35).

Note that to get the correct transform for $t = 0$ (i.e., the $t_2 = t_1$ case in (4.31)) it is necessary to include a convergence factor $\exp(i\omega 0^+)$, where $0^+ \times \infty = \infty$, in carrying out the contour integral. This makes the integral vanish over a half-circle in the upper half-plane. This factor may also be included in $G_0(k,\omega)$ itself. (See Schrieffer (1964), pp. 108–9.)

Appendix J

Electron Propagator in Normal Electron–Phonon System

The Hamiltonian for the coupled electron–phonon system is given in (15.4). In order to calculate the electron propagator for this system, we need to add to the dictionary Table 9.1 expressions for the electron–phonon interaction and the free phonon propagator. In jellium approximation, the

Table J.1　*Diagram dictionary for phonons*

Diagram	Function
free phonon　〰〰〰〰	$iD_0(\mathbf{q}, \omega) = \dfrac{i2\Omega_q}{\omega^2 - \Omega_q^2 + i\delta}$
electron– phonon interaction	$-ig_q = -i\dfrac{4\pi e^2}{q}\sqrt{\left/\left(\dfrac{Z^2 N}{2\Omega_p M}\right)\right.}$

former is given by (15.8). The latter is (see chapter 16, equation just after (16.41)):

$$D_0(\mathbf{q}, \omega) = \frac{1}{\omega - \Omega_q + i\delta} - \frac{1}{\omega + \Omega_q - i\delta} = \frac{2\Omega_q}{\omega^2 - \Omega_q^2 + i\delta} \tag{J.1}$$

which is just like the electron propagator except for the addition of a 'negative frequency' part. (The factor in the numerator of (J.1) is a matter of convention. See Schultz (1964), p. 124.) These quantities may be represented diagrammatically as in Table J.1.

The expansion for G is then given by

$$\tag{J.2}$$

(Note that all bubble diagrams have been omitted because

$$(a) \quad \overset{\mathbf{q}=0}{\rightsquigarrow\!\bigcirc} = 0, \quad (b) \quad \overset{\mathbf{q}=0}{\bullet\!\!\text{\tiny oooo}\!\bigcirc} = 0. \tag{J.3}$$

Equation (a) follows from the statement just after (10.32). Regarding (b), we see that the $\mathbf{q}=0$ phonon has infinite wavelength, meaning a translational movement of the whole crystal. This is eliminated simply by holding the crystal fixed.)

The striking thing about (J.2) is that the phonon lines (including vertex dots) enter the diagrams in exactly the same way as the Coulomb interactions do. This means that we have an effective electron–electron interaction due to the emission and absorption of virtual phonons. This is just the interaction pointed out by Fröhlich, which we discussed in §15.3(a). Stated diagrammatically:

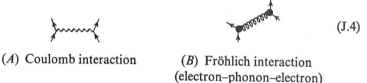 (J.4)

    (A) Coulomb interaction   (B) Fröhlich interaction
                                      (electron–phonon–electron)

Observe that according to Table J.1 (B) involves frequency parameter ω, and is therefore a retarded interaction.

Because of the symmetrical way in which (A), (B) enter series (J.2), we can do a little preliminary partial summation sleight-of-hand to make the series as simple-looking as in the no-phonon case. First, the expansion is rewritten in factored form:

$$\parallel\!\!\uparrow\!\!\parallel = \uparrow + \uparrow \times \left(\bigcirc + \text{\tiny oooo} \right) + \left(\uparrow \, \Diamond \, \uparrow \right) \times$$

$$\times \left(\rightsquigarrow + \bullet\text{\tiny ooooo}\bullet \right) \times \left(\rightsquigarrow + \bullet\text{\tiny ooooo}\bullet \right) + \cdots \tag{J.5}$$

Defining the *combined interaction*

$$\text{\large www} = \rightsquigarrow + \bullet\text{\tiny ooooo}\bullet , \tag{J.6}$$

this reduces to

$$\text{(J.7)}$$

which is just as simple in form as the no-phonon propagator expansion (9.40).

The usual partial summations of chapter 10 can be carried out on (J.7). The Dyson equation is

$$\text{(J.8)}$$

where the irreducible self-energy is given by

$$\text{(J.9)}$$

It is now easy to do all the renormalizations on (J.9) that were done on (10.8). The clothed combined interaction is given by

$$\text{(J.10)}$$

where $\left(\!\Lambda\!\right)$ is the sum over irreducible polarization parts (cannot be broken in two by removing a combined interaction line):

$$\text{(J.11)}$$

similar to (10.54, 52). The free propagator lines may be dressed as in §11.1. The renormalized series then has no insertions in any interaction or propagator, and all interactions and free propagators become clothed:

$$\left(\Sigma\right) = \;\mathrm{[diagram]}\; + \;\mathrm{[diagram]}\; + \;\mathrm{[diagram]}\; + \cdots . \tag{J.12}$$

The solution of (J.10) is evidently

$$\mathrm{[dressed\ line]} = \frac{\mathrm{[wavy\ line]}}{1 - \left(\Lambda\right)\mathrm{[line]}} = \frac{\mathrm{[wavy\ line]}}{1 - \left(\Lambda\right)\mathrm{[line]} - \left(\Lambda\right)\mathrm{[line]}}$$

$$+ \;\frac{\mathrm{[spring\ line]}}{1 - \left(\Lambda\right)\mathrm{[line]} - \left(\Lambda\right)\mathrm{[line]}} \tag{J.13}$$

showing that the dressed combined interaction is the sum of a shielded Coulomb interaction and a shielded Fröhlich interaction. Note that if $\left(\Lambda\right)$ is replaced by the lowest-order pair bubble in (J.11), the shielding term in the denominator is just the sum of a Coulomb and a phonon part.

In the literature (see bibliography at end of chapter 15!), the dressed combined interaction is usually written in another form, more complicated, but having a slightly more direct physical interpretation. We define (1) a dressed Coulomb line by:

$$\mathrm{[dressed\ Coulomb\ line]} = \mathrm{[line]} + \left(\Lambda\right)\mathrm{[line]} + \left(\Lambda\right)\left(\Lambda\right)\mathrm{[line]} + \cdots$$

$$= \frac{\mathrm{[line]}}{1 - \left(\Lambda\right)\mathrm{[line]}} , \tag{J.14}$$

(2) a dressed phonon line by

$$(J.15)$$

or

and (3) a dressed electron–phonon interaction by

$$(J.16)$$

(The line stumps in (J.16) are only to show where electron and phonon lines are to be attached. Note that ⊙ is not an ordinary dressed vertex like (11.30) since it contains insertions only in the phonon branch.) Then it is simple diagram algebra to show that

$$(J.17)$$

That is, the dressed combined interaction is the dressed Coulomb interaction plus dressed phonon interaction.

Let us examine (J.17) in RPA, where $\left(\!\!\begin{array}{c}\wedge\end{array}\!\!\right) = \left(\!\!\begin{array}{c}\vee\end{array}\!\!\right)$. Then, the dressed Coulomb line is just V_{eff} in RPA (10.35):

$$\underset{RPA}{\sim\!\!\sim\!\!\sim\!\!\sim} \equiv -iV_{eff}(\mathbf{q}, \omega) = \frac{-iV_q}{\epsilon_{RPA}(\mathbf{q}, \omega)}. \tag{J.18}$$

The dressed phonon line is (see bibliography at end of chapter 15)

$$\underset{RPA}{\overline{\underline{\text{00000000}}}} \equiv iD(\mathbf{q}, \omega) = \frac{i2\sqrt{\epsilon_{RPA}(\mathbf{q}, 0)}\,\omega_q}{\omega^2 - \omega_q^2 + i\delta} \tag{J.19}$$

where ω_q is the phonon frequency renormalized by the accompanying electron cloud; it is given by

$$\omega_q = \frac{\Omega_q}{\sqrt{\{\epsilon_{RPA}(\mathbf{q}, 0)\}}} = \left(\frac{4\pi Z^2 e^2 N^2/M}{1 + \lambda^2/q^2}\right)^{\frac{1}{2}} \propto q \quad \text{(for small } q\text{)}. \tag{J.20}$$

This is a much more respectable behaviour for a longitudinal phonon than that shown in Fig. 15.1. Note that beside the unimportant $\sqrt{\{\epsilon_{RPA}(\mathbf{q}, 0)\}}$, (J.19) is just the free propagator (J.1) with Ω_q replaced by ω_q. The dressed electron–phonon interaction is given by

$$\underset{RPA}{\bigodot} \equiv -ig_{eff}(\mathbf{q}) = \frac{-ig_q}{\epsilon_{RPA}(\mathbf{q}, 0)}. \tag{J.21}$$

This is a screened interaction and is evidently much smaller than the bare one.

We may use these results to write out the expression for the dressed combined interaction in (J.17):

$$\underset{RPA}{\mathbf{www}} \equiv -i\mathcal{V}_{RPA}(\mathbf{q}, \omega) = -i\left[|g_{eff}(\mathbf{q})|^2 D(\mathbf{q}, \omega) + V_{eff}(\mathbf{q}, \omega)\right]$$

$$= -i\left\{\left(\frac{g_q}{\epsilon_{RPA}(\mathbf{q}, 0)}\right)^2 \times \frac{2\sqrt{\epsilon_{RPA}(\mathbf{q}, 0)}\,\omega_q}{\omega^2 - \omega_q^2 + i\delta} + \right. \tag{J.22}$$

$$\left. + \frac{4\pi e^2}{q^2\,\epsilon_{RPA}(\mathbf{q}, \omega)}\right\}$$

which is evidently a retarded interaction. It is possible at this point to make contact with the electron–phonon–electron interaction (15.13) used by BCS. The BCS interaction is just a static approximation to (J.22) in which $\epsilon_{RPA}(\mathbf{q}, \omega)$ is replaced by the static limit as $\mathbf{q} \to 0$

$$\epsilon_{RPA}(\mathbf{q}, \omega) \to \epsilon(\mathbf{q}) = 1 + \frac{\lambda^2}{q^2} \tag{J.23}$$

and ω in the phonon propagator is set equal to $\epsilon_k - \epsilon_{k-q}$. These replacements in (J.22) yield just (15.13).

Note that the first term in (J.22), i.e., the Fröhlich interaction, is negative for small ω. Thus the criterion for superconductivity using the combined interaction is, again, that the Fröhlich term should dominate the Coulomb term. In the normal case, with which we are dealing in this appendix, the Coulomb term is the dominant one.

Let us now calculate the electron propagator, using the following approximation for the proper self-energy:

$$\bigcirc\!\!\!\!\!\Sigma \approx \begin{array}{c} \mathbf{k}+\mathbf{q}, \\ \omega + \epsilon \end{array} \qquad (J.24)$$

with ⌇⌇⌇⌇ given by (J.22). We have thus

$$\Sigma(\mathbf{k}, \omega) = i \int \frac{d^3 q \, d\epsilon}{(2\pi)^4} G_0(\mathbf{k}+\mathbf{q}, \omega+\epsilon) \left[|g_{\text{eff}}(\mathbf{q})|^2 D(\mathbf{q}, \epsilon) + V_{\text{eff}}(\mathbf{q}, \epsilon) \right]. \qquad (J.25)$$

The evaluation of this is in Schrieffer (1964a), p. 151 ff.; it leads to a quasi particle with an effective mass near the Fermi surface given by

$$m^* = m + \delta m_{\text{Coul.}} + \delta m_{\text{phon.}} \qquad (J.26)$$

where $\delta m_{\text{Coul.}}$ is obtained from (10.47) and

$$\delta m_{\text{phon.}} = m \frac{4}{\pi} \left(\frac{\pi^2}{18} \right)^{\frac{1}{3}} r_s \ln \left[\frac{(2k_F)^2 + \lambda^2}{\lambda^2} \right] \qquad (J.27)$$

where λ is the RPA screening constant in (10.39). This result is just what we would expect: the bare electron has its mass renormalized partly by the accompanying cloud of other electrons, and partly by the accompanying phonon cloud.

When the Fröhlich term dominates the combined interaction, the above calculation is no longer valid, since the system becomes a superconductor. In that case, it is necessary to use the Nambu method to calculate the electron propagator, as described in §15.6.

Appendix K

Spin Wave Functions

Since most people have only seen spin presented in the matrix scheme, we will give a brief review of the less familiar Schrödinger scheme used here. (See Raimes (1961), p. 113 and Margenau (1961), p. 59 for details.) Consider a free fermion. The space part of its wave function, $\phi_k(\mathbf{r})$ is the eigenfunction of the momentum operator, $\mathbf{p} = -i\nabla_r$, i.e.,

$$-i\nabla_r \phi_k(\mathbf{r}) = \mathbf{k}\phi_k(\mathbf{r}), \quad (\hbar = 1), \tag{K.1}$$

where $\mathbf{r}$ is the space coordinate and $\mathbf{k} = (k_x, k_y, k_z)$ are the momentum eigenvalues. Analogously, the spin part of the wave function, $\eta_\sigma(\gamma)$ is the eigenfunction of the z-component of the spin operator, S_z, i.e.,

$$S_z \eta_\sigma(\gamma) = \sigma\eta_\sigma(\gamma), \tag{K.2}$$

where γ is the spin coordinate and σ are the eigenvalues of S_z. σ has the values $\pm\frac{1}{2}\hbar$ or $\pm\frac{1}{2}$ (for $\hbar = 1$). There is no good physical picture of γ, but it may be visualized as the cosine of the angle between the spin axis and the z-axis. However, this angle can only have the values 0 and π so that $\gamma = +1, -1$. The spin wave functions are

$$\eta_{+\frac{1}{2}}(1) = 1, \quad \eta_{+\frac{1}{2}}(-1) = 0, \quad \eta_{-\frac{1}{2}}(1) = 0, \quad \eta_{-\frac{1}{2}}(-1) = 1. \tag{K.3}$$

They thus form a complete orthogonal set, i.e., it is easily checked that

$$\sum_\gamma \eta_\sigma^*(\gamma)\eta_{\sigma'}(\gamma) = \delta_{\sigma\sigma'}. \tag{K.4}$$

The S_x, S_y operators operate as follows on the spin wave functions:

$$S_x \eta_\sigma(\gamma) = \tfrac{1}{2}\eta_{-\sigma}(\gamma), \quad S_y \eta_\sigma(\gamma) = i\sigma\eta_{-\sigma}(\gamma). \tag{K.5}$$

Appendix L

Summary of Different Types of Propagators and their Spectral Representations and Analytic Properties

A. At $T = 0$, with fixed number of particles, N

(For details, see Fetter and Walecka (1971), p. 72 ff., and Abrikosov (1965), p. 49 ff.)

We first give the definitions of the various propagators:

1. *Time-ordered or causal*

$$G(\mathbf{k}, t) = -i\langle \Psi_0 | T\{c_k(t) c_k^\dagger(0)\} | \Psi_0 \rangle. \tag{L.1}$$

2. *Retarded (first type)*

$$G^+(\mathbf{k}, t) = -i\theta_t \langle \Psi_0 | c_k(t) c_k^\dagger(0) | \Psi_0 \rangle. \tag{L.2}$$

3. *Advanced (first type)*

$$G^-(\mathbf{k}, t) = +i\theta_{-t}\langle \Psi_0 | c_k^\dagger(0) c_k(t) | \Psi_0 \rangle. \tag{L.3}$$

4. *Retarded (second type; this is what is usually meant by 'retarded')*

$$G^R(\mathbf{k}, t) = -i\theta_t \langle \Psi_0 | [c_k(t), c_k^\dagger(0)]_+ | \Psi_0 \rangle. \tag{L.4}$$

5. *Advanced (second type)*

$$G^A(\mathbf{k}, t) = +i\theta_{-t} \langle \Psi_0 | [c_k(t), c_k^\dagger(0)]_+ | \Psi_0 \rangle. \tag{L.5}$$

By using the same method as in appendix H, we find the following Lehmann representations for the Fourier transforms of the above propagators:

1. *Causal*

$$G(\mathbf{k}, \omega) = \int_0^\infty d\omega' \left[\frac{A^+(\mathbf{k}, \omega')}{\omega - \omega' - \mu + i\delta} + \frac{A^-(\mathbf{k}, \omega')}{\omega + \omega' - \mu - i\delta} \right]. \tag{L.6}$$

2. *Retarded, first type*

$$G^+(\mathbf{k}, \omega) = \int_0^\infty d\omega' \left[\frac{A^+(\mathbf{k}, \omega')}{\omega - \omega' - \mu + i\delta} \right]. \tag{L.7}$$

3. *Advanced, first type*

$$G^-(\mathbf{k}, \omega) = \int_0^\infty d\omega' \frac{A^-(\mathbf{k}, \omega')}{\omega + \omega' - \mu - i\delta}. \tag{L.8}$$

4. *Retarded, second type*

$$G^R(\mathbf{k}, \omega) = \int_0^\infty d\omega' \left[\frac{A^+(\mathbf{k}, \omega')}{\omega - \omega' - \mu + i\delta} + \frac{A^-(\mathbf{k}, \omega')}{\omega + \omega' - \mu + i\delta} \right]. \tag{L.9}$$

5. *Advanced, second type*

$$G^A(\mathbf{k}, \omega) = \int_0^\infty d\omega' \left[\frac{A^+(\mathbf{k}, \omega')}{\omega - \omega' - \mu - i\delta} + \frac{A^-(\mathbf{k}, \omega')}{\omega + \omega' - \mu - i\delta} \right]. \tag{L.10}$$

The causal propagator, $G(\mathbf{k}, \omega)$ may be calculated diagrammatically. From it we may obtain the spectral functions A^+ and A^- using (3.76) which yields

$$G(\mathbf{k}, \omega) = P \int_0^\infty d\omega' \frac{A^+(\mathbf{k}, \omega')}{\omega - \omega' - \mu} - i\pi \int_0^\infty d\omega' A^+(\mathbf{k}, \omega') \delta(\omega - \omega' - \mu)$$

$$+ P \int_0^\infty d\omega' \frac{A^-(\mathbf{k}, \omega')}{\omega + \omega' - \mu} + i\pi \int_0^\infty d\omega' A^-(\mathbf{k}, \omega') \delta(\omega + \omega' - \mu) \tag{L.11}$$

from which we immediately obtain from the imaginary parts:

$$A^+(\mathbf{k}, \omega - \mu) = -\frac{1}{\pi}\operatorname{Im} G(\mathbf{k}, \omega), \quad \omega > \mu$$

$$A^-(\mathbf{k}, \mu - \omega) = +\frac{1}{\pi}\operatorname{Im} G(\mathbf{k}, \omega), \quad \omega < \mu. \tag{L.12}$$

All the other propagators can be calculated from A^+ and A^-.

B. At $T = 0$ with fixed chemical potential, μ

The above $T = 0$ representation is for a system with a fixed number of particles, N. For a system with fixed chemical potential (see §9.7) we simply replace ω by $\omega + \mu$ everywhere in (L.6)→(L.10) (see (9.47), also Abrikosov (1965), p. 63). If in addition we define a new spectral density function:

$$\begin{aligned} A(\mathbf{k}, \omega') &= A^+(\mathbf{k}, \omega'), \quad \omega' > 0 \\ &= A^-(\mathbf{k}, -\omega'), \quad \omega' < 0 \end{aligned} \tag{L.13}$$

then (L.6)→(L.10) assume the simple forms (note: $\bar{G}(\mathbf{k}, \omega) \equiv G(\mathbf{k}, \omega + \mu)$ is the propagator for fixed μ):

1. *Causal*

$$\bar{G}(\mathbf{k}, \omega) = \int_{-\infty}^{+\infty} d\omega' \, A(\mathbf{k}, \omega') \left[\frac{\theta_{\omega'}}{\omega - \omega' + i\delta} + \frac{\theta_{-\omega'}}{\omega - \omega' - i\delta} \right]. \tag{L.14}$$

2. *Retarded, first type*

$$\bar{G}^+(\mathbf{k}, \omega) = \int_{-\infty}^{+\infty} d\omega' \, \frac{A(\mathbf{k}, \omega')\,\theta_{\omega'}}{\omega - \omega' + i\delta}. \tag{L.15}$$

3. *Advanced, first type*

$$\bar{G}^-(\mathbf{k}, \omega) = \int_{-\infty}^{+\infty} d\omega' \, \frac{A(\mathbf{k}, \omega')\,\theta_{-\omega'}}{\omega - \omega' - i\delta}. \tag{L.16}$$

4. *Retarded, second type*

$$\bar{G}^R(\mathbf{k}, \omega) = \int_{-\infty}^{+\infty} d\omega' \, \frac{A(\mathbf{k}, \omega')}{\omega - \omega' + i\delta}. \tag{L.17}$$

5. *Advanced, second type*

$$\bar{G}^A(\mathbf{k}, \omega) = \int_{-\infty}^{+\infty} d\omega' \, \frac{A(\mathbf{k}, \omega')}{\omega - \omega' - i\delta}. \tag{L.18}$$

C. Analytic properties of propagators at $T=0$

These expressions may be used to study the analytic properties of the various propagators (Abrikosov (1965), pp. 55, 56). The argument and results are the same for fixed N or fixed μ; we use fixed μ since the expressions are simpler. Consider $\bar{G}^R$ first, and apply (3.76):

$$\bar{G}^R(\mathbf{k}, \omega) = P \int_{-\infty}^{+\infty} d\omega' \frac{A(\mathbf{k}, \omega')}{\omega - \omega'} - i\pi A(\mathbf{k}, \omega). \qquad (L.19)$$

Hence

$$\text{Re } \bar{G}^R(\mathbf{k}, \omega) = \frac{P}{\pi} \int_{-\infty}^{+\infty} d\omega' \frac{\text{Im } \bar{G}^R(\mathbf{k}, \omega')}{\omega' - \omega}. \qquad (L.20)$$

A function which obeys this 'dispersion relation' between its real and imaginary parts is analytic in the upper half ω-plane (UHP) but not necessarily in the lower half ω-plane (LHP) where it may have poles (Titchmarsh (1948)). A simple example is when A is the Lorentzian:

$$A(\mathbf{k}, \omega) = \frac{\Delta/\pi}{[\omega - (\epsilon_k - \mu)]^2 + \Delta^2}. \qquad (L.21)$$

This yields

$$\bar{G}^R(\mathbf{k}, \omega) = \frac{\omega - (\epsilon_k - \mu)}{[\omega - (\epsilon_k - \mu)]^2 + \Delta^2} - i \frac{\Delta}{[\omega - (\epsilon_k - \mu)]^2 + \Delta^2} = \frac{1}{\omega - (\epsilon_k - \mu) + i\Delta} \qquad (L.22)$$

which is analytic in the UHP but has a simple pole at $\omega = (\epsilon_k - \mu) - i\Delta$ in the LHP.

Similarly, $\bar{G}^A$ obeys the dispersion relation

$$\text{Re } \bar{G}^A(\mathbf{k}, \omega) = -\frac{P}{\pi} \int_{-\infty}^{+\infty} d\omega' \frac{\text{Im } \bar{G}^A(\mathbf{k}, \omega')}{\omega' - \omega} \qquad (L.23)$$

which is characteristic for functions analytic in the LHP, with poles in the UHP. If A in (L.21) is used, we find an example of this:

$$\bar{G}^A(\mathbf{k}, \omega) = \frac{1}{\omega - (\epsilon_k - \mu) - i\Delta} \qquad (L.24)$$

which has a pole in the UHP.

The causal propagator obeys

$$\text{Re}\, \bar{G}(\mathbf{k}, \omega) = \frac{1}{\pi} P \int\limits_{-\infty}^{+\infty} d\omega' \, \frac{\text{sgn}\,(\omega')\,\text{Im}\,\bar{G}(\mathbf{k}, \omega')}{\omega' - \omega} \qquad (L.25)$$

which is not analytic in either half-plane. It is thus advantageous to use $\bar{G}^R$, $\bar{G}^A$ rather than $\bar{G}$ because of their simpler analytic properties.

By applying (3.76) to $\bar{G}$, $\bar{G}^R$, and $\bar{G}^A$ it is easy to show that for real ω:

$$\text{Re}\, \bar{G}(\mathbf{k}, \omega) = \text{Re}\, \bar{G}^R(\mathbf{k}, \omega)$$
$$\text{Im}\, \bar{G}(\mathbf{k}, \omega) = \text{Im}\, \bar{G}^R(\mathbf{k}, \omega) \quad \text{for } \omega > 0$$
$$= -\text{Im}\, \bar{G}^R(\mathbf{k}, \omega) \quad \text{for } \omega < 0 \qquad (L.26)$$

$$\text{Re}\, \bar{G}(\mathbf{k}, \omega) = \text{Re}\, \bar{G}^A(\mathbf{k}, \omega)$$
$$\text{Im}\, \bar{G}(\mathbf{k}, \omega) = -\text{Im}\, \bar{G}^A(\mathbf{k}, \omega) \quad \text{for } \omega > 0$$
$$= \text{Im}\, \bar{G}^A(\mathbf{k}, \omega) \quad \text{for } \omega < 0 \qquad (L.27)$$

whence

$$\bar{G}(\mathbf{k}, \omega) = \bar{G}^R(\mathbf{k}, \omega) \quad \text{for } \omega > 0$$
$$= \bar{G}^A(\mathbf{k}, \omega) \quad \text{for } \omega < 0. \qquad (L.28)$$

D. Finite temperature, fixed μ

The real time propagators at finite T are defined just as the $T=0$ ones (L.1)→(L.5) except that the average is now over a grand canonical ensemble (see, e.g., (14.18)). The Lehmann representations of their Fourier transforms are just like those at $T=0$ except that θ's are replaced by Fermi functions, $f^-(\omega) = [\exp(\beta\omega)+1]^{-1}$, $f^+ = 1 = f^-$. They may be found by the finite T analogue of the method in Appendix H (see Fetter and Walecka (1971), p. 292 ff., Abrikosov (1965), p. 141 ff.).

1. *Causal*

$$G(\mathbf{k}, \omega) = \int\limits_{-\infty}^{+\infty} d\omega' \, A(\mathbf{k}, \omega') \left[\frac{f^+(\omega')}{\omega - \omega' + i\delta} + \frac{f^-(\omega')}{\omega - \omega' - i\delta} \right]. \qquad (L.29)$$

2. *Retarded, first type*

$$\bar{G}^+(\mathbf{k}, \omega) = \int\limits_{-\infty}^{+\infty} d\omega' \, \frac{A(\mathbf{k}, \omega')f^+(\omega')}{\omega - \omega' + i\delta}. \qquad (L.30)$$

3. *Advanced, first type*

$$\bar{G}^-(\mathbf{k}, \omega) = \int\limits_{-\infty}^{+\infty} d\omega' \, \frac{A(\mathbf{k}, \omega')f^-(\omega')}{\omega - \omega' - i\delta}. \qquad (L.31)$$

4. *Retarded, second type*

$$\bar{G}^R(\mathbf{k}, \omega) = \int\limits_{-\infty}^{+\infty} d\omega' \, \frac{A(\mathbf{k}, \omega')}{\omega - \omega' + i\delta}. \tag{L.32}$$

5. *Advanced, second type*

$$\bar{G}^A(\mathbf{k}, \omega) = \int\limits_{-\infty}^{+\infty} d\omega' \, \frac{A(\mathbf{k}, \omega')}{\omega - \omega' - i\delta}. \tag{L.33}$$

None of the above propagators can be calculated diagrammatically. But we can use diagrams to calculate the related 'imaginary time' propagator, from which we may obtain the above propagators. The imaginary time propagator is:

$$\mathscr{G}(\mathbf{k}, \tau) = -\langle T\{c_k(\tau) c_k^\dagger(0)\}\rangle. \tag{L.34}$$

Its Fourier transform $\mathscr{G}(\mathbf{k}, i\omega_n)$ (the 'imaginary frequency' propagator), has the Lehmann representation (see Fetter and Walecka (1971), p. 297):

$$\mathscr{G}(\mathbf{k}, i\omega_n) = \int\limits_{-\infty}^{+\infty} d\omega' \, \frac{A(\mathbf{k}, \omega')}{i\omega_n - \omega'} \qquad \omega_n = \frac{(2n+1)\pi}{\beta} \tag{L.35}$$

where $A(\mathbf{k}, \omega')$ is the same as that in (L.29)→(L.33). This may be analytically continued from the points $i\omega_n$ to the whole complex z-plane by replacing $i\omega_n$ by z (Fetter and Walecka (1971), p. 297–8). We may then obtain A from

$$\mathscr{G}(\mathbf{k}, z \to \omega + i\delta) = \int\limits_{-\infty}^{+\infty} d\omega' \, \frac{A(\mathbf{k}, \omega')}{\omega - \omega' + i\delta} \tag{L.36}$$

by applying (3.76) which yields

$$A(\mathbf{k}, \omega) = -\frac{1}{\pi} \operatorname{Im} \mathscr{G}(\mathbf{k}, \omega + i\delta). \tag{L.37}$$

Similarly:

$$A(\mathbf{k}, \omega) = +\frac{1}{\pi} \operatorname{Im} \mathscr{G}(\mathbf{k}, \omega - i\delta).$$

Also:

$$A(\mathbf{k}, \omega) = \frac{1}{2\pi i} [\mathscr{G}(\mathbf{k}, \omega - i\delta) - \mathscr{G}(\mathbf{k}, \omega + i\delta)]. \tag{L.38}$$

Having obtained A from the imaginary frequency propagator, we can place it in (L.29)→(L.33) to find all the real frequency propagators. Note that

$$\bar{G}^R(\mathbf{k}, \omega) = \mathcal{G}(\mathbf{k}, \omega + i\delta) \tag{L.39}$$

$$\bar{G}^A(\mathbf{k}, \omega) = \mathcal{G}(\mathbf{k}, \omega - i\delta). \tag{L.40}$$

Diligent use of (3.76) yields that $\bar{G}^R$ and $\bar{G}^A$ at finite T obey the same dispersion relations as at $T = 0$ (see (L.19), (L.23)) so their analytic properties are the same as at $T = 0$. $\bar{G}$ is easily shown to obey (cf. (L.25))

$$\mathrm{Re}\,\bar{G}(\mathbf{k}, \omega) = \frac{P}{\pi} \int_{-\infty}^{+\infty} d\omega' \frac{\coth(\tfrac{1}{2}\beta\omega')\,\mathrm{Im}\,\bar{G}(\mathbf{k}, \omega)}{\omega' - \omega} \tag{L.41}$$

so it is not analytic. We also find for real ω:

$$\bar{G}(\mathbf{k}, \omega) = f^+(\omega)\,\bar{G}^R(\mathbf{k}, \omega) + f^-(\omega)\,\bar{G}^A(\mathbf{k}, \omega) \tag{L.42}$$

which is the finite T version of (L.28).

E. Two-particle propagators

The above discussion has been restricted to single particle propagators. Two-particle propagators (or the related scattering amplitudes) are much more complicated since they involve in general three independent frequencies. However, in many cases of interest, such as the polarization propagator (13.12) or the K-matrix in Exercises 10.4, 10.7, only one frequency occurs. In such cases, the analytic properties are the same as those for the single particle propagators. (Abrikosov (1965), p. 148.)

Exercise

L.1 Show with the aid of (L.17) that the Lorentzian spectral density (L.21) yields the propagator (L.22) (use contour integration).

Appendix M

The Decoupled Equations of Motion for the Green's Function Expressed as a Partial Sum of Feynman Diagrams

Throughout this book we have calculated the Green's function by expanding it in a perturbation series with the aid of Feynman diagrams, then partially summing over a selected set of diagrams. An alternative method is to write out the hierarchy of differential equations satisfied by the Green's function and solve these equations approximately by a 'decoupling' procedure. We will briefly indicate how this decoupling procedure works, and show that any decoupling is equivalent to a partial sum. For details, see Mattuck and Theumann (1971), abbreviated MT.

Let us write the general n-particle propagator in the form

$$G_n(12,\ldots,n;1'2',\ldots,n') = (-i)^n \langle T\{\psi(1)\psi(2),\ldots,\psi(n)\,\psi^\dagger(n'),\ldots,\psi^\dagger(1')\}\rangle.$$
(M.1)

The equation of motion for G_n may be found with the aid of the equation of motion for the ψ operators in Heisenberg picture:

$$\left. \begin{aligned} \left(i\frac{\partial}{\partial t} + \frac{\nabla^2}{2m}\right)\psi(\mathbf{r},t) &= \int d\mathbf{r}'\, v(\mathbf{r}-\mathbf{r}')\,\psi^\dagger(\mathbf{r}',t)\,\psi(\mathbf{r}',t)\,\psi(\mathbf{r},t), \\ \left(i\frac{\partial}{\partial t} - \frac{\nabla^2}{2m}\right)\psi^\dagger(\mathbf{r},t) &= -\int d\mathbf{r}'\, v(\mathbf{r}-\mathbf{r}')\,\psi^\dagger(\mathbf{r},t')\,\psi^\dagger(\mathbf{r}',t)\,\psi(\mathbf{r}'\,t) \end{aligned} \right\}.$$
(M.2)

Differentiating both sides of (M.1) with respect to any of the times t_j, and using (M.2) yields an equation of motion expressing G_n in terms of G_{n-1} and G_{n+1}:

$$\left(i\frac{\partial}{\partial t_j} + \frac{\nabla_j^2}{2m}\right)G_n(1,\ldots,n;1',\ldots,n')$$

$$= \sum_{l'=1'}^{n'} \delta(j-l')(-1)^{j+l'}\, G_{n-1}(1,\ldots,j-1,j+1,\ldots,n;1',\ldots l'-1,l'+1,\ldots,n')$$

$$\qquad -i\int d\mathbf{r}\, v(\mathbf{r}_j-\mathbf{r})\, G_{n+1}(\mathbf{r}t_j,1,\ldots,j,\ldots,n;\mathbf{r}t_j^+,1',\ldots,n').$$
(M.3)

Note that in the expression: $(-1)^{j+l'}$, l' means the *number l*. Also, $1, \ldots, j-1, j+1, \ldots, n$, means $1, 2, 3, \ldots, n$ with j omitted, and $\delta(j-l)=\delta(\mathbf{r}_j-\mathbf{r}_l)\delta(t_j-t_l)$. Also $G_{1-1}=1$. There is a similar equation when we differentiate with respect to $t_{j'}$.

(M.3) constitutes a 'hierarchy' of equations for G. I.e., when $n=1$ we find an equation for $G\,(\equiv G_1)$ in terms of G_2:

$$\left(i\frac{\partial}{\partial t_1} + \frac{\nabla_1^2}{2m}\right)G(\mathbf{r}_1,t_1;\mathbf{r}_1',t_1') = \delta(\mathbf{r}_1-\mathbf{r}_1')\,\delta(t_1-t_1')$$

$$\qquad\qquad -i\int d^3\mathbf{r}\, v\,(\mathbf{r}_1-\mathbf{r})\, G_2(\mathbf{r}t_1,\mathbf{r}_1\,t_1;\mathbf{r}t_1^+,\mathbf{r}_1'\,t_1').$$
(M.4)

When $n=2$ we get an equation for G_2 in terms of G and G_3, etc. To solve this hierarchy, the usual thing is to express say G_{l+1} in the equation for G_l approximately in terms of products of lower order G's so that the equations for $n > l$ become decoupled from those for $n \leqslant l$ and the hierarchy becomes finite. For example, if in (M.4) we set

$$G_2(12; 1'2') \simeq G(1; 1') G(2; 2') - G(1; 2') G(2; 1'). \qquad (M.5)$$

we have just a single differential equation in the hierarchy, to be solved for G. Or we could decouple G_3 in the form

$$G_3(123; 1'2'3') \approx G(1; 3') G_2(23; 1'2'), \qquad (M.6)$$

in which case there would be two equations in the hierarchy, to be solved for G, G_2.

In order to see the relation between the decoupling and partial sum methods, we first re-write the hierarchy (M.3) in integral form. The integral form of the first equation in the hierarchy, (M.4) is

$$G(1; 1') = G_0(1; 1') - i \iint d2\,d3\,v(2; 3)\,G_0(1; 2)\,G_2(3, 2; 3^+, 1') \qquad (M.7a)$$

where $v(s; s') \equiv v(\mathbf{r}_s - \mathbf{r}_{s'}) \delta(t_s - t_{s'})$. This may be checked by differentiating both sides with respect to t_1 and noting that for the free propagator we have:

$$(i\,\partial/\partial t_1 + \nabla_1^2/2m)\,G_0(1, 1') = \delta(1 - 1'). \qquad (M.7b)$$

Using the 'stretched skin' representation of G_2 introduced in (13.1'), (13.2), (M.7a) may be drawn as follows:

$$(M.7c)$$

Similarly, the $G_2 - G_3$ equation in integral form is:

$$(M.8)$$

That is, (M.7) is the integral form of (M.4), etc. Note that there are actually *two* possible forms for the $G - G_2$ equation, of which (M.7) is one, and *four* possible forms of the $G_2 - G_3$ equation (M.8). (See MT, p. 730.)

Consider now the decoupling (M.5). In diagram form this is

$$\text{(M.9)}$$

Placing this in the integral equation of motion (M.7) we find

$$\text{(M.10)}$$

which is just the self-consistent Hartree–Fock approximation.

A similar procedure can be used to reveal the secret meaning of the decoupling of G_3 in (M.6). In diagram form (M.6) is

$$\text{(M.11)}$$

If this is substituted into the 'fish' part of (M.8), we find the result in the second diagram of (M.12):

$$\text{(M.12)}$$

This may be distorted into the third diagram of (M.12), which is topologically equivalent to (and therefore the same as) the second diagram. Substituting this result into the whole equation in (M.8) yields

$$\text{(M.13)}$$

Iterating (M.13) by substituting the whole right-hand side of this equation for the G_2 on the right, etc., generates immediately the following series for G_2:

$$(M.14)$$

We now follow the usual procedure in the decoupling method and replace the clothed propagators in (M.14) by bare ones, giving us the ladder sum for G_2:

$$(M.15)$$

which shows that decoupling (M.6) is equivalent to ladder approximation.

These two examples illustrate that any given decoupling of the equations of motion for the propagators is equivalent to a particular partial sum of Feynman diagrams. (The reverse correspondence does not appear to be true, i.e., to any given partial sum there does not necessarily correspond a decoupling. See MT, p. 740.)

Appendix N

The Reduced Graph Expansion

The reduced graph method originated in quantum field theory where it is called the 'unitarity expansion', and was first used in many-body theory by J. S. Langer (1961, 1962). In this brief introduction, we follow the unpublished notes of U. Larsen (1973).

Reduced graphs may be used to calculate the imaginary part of any field-theoretic quantity, e.g., the self-energy or the scattering amplitude. The real part may then be found using dispersion relations such as (L.20). Reduced graphs are not ordinary Feynman diagrams, although they do resemble them. The idea is to notice that the imaginary part of the value of any given Feynman diagram may be broken up into a sum of contributions, one from each of the poles associated with the intermediate states of the diagram. Each of these contributions is expressed in the form of a diagram called a *reduced graph*, which may be evaluated by certain rules. We will show in detail how this is done in a couple of simple cases then generalize.

By (L.37, 38), the imaginary part of a quantity Q may be obtained from the discontinuity of Q across the real axis:

$$\text{Im } Q(x+i\delta) = \frac{1}{2i}[Q(x+i\delta) - Q(x-i\delta)] \equiv \delta Q.$$

Let us first consider the reduced graphs for the imaginary part, $\delta\Sigma$, of the self-energy, Σ. To introduce the basic concepts, we will calculate the contribution to the discontinuity from two of the lowest-order self-energy diagrams at finite T. Note that all propagator lines in the diagrams are clothed, and the spectral representation is used, where $\rho(x)$ is the spectral density. For simplicity, the interaction is assumed independent of wave number, and equal to a constant, V, represented diagrammatically by a single point. The first diagram is (we sometimes use the abbreviated notation $1 \equiv x_1$, $2 \equiv x_2$, $3 \equiv x_3$, $i\omega_n \equiv n$, etc.):

$$\equiv (-1) V^2 \int d1 \rho(1) \int d2 \rho(2) \int d3 \rho(3) F(n, 1, 2, 3)$$

where

$$F(n; 1, 2, 3) = \frac{1}{\beta} \sum_m \frac{1}{x_1 - i\omega_m} \cdot \frac{1}{\beta} \sum_l \frac{1}{x_2 - i\omega_n + i\omega_m - i\omega_l} \cdot \frac{1}{x_3 - i\omega_l}. \quad \text{(N.1)}$$

The sum over ω_l is carried out as in §14.5 and yields

$$\frac{1}{\beta} \sum_l = \frac{f(2) - f(3)}{i\omega_m - 3 + 2 - i\omega_n} \quad \text{(N.2)}$$

where we have used $f(2 - i\omega_n + i\omega_m) = f(2)$, since $\omega_n + \omega_m$ is even. Placing this in the sum over ω_m and evaluating this sum by the same technique gives

$$F(n, 1, 2, 3) = \frac{[f(3) - f(2)][f(1) + g(3 - 2)]}{1 + 2 - 3 - i\omega_n} \quad \text{(N.3)}$$

where we have used that

$$f(3-2+i\omega_n) = -g(3-2) = -[e^{\beta(3-2)}-1]^{-1} \quad \text{(Bose function)},$$

since ω_n is odd. Analytically continuing $i\omega_n \to z$, and juggling the Fermi and Bose functions a bit, F may be written in the form

$$F(z,1,2,3) = \frac{R(1,2,3)}{1+2-3-z}, \quad R(1,2,3) = \frac{-f(1)f(2)f(-3)}{f(1+2-3)}. \quad \text{(N.4)}$$

Hence

$$\Sigma^{(1)}(z) = (-1)V^2 \int d1\rho(1) \int d2\rho(2) \int d3\rho(3) \frac{R(1,2,3)}{1+2-3-z}. \quad \text{(N.5)}$$

The discontinuity, $\delta\Sigma^{(1)}$, across the real axis is given by (abbreviate $\int d1\rho(1)$ by $\int_1$, etc.)

$$\delta\Sigma^{(1)} = \frac{1}{2i}\{\Sigma^{(1)}(\omega+i\delta) - \Sigma^{(1)}(\omega-i\delta)\} = \iiint_{1\ 2\ 3} \delta F$$

where, using (3.76),

$$\delta F = \frac{1}{2i}\{F(\omega+i\delta) - F(\omega-i\delta)\}$$

$$= \frac{R(1,2,3)}{2i}\left[\frac{1}{1+2-3-\omega-i\delta} - \frac{1}{1+2-3-\omega+i\delta}\right]$$

$$= \pi R(1,2,3)\,\delta(1+2-3-\omega). \quad \text{(N.6)}$$

Hence, Im $\Sigma^{(1)}$ is given by

$$\text{Im}\,\Sigma^{(1)} \equiv \delta\Sigma^{(1)} = \pi \iiint_{1\ 2\ 3} R(1,2,3)\,\delta(1+2-3-\omega)$$

$$= -\pi f^{-1}(\omega)\iiint_{1\ 2\ 3} f(1)f(2)f(-3)\,\delta(1+2-3-\omega) \quad \text{(N.7)}$$

where we have used the fact that because of the δ-function, $f(\omega)=f(1+2-3)$.

From this result, we can see that the discontinuity arises from the pole of F, which occurs at $z=1+2-3$. Note that the location of this pole may be obtained directly from the original graph by drawing a dotted line cutting the graph into two pieces:

$$\text{(N.8)}$$

and taking particle lines as positive, holes negative: $z = 1 + 2 - 3$. Furthermore, note that each particle, p, has a factor $f(p)$, and each hole, h, a factor $f(-h)$. The -1 is for the fermion loop, and the delta function gives frequency conservation.

For the second example, consider the self-energy diagram given by

$$\Sigma^{(2)}(n) \equiv \underbrace{\begin{array}{c} n+s-l \quad n+t-l \\ \vdots \end{array}}_{} \equiv (-1)^2 V^3 \int\int\int\int\int_{1\,2\,3\,4\,5} F(n; 1, 2, 3, 4, 5) \quad \text{(N.9)}$$

where

$$F(n; 1, 2, 3, 4, 5) = \frac{1}{\beta} \sum_l \frac{1}{5-l} \cdot \frac{1}{\beta} \sum_s \frac{1}{2-s} \cdot \frac{1}{1-s-n+l} \times$$

$$\times \frac{1}{\beta} \sum_t \frac{1}{4-t} \cdot \frac{1}{3-t-n+l}. \quad \text{(N.10)}$$

Carrying out the frequency sums yields, after continuing $i\omega_n \rightarrow z$

$$F(z) = -[f(2) - f(1)][f(4) - f(3)]$$

$$\times \left\{ \frac{f(5)[2-1-4+3] + g(2-1)[5-4+3-z] - g(4-3)[5-2+1-z]}{(5-2+1-z)(5-4+3-z)(2-1-4+3)} \right\}. \quad \text{(N.11)}$$

This expression has two poles, in contrast to the first example, where there was just one.

The idea now is to evaluate δF expressing it (hence $\delta \Sigma^{(2)}$) as the sum of two contributions, one from each pole. There is a technical problem here, since both poles lie on the real axis (because $5-2+1$, and $5-4+3$ are real). This means that when the integrals over 1, 2, 3, 4, 5, are carried out, these poles give rise to 'overlapping branch cuts', i.e., branch cuts lying on top of each other, on the real axis. This leads to ambiguities and indeterminate integrals.

To avoid such ugly things, it is necessary to displace the overlapping cuts from each other. This is done by adding infinitesimal imaginary parts to the integration variables, i.e., $1 \rightarrow \bar{1} \equiv 1 + i\delta_1$, $2 \rightarrow \bar{2} \equiv 2 + i\delta_2$, etc., choosing the $i\delta$'s in such a way that the first pole, $x_{1p} = 5-2+1$, becomes $\bar{x}_{1p} = x_{1p} + i\eta$, the second pole, $x_{2p} = 5-4+3$ becomes $\bar{x}_{2p} = x_{2p} + 2i\eta$, etc. When this is done, we find

$$F(z; \bar{1}, \bar{2}, \bar{3}, \bar{4}, \bar{5}) = \frac{R(z; \bar{1}, \ldots, \bar{5})}{(5-2+1-z+i\eta)(5-4+3-z+2i\eta)} \quad \text{(N.12)}$$

where

$$R(z; \bar{1}, \ldots, \bar{5}) = -[f(2)-f(1)][f(4)-f(3)]$$

$$\times \left\{ \frac{\begin{array}{l} f(5)[2-1-4+3+i\eta]+g(2-1) \\ \times [5-4+3-z+2i\eta]-g(4-3)[5-2+1-z+i\eta] \end{array}}{2-1-4+3+i\eta} \right\}.$$

$$\text{(N.13)}$$

The total discontinuity across *all* the cuts is given by

$$\eth \Sigma^{(2)} = \frac{1}{2i} \{ \Sigma^{(2)}(\omega+i\delta) - \Sigma^{(2)}(\omega-i\delta) \} = \iiiint\limits_{1\,2\,3\,4\,5} \eth F. \quad \text{(N.14)}$$

Here, $\delta \ll \eta$ and

$$\eth F = \frac{1}{2i} \{ F(\omega+2i\eta+i\delta) - F(\omega+i\eta-i\delta) \}$$

$$= \frac{1}{2i} \{ [F(\omega+2i\eta+i\delta) - F(\omega+2i\eta-i\delta)] + [F(\omega+i\eta+i\delta) - F(\omega+i\eta-i\delta)] \},$$

$$\text{(N.15)}$$

where we have used that $F(\omega+2i\eta-i\delta)=F(\omega+i\eta+i\delta)$ since $\delta \ll \eta$ and there are no cuts in between that at $z=\omega+2i\eta-i\delta$ and that at $z=\omega+i\eta+i\delta$. With the aid of (N.12), (N.15) becomes:

$$\eth F = \pi\delta(5-2+1-\omega)\frac{R(\omega+i\eta; \bar{1}, \ldots, \bar{5})}{5-4+3-\omega-i\eta+2i\eta}$$

$$+ \pi\delta(5-4+3-\omega)\frac{R(\omega+2i\eta; \bar{1}, \ldots, \bar{5})}{5-2+1-\omega-2i\eta+i\eta}. \quad \text{(N.16)}$$

Placing the expression for R in (N.13) into (N.16), simplifying with the aid of the δ-functions, and placing the result for δF into (N.14) yields:

$$\delta \Sigma^{(2)} = \pi V^3 f^{-1}(\omega) \iiint\limits_{1\,2\,5} \delta(5-2+1-\omega)(-1)f(5)f(1)f(-2)$$

$$\times \iint\limits_{3\ 4} \frac{f(3)-f(4)}{3-4-(\omega-5-i\eta)} +\pi V^3 f^{-1}(\omega) \iiint\limits_{3\,4\,5} \delta(5-4+3-\omega)$$

$$\times (-1)f(5)f(3)f(-4) \iint\limits_{1\ 2} \frac{f(1)-f(2)}{1-2-(\omega-5-i\eta)} . \tag{N.17}$$

We can now extend the rules used to get (N.7) directly from graph (N.8), and use these extended rules to get (N.17) directly from graphs. This is done as follows: We cut graph (N.9) into two internally connected pieces in all ways such that the incoming line enters in the left-hand piece and the outgoing line exits from the right-hand piece:

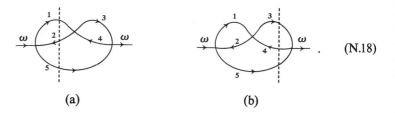

(a)   (b)   (N.18)

These graphs, as well as (N.8), have the general form:

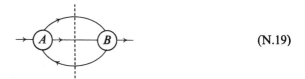

(N.19)

where A and B are called 'endpieces' and the set of particle and hole lines cut by the dotted line are called the 'intermediate state'. Such graphs are called 'reduced graphs'. We see that the two terms in (N.17) can be obtained from the reduced graphs in (N.18) by the following rules:

(1) A factor for each endpiece given by evaluating the endpiece as an ordinary Feynman diagram. (For example, in (N.18a), the endpiece A gives just a factor V, while endpiece B produces V^2 times the double integral over 3 and 4 in the first term of (N.17).)

(2) A factor $f(x_i)$ for each particle in the intermediate state, and $f(-x_i)$ for each hole.

(3) A factor $\delta(\Sigma x_i \text{ (particle)} - \Sigma x_i \text{ (hole)})$ for the intermediate state.

(4) $\int dx_i \rho(x_i)$ for each x_i in the intermediate state.

(5) Overall factor $[f(\omega)]^{-1}$.

(6) Factor (-1) for each fermion loop not already included in evaluating the endpieces.   (N.20)

In the case of a general Feynman diagram, to find its contribution to the imaginary part of the self energy, ImΣ, we first separate it into two pieces in all possible ways such that the incoming particle enters the left-hand piece and the outgoing particle leaves the right-hand piece. The resulting diagrams all have the general form

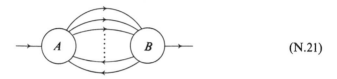

(N.21)

with n particles in the intermediate state, and $n-1$ holes. The contribution of each of these reduced diagrams to ImΣ is given by the rules in (N.20).

Finally, to get the *reduced graph expansion* for the imaginary part of the self-energy, we decompose every Feynman diagram into a sum of reduced graphs, and regroup all the resulting reduced graphs according to the number of lines in the intermediate states. This yields

$$\text{Im} \to \Sigma \to \quad = \quad \to \bigcirc \longrightarrow \bigcirc \to \quad + \quad \to \bigcirc = \bigcirc \to \quad + \cdots \quad \text{(N.22)}$$

where the endpieces are the sum over all possible Feynman diagrams with the indicated number of lines entering and leaving.

A similar expansion exists for the imaginary part of the scattering amplitude (§10.6):

$$\text{Im} \; \Gamma \quad = \quad + \quad + \quad + \cdots ,$$

(N.23)

and also for the higher-order field theoretic quantities, with many particles entering and leaving.

Answers to Exercises

Chapter 2

2.1

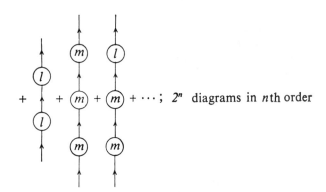

$+$ ⎡... ⎤ $+ \cdots$; 2^n diagrams in nth order

2.2 $P(\mathbf{r}_2, \mathbf{r}_1) = P_0(\mathbf{r}_2, \mathbf{r}_1) + P_0(\mathbf{r}_m, \mathbf{r}_1) P(m) P_0(\mathbf{r}_2, \mathbf{r}_m) + P_0(\mathbf{r}_l, \mathbf{r}_1) P(l) P_0(\mathbf{r}_2, \mathbf{r}_l)$

$\qquad + P_0(\mathbf{r}_m, \mathbf{r}_1) P(m) P_0(\mathbf{r}_m, \mathbf{r}_m) P(m) P_0(\mathbf{r}_2, \mathbf{r}_m) + \dots.$

2.3 $P(\mathbf{r}_2, \mathbf{r}_1) = c + c^2[P(m) + P(l)] + c^3[P(m)^2 + P(m) P(l) + P(l) P(m) + P(l)^2] + c^4\dots$

$\qquad = c\{1 + c[P(m) + P(l)] + c^2[P(m) + P(l)]^2 + \dots\} = \dfrac{c}{1 - c[P(m) + P(l)]}.$

2.4 $P(\mathbf{r}_2, \mathbf{r}_1) = \dfrac{c}{1 - c\sum\limits_a P(a)}.$

402

Chapter 3

3.1
$$\phi_n(x) = \sqrt{\frac{2}{a}} \sin\left(\frac{\pi n x}{a}\right); \quad \epsilon_n = \frac{\pi^2 n^2}{2ma^2}, \quad n = 1, 2, 3, \ldots$$

$$G_0^+(n, t_2 - t_1) = - i\theta_{t_2 - t_1} \exp\left[-i\left(\frac{\pi^2 n^2}{2ma^2}\right)(t_2 - t_1)\right]$$

$$G_0^+(n, \omega) = \left[\omega - \frac{\pi^2 n^2}{2ma^2} + i\delta\right]^{-1}.$$

3.2 We have
$$H = p^2/2m + U(x), \quad H\phi_n = \frac{\pi^2 n^2}{2ma^2} \phi_n,$$

thus:
$$V_{mn} = \int dx \phi_m^*(x) \underbrace{B(p^2/2m + U(x))^3 \phi_n(x)}_{= B\left(\frac{\pi^2 n^2}{2ma^2}\right)^3 \phi_n}$$

$$= B\left(\frac{\pi^2 n^2}{2ma^2}\right)^3 \delta_{mn}.$$

3.3 The diagram series here has just the form (3.33) except that the $\approx$ is replaced by $=$. The (M) diagram is equal to
$$- iV_{nn} = - iB\left(\frac{\pi^2 n^2}{2ma^2}\right)^3$$

and the directed lines are just the propagators of Ex. 3.1. Thus
$$G^+(n, \omega) = \left[\omega - \frac{\pi^2 n^2}{2ma^2} + i\delta - B\left(\frac{\pi^2 n^2}{2ma^2}\right)^3\right]^{-1}.$$

3.4
$$\epsilon_n' = \frac{\pi^2 n^2}{2ma^2} + B\left(\frac{\pi^2 n^2}{2ma^2}\right)^3; \quad \tau_n = \infty.$$

3.5 See Eq. (2.23).

3.7 Use (3.77), (3.78). $\quad m^* = m\left[1 + \frac{2Nm^2 W^2}{\Omega \pi^2 a}\right]^{-1}.$

3.9

$$= (i/\Omega^4) G_0^+(s) G_0^+(k) \sum_{r, q, p} G_0^+(r) G_0^+(q) G_0^+(p)$$

$$\times W_{sr} W_{pk} W_{rq} W_{qp} \sum_j e^{-i(r-p)\cdot R_j} \sum_l e^{-i(s-r+p-k)\cdot R_l}$$

$$= iG_0^+(k)^2 \left(\frac{N}{\Omega}\right)^2 \int \frac{d^3 p}{(2\pi)^3} \int \frac{d^3 q}{(2\pi)^3} G_0^+(p)^2 G_0^+(q) W_{pk} W_{kp} W_{pq} W_{qp} \delta_{ks}.$$

Chapter 4

4.1 $-\mathbf{k}$.

4.2 (a) 0.

(b) $|1_3^h, 1_8^p\rangle$.

(c) $c_k^\dagger c_k|11111000...\rangle = |11111000...\rangle$, for $k < k_F$, and $= 0$, for $k > k_F$.
Hence $\sum_k \epsilon_k c_k^\dagger c_k|11111000...\rangle = (\epsilon_1 + \epsilon_2 + \epsilon_3 + \epsilon_4 + \epsilon_5)|11111000...\rangle$.

4.3 Equation (4.39) with extra term, $-|V_{kl}|^2 (\omega - \epsilon_l + i\delta)^{-1}$, in denominator.

4.4 See (9.33i, h) and after (9.32).

4.5 (a)

(b) Energy is not conserved:

$$k^2/2m + p^2/2m \neq (k-q)^2/2m + (p+q)^2/2m.$$

4.6 $[iG_0^+(\mathbf{k}, \omega)]^2 \times F(\mathbf{k}, \omega)$ where:

$$F(\mathbf{k}, \omega) = \sum_{q,l} \int \frac{d\epsilon}{2\pi} \int \frac{d\beta}{2\pi} iG_0^+(\mathbf{k} - \mathbf{q}, \omega - \epsilon) \times (-iV_q)^2$$

$$\times (-1) \times iG_0^+(\mathbf{l} + \mathbf{q}, \beta + \epsilon) \times iG_0^-(\mathbf{l}, \beta)$$

A function of **k** and ω.

4.7 By conservation of momentum and frequency, the directed line joining the two

parts must have the label **k**, ω. The result then follows immediately.

4.8 See (10.5), together with (4.76, 77) and $F(\mathbf{k}, \omega)$ in Ex. 4.6.

4.11(a)

or $G^+(\mathbf{q},\mathbf{p},\omega) = G_0^+(\mathbf{p},\omega)\, \delta_{pq} + \dfrac{G_0^+(\mathbf{q},\omega)\, A\, G_0^+(\mathbf{p},\omega)}{1-A\left[\displaystyle\sum_{l>k_F} G_0^+(l,\omega) + \sum_{m<k_F} G_0^-(\mathbf{m},\omega)\right]}$

(b) $G^+(\mathbf{q},\mathbf{p},\omega) = G_0^+(\mathbf{p},\omega)\, \delta_{pq} + \dfrac{G_0^+(\mathbf{q},\omega)\, A f_p f_q\, G_0^+(\mathbf{p},\omega)}{1-A\left[\displaystyle\sum_{l>k_F} G_0^+(l,\omega) f_l^2 + \sum_{m<k_F} G_0^-(\mathbf{m},\omega) f_m^2\right]}$

Chapter 5

5.1 $-\displaystyle\int_0^t dt_4 \int_0^t dt_3 \int_0^t dt_2 \int_0^t dt_1 \sum_{p,q} iG_0^+(p, t_3-t_1) \times (-i)\, V_{1p} \times iG_0^-(1, t_2-t_3)$

$t_4 > t_3 > t_2 > t_1$

$\times (-i)\, V_{q1} \times iG_0^+(q, t_4-t_2) \times (-i)\, V_{1q} \times iG_0^-(1, t_1-t_4) \times (-i)\, V_{p1}.$

5.2 (a) Equivalent.
    (b) Distinct.
    (c) Distinct.

5.3 The diagrams are exactly the same as in (5.12, 13) except that all 1's are replaced by m, where m is such that $\epsilon_m < \epsilon_F$, ϵ_F being the Fermi energy of the system.

5.5 Both have a particle and a hole in the same state, which is impossible.

Chapter 7

7.1 $(-1)|0010100...\rangle$.

7.2 $(A^*\langle 100...| + B^*\langle 11100...|)\, \hat{N}(A|100...\rangle + B|11100...\rangle) = A^2 + 3B^2.$

7.3 $c_1 c_2^\dagger|\Psi\rangle + c_2^\dagger c_1|\Psi\rangle = -A|0100...\rangle + A|0100...\rangle = 0.$

7.4 Using (7.47) and the ϵ_n from the answer to Ex. 3.1, gives

$$H_0 = \sum_n \left(\frac{\pi^2 n^2}{2ma^2}\right) c_n^\dagger c_n.$$

7.5 Using (7.50) plus the answer to Ex. 3.2 yields

$$H_2 = \sum_{mn} B\left(\frac{\pi^2 n^2}{2ma^2}\right)^3 \delta_{mn} c_m^\dagger c_n = \sum_n B\left(\frac{\pi^2 n^2}{2ma^2}\right)^3 c_n^\dagger c_n$$

7.6 Using (7.69), (7.69'), yields:

$$V_{klmn} = \frac{A}{\Omega}\, \delta(\mathbf{k}+\mathbf{l}, \mathbf{m}+\mathbf{n})$$

Hence,

$$V^{\text{occ}} = \frac{A}{\Omega} \sum_{m,n,q} c_{n-q}^\dagger c_{m+q}^\dagger c_m c_n.$$

7.11
$$S_x^{occ} = \sum_{k\sigma, l\sigma'} \langle k\sigma|S_x|l\sigma'\rangle \, c_{k\sigma}^\dagger c_{l\sigma'},$$

where
$$\langle k\sigma|S_x|l\sigma'\rangle = \delta_{kl}\langle\sigma|S_x|\sigma'\rangle.$$

Using appendix (K.5), (K.4),

$$\langle\sigma|S_x|\sigma'\rangle = \sum_\gamma \eta_\sigma^*(\gamma)\, S_x\, \eta_{\sigma'}(\gamma) = \tfrac{1}{2}\sum_\gamma \eta_\sigma^*(\gamma)\, \eta_{-\sigma'}(\gamma) = \tfrac{1}{2}\delta_{\sigma,-\sigma'}.$$

Hence
$$S_x^{occ} = \tfrac{1}{2}\sum_{k\sigma, l\sigma'} \delta_{kl}\delta_{\sigma,-\sigma'}\, c_{k\sigma}^\dagger c_{l\sigma'} = \tfrac{1}{2}\sum_k [c_{k\uparrow}^\dagger c_{k\downarrow} + c_{k\downarrow}^\dagger c_{k\uparrow}].$$

Chapter 8

8.1 $\mathbf{k} = [iG_0^-(\mathbf{k}, \omega)]^2(-iV_{kkkk})(-1)(-1),$

$= [iG_0^-(\mathbf{k}, \omega)]^2(-iV_{kkkk})(-1).$

The second pair of diagrams cancels in a similar way.

8.3
$$\epsilon_k' = \frac{k^2}{2m} - 0.03\frac{kk_F}{2m} \approx \frac{k^2}{2m}, \quad \epsilon_F' \approx \frac{k_F^2}{2m}$$

$$\frac{\tau_k^{-1}}{\epsilon_k' - \epsilon_F'} = \frac{0.25(k-k_F)^2}{k^2 - k_F^2} = 0.25\left(\frac{k/k_F - 1}{k/k_F + 1}\right).$$

This ratio is small (< 0.1) provided $k/k_F < 2$.

Chapter 9

9.1 $(-1)\times(-1)\times c_l^\dagger(t_2)\, c_m^\dagger(t_3)\, c_k(t_1).$

9.5 (a) and (c).

9.6

$\equiv [iG_0(\mathbf{k}, \omega)]^2 \displaystyle\int \frac{d^3\mathbf{q}}{(2\pi)^3} \int \frac{d^3\mathbf{p}}{(2\pi)^3} \int \frac{d\epsilon}{2\pi} \int \frac{d\beta}{(2\pi)}$

$\times (-iV_q)(-iV_{k-p-q}) \times iG_0(\mathbf{k}-\mathbf{q}, \omega-\epsilon)$

$\times iG_0(\mathbf{p}, \beta) \times iG_0(\mathbf{p}+\mathbf{q}, \beta+\epsilon).$

9.9 Because they violate rule (2) in §9.6.

Chapter 10

10.1 b, c are self-energy parts. b is proper, c improper.

10.2

$$iG(k_2\,k_1, \omega) = iG_0(k_1, \omega)\,\delta_{k_2 k_1} + \sum_{l,m} iG_0(k_2, \omega)\,\delta_{k_2 m} \times [-iV_{ml}] \times iG(lk_1, \omega).$$

10.3 See (4.96).

10.4 Using (10.21), with $V_k = A$, we find

$$K(\mathbf{q}, \omega) = A\left[1 - Ai \int \frac{d^3\mathbf{p}\,d\epsilon}{(2\pi)^4}\, G_0^+(\mathbf{p}, \epsilon)\, G_0^+(\mathbf{q}-\mathbf{p}, \omega-\epsilon)\right]^{-1}$$

as can be verified by substitution in (10.21).

10.5 a, d are polarization parts. a is proper, d improper.

10.6

$$-i\pi(\mathbf{q}, \omega) = -i\pi_0(\mathbf{q}, \omega)$$
$$+ 4\int \frac{d^3\mathbf{k}'}{(2\pi)^3} \int \frac{d\gamma'}{2\pi} \int \frac{d^3\mathbf{k}}{(2\pi)^3} \int \frac{d\gamma}{2\pi} iG_0(\mathbf{k}', \gamma') \times iG_0(\mathbf{k}, \gamma)$$
$$\times iG_0(\mathbf{k}+\mathbf{q}, \gamma+\omega) \times iG_0(\mathbf{k}'+\mathbf{q}, \gamma'+\omega) \times (-i)\,K(\mathbf{k}', \gamma', \mathbf{k}, \gamma; \mathbf{q}, \omega)$$

(factor of 4 for sum over spins).

10.7 (a) $K(\mathbf{p}', \epsilon', \mathbf{p}, \epsilon; \mathbf{q}, \omega) = Au_p u_{p'}[1 - AI(\mathbf{q}, \omega)]^{-1}$,

where $I(\mathbf{q}, \omega) = i \int \frac{d^3\mathbf{p}''\,d\epsilon''}{(2\pi)^4}\, G_0(\mathbf{p}'', \epsilon'')\, G_0(\mathbf{q}-\mathbf{p}'', \omega-\epsilon'')\, u_{p''}^2$

(b) $I^+(\mathbf{q}, \omega) = \int \frac{d^3\mathbf{p}}{(2\pi)^3}\, \frac{u_{p''}^2}{\omega - \epsilon_{p''} - \epsilon_{q-p''} + i\delta}$

$I^-(\mathbf{q}, \omega) = -\int \frac{d^3\mathbf{p}}{(2\pi)^3}\, \frac{u_{p''}^2}{\omega - \epsilon_{p''} - \epsilon_{q-p''} - i\delta}$

(c) $I^+(0, 0) = -\frac{m}{2\pi^2}(w - k_F);\quad I^-(0, 0) = \frac{mk_F}{2\pi^2}$

(d) At low density, $k_F a \ll 1$ (see beginning of §10.3). But, since V_p has range w, Fourier transforming shows $V(\mathbf{r})$ has range $\sim 1/w$, so $a \approx 1/w$. Hence $k_F \ll w$, which means $|I^+(0,0)| \gg I^-(0,0)$.

10.8 $KE = \dfrac{3\pi^2 \hbar^2}{mL^2}$ for two non-interacting electrons in ground state in a three-dimensional box. $PE = e^2/a_0 r_s$. From (10.24), $L \sim 2a_0 r_s$.

Hence

$$KE/PE = \frac{3\pi^2 \hbar^2}{2me^2 L} \sim \frac{3\pi^2 a_0}{4a_0 r_s} \sim \frac{8}{r_s}.$$

10.9 For $q \ll 2$: $V_{eff}(q,0) \approx (4\pi e^2/k_F^2) \left[\dfrac{\lambda^2}{k_F^2} + q^2 \left(1 - \dfrac{\lambda^2}{12k_F^2} \right) \right]^{-1}$

For $q \gg 2$: $V_{eff}(q,0) \approx (4\pi e^2/k_F^2) \left[q^2 + \dfrac{4}{3} \dfrac{\lambda^2}{k_F^2} q^{-2} \right]^{-1}$

at $q = 2$: $\dfrac{dV_{eff}}{dq} = +\infty.$

Chapter 11

11.1 Diagram (b) should not be included, since it is already included in diagram 4 of (11.3). None of them should be in (11.5) since (a) and (b) are already included in the third diagram on the right of (11.5) and (c) is included in the second diagram on the right of (11.5).

11.2 $G(\mathbf{k}, \omega)$ as in (11.21) with $\epsilon'_k = \epsilon_k(1 - A)^{-1}$,

$$Z_k = (1-A)^{-1}, \quad \tau_k^{-1} = \frac{\text{sgn}\,[\epsilon_k(1-A)^{-1} - \mu]\, B[\epsilon_k(1-A)^{-1} - \mu]^2}{(1-A)}.$$

11.3 $\equiv -i \sum_l [V_{klkl} - V_{lkkl}] \int_{-\infty}^{+\infty} \frac{d\beta}{2\pi} \frac{e^{i\beta\eta}}{\beta - \epsilon'_l + i(\beta - \mu)\delta}.$

Poles are at $\omega = \epsilon'_l - i(\epsilon'_l - \mu)\delta$. Close contour in upper half β-plane. By residue theorem, $\int d\beta = 0$ for $\epsilon'_l > \mu$, and $= +i$ for $\epsilon'_l < \mu$, from which the result follows immediately.

Chapter 12

12.2

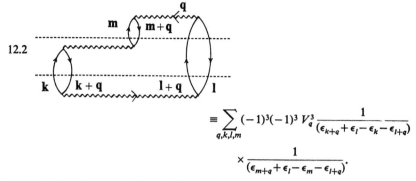

$$\equiv \sum_{q,k,l,m} (-1)^3(-1)^3 V_q^3 \frac{1}{(\epsilon_{k+q}+\epsilon_l-\epsilon_k-\epsilon_{l+q})}$$

$$\times \frac{1}{(\epsilon_{m+q}+\epsilon_l-\epsilon_m-\epsilon_{l+q})}.$$

12.3 The first diagram is given in (12.7). The second is

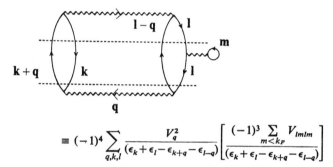

$$\equiv (-1)^4 \sum_{q,k,l} \frac{V_q^2}{(\epsilon_k+\epsilon_l-\epsilon_{k+q}-\epsilon_{l-q})} \left[\frac{(-1)^3 \sum\limits_{m<k_F} V_{lmlm}}{(\epsilon_k+\epsilon_l-\epsilon_{k+q}-\epsilon_{l-q})} \right]$$

The diagram with n bubbles contains the factor in brackets to the nth power. Hence we have a geometric series which may be summed to yield

$$\mathcal{E} = (-1)^4 \sum_{qkl} \left\{ \frac{-\tfrac{1}{2}V_q^2}{\epsilon_k+\epsilon_l-\epsilon_{k+q}-\epsilon_{l-q}} + \frac{V_q^2}{\epsilon_k+(\epsilon_l+\beta_l)-\epsilon_{k+q}-\epsilon_{l-q}} \right\}$$

where

$$\beta_l = \sum_{m<k_F} V_{lmlm}.$$

Chapter 13

13.1

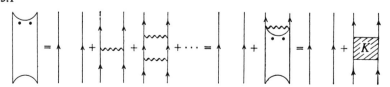

(c) Substituting for $\overset{i}{\diagup}$ from Ex. 13.2 yields (a).

13.7 We have

$$-iG_2(4,3,2,1) = -\langle 0|T\{\psi(4)\,\psi^\dagger(3)\,\psi(2)\,\psi^\dagger(1)\}|0\rangle - \langle 0|T\{\psi(4)\,\psi^\dagger(3)\,\psi(2)\,\psi^\dagger(1)\}|0\rangle$$

$$= iG_0(4,1) \times iG_0(2,3) - iG_0(4,3) \times iG_0(2,1),$$

which may immediately be translated into the zeroth-order diagrams of (13.2).

Chapter 14

14.1
$$\bar{N} = \frac{\sum_i N_i e^{-\beta(E_i - \mu N_i)}}{\sum_i e^{-\beta(E_i - \mu N_i)}} = \left[\frac{\partial \ln Z}{\partial(\beta\mu)}\right]_\beta.$$

14.2 By (14.12), $\ln Z_0 = \sum_k \ln(1 + e^{-\beta(\epsilon_k - \mu_0)})$.

Hence $\bar{N}_0 = \sum_k (e^{+\beta(\epsilon_k - \mu_0)} + 1)^{-1}$.

14.5

$$= (+1)[i\mathscr{G}_0(\mathbf{k},\omega_n)]^2$$

$$\times \sum_{p,q} \frac{1}{\beta^2} \sum_{m,i=-\infty}^{+\infty} i\mathscr{G}_0(\mathbf{k}-\mathbf{q},\omega_n-\epsilon_m)$$

$$\times i\mathscr{G}_0(\mathbf{p},\beta_i) \times i\mathscr{G}_0(\mathbf{p}+\mathbf{q},\beta_i+\epsilon_m)$$

$$\times (-V_q)^2.$$

Chapter 15

15.1 $\displaystyle 2\sum_k \epsilon_k b_k^\dagger b_k |\Psi_0\rangle = 2\sum_k \epsilon_k c_{k\uparrow}^\dagger c_{k\uparrow} c_{-k\downarrow}^\dagger c_{-k\downarrow}|\ldots n_{k\uparrow}\ldots n_{-k\downarrow}(=n_{k\uparrow})\ldots\rangle$

$\displaystyle \qquad = 2\sum_k \epsilon_k n_{k\uparrow} n_{-k\downarrow}|\Psi_0\rangle$

$\displaystyle \qquad = 2\sum_k \epsilon_k n_{k\uparrow}|\Psi_0\rangle \quad \text{(since } n_{k\uparrow}=n_{-k\downarrow}=0 \text{ or } 1)$

$\displaystyle \qquad = \sum_{k\sigma} \epsilon_k n_{k\sigma}|\Psi_0\rangle.$

15.2 $[\alpha_k, \alpha_{k'}^\dagger] = u_k u_{k'}[c_{k\uparrow}, c_{k'\uparrow}^\dagger] - u_k v_{k'}[c_{k\uparrow}, c_{-k'\downarrow}] - \ldots, \text{ etc.}, \ldots$

$\qquad = \delta_{kk'}[u_k u_{k'} + v_k v_{k'}] = \delta_{kk'}.$

15.3 $E = E_0 + E_{k_1} + 2E_{k_2}.$

15.5 $\displaystyle \sum_k \epsilon_k (c_{k\uparrow}^\dagger, c_{-k\downarrow})\begin{pmatrix}1 & 0\\ 0 & -1\end{pmatrix}\begin{pmatrix}c_{k\uparrow}\\ c_{-k\downarrow}^\dagger\end{pmatrix} = \sum_k \epsilon_k(c_{k\uparrow}^\dagger c_{k\uparrow} + c_{-k\downarrow}^+ c_{-k\downarrow}) - \sum_k \epsilon_k.$

15.6

$$\equiv (-1)\left[i\mathbb{G}_0(\mathbf{k},\omega)\right]^2$$

$$\times \int \frac{d^3\mathbf{k}'\,d\omega'}{(2\pi)^4}\,\tau_3\,(-i\mathscr{V}_{k,o})\,\mathrm{Tr}\left[\tau_3\times i\mathbb{G}(\mathbf{k}',\omega')\right]e^{i\omega'0^+}$$

15.12

$$F(k,0^+) = -i\langle\Psi_0|c_{k\uparrow} c_{-k\downarrow}|\Psi_0\rangle$$

$$= v_k\sqrt{1-v_k^2}\langle\phi|\phi\rangle = v_k\sqrt{1-v_k^2}$$

where $|\phi\rangle = \prod_{l\neq k}(u_l + v_l b_l^\dagger)|0\rangle$. In normal state, $v_k = 1 - \theta_{\epsilon_k}$ so $F=0$. In super-conducting state $F\neq 0$.

Chapter 17

17.1 (b) Let T be the global gauge transformation operator. Then, $H|\Psi_0\rangle = E_0|\Psi_0\rangle$, so $THT^{-1}T|\Psi_0\rangle = E_0 T|\Psi_0\rangle$. But, by 17.1 (a), $THT^{-1} = H$. Hence $HT|\Psi_0\rangle = E_0 T|\Psi_0\rangle$. But since $|\Psi_0\rangle$ is non-degenerate, this means $T|\Psi_0\rangle = |\Psi_0\rangle$, Q.E.D.

Appendix 𝒜

𝒜.1 We have:

$$[b_q, b_{q'}^\dagger] = b_q b_{q'}^\dagger - b_{q'}^\dagger b_q = \frac{1}{2m\hbar\sqrt{(\omega_q \omega_{q'})}}[\mathscr{P}_q, \mathscr{P}_{q'}] + \frac{i}{2\hbar}\sqrt{\frac{\omega_{q'}}{\omega_q}}[\mathscr{P}_q, \mathscr{U}_{q'}]$$

$$-\frac{i}{2\hbar}\sqrt{\frac{\omega_q}{\omega_{q'}}}[\mathscr{U}_q, \mathscr{P}_{q'}] + \frac{m}{2\hbar}\sqrt{\omega_q \omega_{q'}}[\mathscr{U}_q, \mathscr{U}_{q'}].$$

Using ($\mathscr{A}$.32) yields ($\mathscr{A}$.37a) immediately.

$\mathscr{A}$.3 $\sqrt{(4)}\,\psi_{4_{q_3}}$.

$\mathscr{A}$.5 (a) $= b_{q_3}^\dagger\,\psi_{0_{q_1}}\,\psi_{3_{q_3}}\,\psi_{0_{q_3}}\ldots = \psi_{0_{q_1}}\sqrt{(4)}\,\psi_{4_{q_3}}\,\psi_{0_{q_3}}\ldots = \sqrt{(4)}\,\Psi_{0_{q_1},\,4_{q_3},\,0_{q_3},\,\ldots}$.

(b) $= 0$.

(c) $\sqrt{(3)}\,\sqrt{(5)}\,\sqrt{(2)}\,\Psi_{3_{q_1},\,4_{q_3},\,0_{q_3},\,0,\,0,\,\ldots}$

$\mathscr{A}$.6 Five phonons.

$\mathscr{A}$.7 $E = \tfrac{1}{2}\sum_{q}\hbar\omega_q + 2\hbar\omega_{q_3} + 3\hbar\omega_{q_5}$.

Appendix G

G.3, G.4 The non-vanishing contracted elements and corresponding diagrams are (omit time for brevity):

$$\sum_{klmn}(-iV_{klmn}\langle 0|a_{k_2}a_l^\dagger\,b_k\,a_m\,b_n^\dagger\,a_{k_1}^\dagger|0\rangle \equiv$$

$$\sum_{klmn}(-iV_{klmn})\langle 0|a_{k_2}a_l^\dagger\,b_k\,b_m^\dagger\,a_n\,a_{k_1}^\dagger|0\rangle \equiv$$

$$\sum_{klmn}(-iV_{klmn})\langle 0|a_{k_2}b_l\,a_k^\dagger\,b_m^\dagger\,a_n\,a_{k_1}^\dagger|0\rangle \equiv$$

$$\sum_{klmn}(-iV_{klmn})\langle 0|a_{k_2}b_l\,a_k^\dagger\,a_m\,b_n^\dagger\,a_{k_1}^\dagger|0\rangle \equiv$$

$$\sum_{klmn}(-iV_{klmn})\langle 0|a_{k_2}b_l\,b_k\,b_m^\dagger\,b_n^\dagger\,a_{k_1}^\dagger|0\rangle \equiv$$

$$\sum_{klmn}(-iV_{klmn})\langle 0|a_{k_2}b_l\,b_k\,b_m^\dagger\,b_n^\dagger\,a_{k_1}^\dagger|0\rangle \equiv$$

Appendix H

H.3

(a) $\omega_{n0} = \dfrac{\hbar^2}{2m^*L^2}[n_x^2 + n_y^2 + n_z^2],\ (n_i = 0, 1, 2, \ldots)$

(b) $\Delta E = \hbar^2/2m^*L^2,\ t\ \text{(build-up)} = 2m^*L^2/\hbar$

(c) velocity at Fermi surface $= v_F = \sqrt{\dfrac{2\epsilon_F}{m^*}} = \dfrac{\hbar}{m^*}\dfrac{(3\pi^2 N)^{\frac{1}{3}}}{L}$

so t (traverse) $= L/v_F = m^* L^2/\hbar(3\pi^2 N)^{\frac{1}{3}}$

(d) t (traverse)/t (build-up) $= 2/(3\pi^2 N)^{\frac{1}{3}}$. This is $\sim 10^{-8}$ for a metal and $\sim 0\cdot 2$ for a heavy nucleus.

References

Note: (1) Textbooks devoted wholly or largely to the many-body problem are denoted (T).
 (2) Review articles are denoted (R).
 (3) An article appearing in a collection of articles on the many-body problem is denoted (C).

 Abrikosov, A. A., 'Electron Scattering on Magnetic Impurities in Metals', *Physics*, **2**, 5 (1965).

 Abrikosov, A. A. (with A. A. Migdal), 'On the Theory of the Kondo Effect', *J. Low Temp. Phys.*, **3**, 519 (1970).

(T) Abrikosov, A. A. (with L. P. Gorkov and I. E. Dzyaloshinski), *Quantum Field Theoretical Methods in Statistical Physics*, Pergamon Press, Oxford (1965).

(C) Ambegaokar, V., 'Green's Functions in Many-Body Problems', in *Astrophysics and the Many-Body Problem*, 1962 Brandeis Lectures, Vol. 2, W. A. Benjamin, New York (1963).

(T) Anderson, P. W., *Concepts in Solids*, Benjamin, New York (1963).

 Anderson, P. W. (with W. F. Brinkman), 'Anisotropic Super-fluidity in ^{3}He: A possible interpretation of its stability as a Spin Fluctuation Effect', *Phys. Rev. Lett.*, **30**, 1108 (1973).

(T) Atkins, P. W., *Diagrammatic Perturbation Theory* (notes, especially for chemists) Lincoln College, Oxford (1972).

 Baym, G. (with C. Pethick, D. Pines, M. Ruderman), 'Spin Up in Neutron Stars: the Future of the Vela Pulsar', *Nature*, **224**, 872 (1969).

(C) Beeby, J. L., 'Electrons in Disordered Systems', in *Lectures on the Many-Body Problem*, Vol. 2, Ed. E. R. Caianiello, Academic Press, New York (1964).

(C) Beliaev, S. T., 'Introduction to the Bogliubov Canonical Transformation Method', p. 343 of *The Many-Body Problem (Les Houches)*, Wiley-Dunod, New York—Paris (1959).

 Bjorken, J. D. (with S. D. Drell), *Relativistic Quantum Mechanics*, McGraw-Hill, New York (1964).

 Bjorken, J. D. (with S. D. Drell), *Relativistic Quantum Fields*, McGraw-Hill, New York (1965).

(C) Bloch, C., 'General Perturbation Formalism for the Many-Body Problem at Nonzero Temperatures', in *Lectures on the Many-Body Problem*, Vol. 1, p. 31, Ed. E. R. Caianiello, Academic Press, New York (1962).

Bogoliubov, N. N. (with D. V. Shirkov), *Introduction to the Theory of Quantized Fields*, Chapter VIII, Interscience. New York (1959).

Bohr, A. (with B. R. Mottelson, D. Pines), 'Possible Analogy Between the Excitation Spectra of Nuclei and those of the Superconducting Metallic State,' *Phys. Rev.*, 110, 936 (1958).

Bohr, A. (with B. R. Mottelson), *Nuclear Structure*, vol. I, Benjamin, New York (1969).

(T) Bonch Bruevich, V. (with S. Tyablikov), *The Green Function Method in Statistical Mechanics*, North-Holland, Amsterdam (1962).

(T) Brout, R. (with P. Carruthers), *Lectures on the Many Electron Problem*, Wiley, New York (1963).

(T) Brout, R., *Phase Transitions*, Benjamin, New York (1965).

(R) Brout, R., 'Theoretical Methods in the Critical Region of a Phase Transition', *Phys. Rpts.*, 10C, 1 (1974).

(T) Brown, G. E., *Many-Body Problems*, North-Holland (1972) Amsterdam.

(R) Brown, G. E., 'Collective Motion and the Application of Many-Body Techniques', in *Nuclear Physics* (Proc. Int. School of Physics, Varenna—course 23), p. 99, Academic Press, New York (1963).

Cheung, C. Y. (with R. D. Mattuck), 'Removing the Divergence at the Kondo Temperature by means of Self-consistent Perturbation Theory', *Phys. Rev. B*, 2, 2735 (1970).

Daniel, E. (with S. H. Vosko), 'Momentum Distribution of an Interacting Electron Gas', *Phys. Rev.*, 120, 2041 (1960).

Dennery, P. (with A. Krzywicki), *Mathematics for Physicists*, Harper & Row, New York (1967).

Dicke, R. (with J. Wittke), *Introduction to Quantum Mechanics*, Addison-Wesley, Reading (1960).

Dirac, P. A. M., *The Principles of Quantum Mechanics*, 3rd Ed., 4th Ed., Oxford, Clarendon Press (1947, 1958).

(R) Falicov, L. (with V. Heine), 'The Many-Body Theory of Electrons in Metals or Has a Metal Really Got a Fermi Surface?', *Adv. Phys.*, 10, 57 (1961).

(R), (C) Falkoff, D., 'The *N*-body Problem', in *The Many-Body Problem* (Bergen School Lectures), Ed. C. Fronsdal, Benjamin, New York (1962).

(T)   Fetter, A. L. (with J. D. Walecka), *Quantum Theory of Many-Particle Systems*, McGraw-Hill, New York (1971).

Feynman, R. P., *Quantum Electrodynamics*, Benjamin, New York (1962).

Feynman, R. P. (with A. R. Hibbs), *Quantum Mechanics and Path Integrals*, McGraw-Hill, New York (1965).

Fishlock, T. P. (with J. B. Pendry), 'Electron Correlation at Metallic Densities. II', *J. Phys. C*, **6**, 1909 (1973).

Fowler, M. (with A. Zawadowski), 'Scaling and Renormalization Group in the Kondo Effect,' *Solid State Commun.*, **9**, 471 (1971).

Fowler, M., 'Renormalization Group Techniques in the Kondo Effect', *Phys. Rev.*, *B*, **6**, 3422 (1972).

Fröhlich, H., 'Interaction of Electrons with Lattice Vibrations', *Proc. Roy. Soc.* **A215**, 291 (1952).

(C)   Galitski, V. M., 'Collective Excitations in Fermi Systems', in Report of The International Congress on Many-particle physics (June 1960), *Physica* (supplement), **26**, S–174 (1960).

Galitski, V. M. (with A. B. Migdal), *Soviet Phys. JETP*, **34** (7), 96 (1958).

Gell-Mann, M. (with F. E. Low), 'Quantum Electrodynamics at Small Distances', *Phys. Rev.*, **95**, 1300 (1954).

Gell-Mann, M. (with K. A. Brueckner), *Phys. Rev.*, **106**, 364 (1957).

(R)   Ginzburg, V. L., 'The Problem of High Temperature Superconductivity', *Contemp. Phys.*, **9**, 355 (1968).

Hatano, A. (with H. Kanazawa and Y. Mizuno), 'Electron Interaction and Positron Annihilation in Electron Gases', *Progr. Theor. Phys.*, **34**, 875 (1965).

Hedin, L., 'New Method for Calculating the One-Particle Green's Function with Application to the Electron Gas Problem', *Phys. Rev.*, **139**, A796 (1965).

(R)   Hedin, L. (with S. Lundqvist), 'Effects of Electron-Electron and Electron-Phonon Interaction on the One-electron States of Solids', in *Solid State Physics*, Ed. Seitz, **23** (1969).

(R)   Heeger, A. J., 'Localized Moments and Non-Moments in Metals: The Kondo Effect', in *Solid State Physics*, Ed. Seitz, **23** (1969).

Heisenberg, W., *Ann. Physik*, **10**, 888 (1931).

Højgaard Jensen, H., 'The Free Phonon Field', in *Phonons and Phonon Interactions* (Århus Lectures, 1963), Benjamin, New York (1964).

(R)   Hugenholtz, N. M., 'Quantum Theory of Many-Body Systems', *Reports on Progress in Physics*, **28**, 201 (1965).

Johansson, B., 'Symmetry-breaking Average and Field Theoretic Method in Superconductivity', *Physica*, **32**, 2164 (1966).

(T)   Kadanoff, L. (with G. Baym), *Quantum Statistical Mechanics*, Benjamin, New York (1962).

Kaempffer, F. A., *Concepts in Quantum Mechanics*, Academic Press, New York (1965).

(C)   Katz, A., 'The Analytic Structure of Many-Body Perturbation Theory', p. 223 of *Lecture Notes on the Many-Body Problem* (Bergen School), Benjamin, New York (1962).

(T)   Kirzhnits, D. A., *Field Theoretical Methods in Many-Body Systems*, Pergamon, Oxford (1967).

(T)   Kittel, C., *Quantum Theory of Solids*, Wiley, New York (1963).

(R), (C)   Klein, A., 'Theory of Normal Fermion Systems', p. 279 of *Lectures on the Many-Body Problem*, Vol. 1, Ed. E. R. Caianiello, Academic Press, New York (1962).

Kohn, W. (with J. M. Luttinger), 'Quantum Theory of Electrical Transport Phenomena', *Phys. Rev.* **108**, 590 (1957).

Kondo, J., 'Theory of Dilute Magnetic Alloys', in *Solid State Physics*, Ed. Seitz, **23** (1969).

Kurki-Suonio, K., 'A Note on the Partial Summation of Graphs in Many-Body Theory', *Phys. Letters*, **14**, 298 (1965).

Landau, L. (with E. Lifshitz), *Quantum Mechanics*, Pergamon, Oxford (1958).

Landau, L. (with E. Lifshitz), *Statistical Physics*, Pergamon, London (1959).

Larsen, U. (with Mattuck, R. D.), 'Impurity Spin Lifetime and Electrical Resistivity in Dilute Magnetic Alloys', in *Proceedings of the International Conference of Magnetism ICM-73* volume V, p. 88, Nauka, Moscow (1974).

Larsen, U., *On the Theory of the Kondo Effect*, unpublished report, Physics Laboratory I, H. C. Ørsted Institute, University of Copenhagen, 2100 Copenhagen, Denmark.

Larsen, U., 'A Note on the Pseudo-Fermion Representation of a Spin $\frac{1}{2}$ or 1', *Z. Physik*, **256**, 65 (1972).

Layzer, A. (with D. Fay), 'Spin Fluctuation Exchange: Mechanism for a Superfluid transition in Liquid ^{3}He', *Solid State Commun.* (1974).

Leath, P. L., 'Equivalence of Expanding in Localized or Bloch States in Disordered Alloys', *Phys. Rev.* **B2**, 3078 (1970).

(T)   Lefebre, R. (with C. Moser), 'Correlation Effects in Atoms and Molecules', in series *Adv. in Chem. Phys.*, **XIV**, London, Interscience (1969).

Leggett, A. J., 'Microscopic Theory of NMR in an Anisotropic Superfluid (^{3}He A)', *Phys. Rev. Lett.* **31**, 352 (1973).

Lundqvist, B. I. (with C. Lydén), 'Calculated Momentum Distributions and Compton Profiles of Interacting Conduction Electrons in Lithium and Sodium', *Phys. Rev.* **B, 4**, 3360 (1971).

Luttinger, J. M. (with J. C. Ward), 'Ground State Energy of a Many-Fermion System II', *Phys. Rev.*, **118**, 1417 (1960a).

Luttinger, J. M., 'Fermi Surface and Some Simple Equilibrium Properties of a System of Interacting Fermions', *Phys. Rev.*, **119**, 1153 (1960b).

Luttinger, J. M., 'Theory of the Fermi Surface', in *The Fermi Surface*, Eds. W. Harrison and M. B. Webb, Wiley, New York (1960c).

Luttinger, J. M., 'Analytic Properties of Single Particle Propagators for Many-Fermion Systems', *Phys. Rev.* **121**, 942 (1961).

Lynton, E. A., *Superconductivity*, Methuen-Wiley, London, New York (1962).

March, N. H. (with W. H. Young and S. Sampanthar), *The Many-Body Problem in Quantum Mechanics*, Cambridge University Press (1967).

Margenau, H., in *Quantum Theory. I. Elements*, Ed. Bates, Acad. Press, New York (1961).

Mattuck, R. D., and Johansson, B., 'Quantum Field Theory of Phase Transitions in Fermi Systems', *Advances in Physics*, **17**, 509 (July 1968).

Mattuck, R. D., 'Lifetime of Quasi Particles in Fermi System', *Phys. Letters*, **11**, 29 (1964).

Mattuck, R. D. (with Alba Theumann), 'Expressing the Decoupled Equations of Motion for the Green's Function as a Partial Sum of Feynman Diagrams', *Adv. Phys.*, **20**, 721 (1971).

Mattuck, R. D. (with L. Hansen and C. Y. Cheung), 'Removing the Divergence at the Kondo Temperature by means of Self-Consistent Perturbation Theory', *Journ. de Phys.*, Supplement to vol. **32**, P. C1-432 (1971).

Merzbacher, E., *Quantum Mechanics*, 2nd Edn, Wiley, New York (1970).

Migdal, A. B., *Soviet Phys. JETP*, **34**, (7) 996 (1958).

(T) Mills, R., *Propagators for Many-Particle Systems*, Gordon and Breach, New York (1969).

Murata, K. K., *Potential and Correlation Effects in Dilute Localized Moment Systems*, Ph.D. Thesis, Cornell University (1971).

Nambu, Y., 'Quasi particles and Gauge Invariance in the Theory of Superconductivity', *Phys. Rev.*, **117**, 648 (1960).

(T) Nozières, P., *Theory of Interacting Fermi Systems*, Benjamin, New York (1964).

Nozières, P. (with J. Gavoret and B. Roulet), 'Singularities in the X-Ray Absorption and Emission of Metals. II. Self-Consistent Treatment of Divergences', *Phys. Rev.*, **178**, 1084 (1969).

Nozières, P. (with C. T. De Dominicis), 'Singularities in the X-Ray Absorption and Emission of Metals. III. One-Body Theory Exact Solution', *Phys. Rev.*, **178**, 1097 (1969).

Osheroff, D. D. (with W. J. Gully, R. C. Richardson, and D. M. Lee), 'New Magnetic Phenomenon in Liquid He^3 Below 3 mK', *Phys. Rev. Lett.*, **29**, 920 (1972).

Park, D., *Introduction to the Quantum Theory*, McGraw-Hill, New York (1964).

(R) Parry, W. (with R. Turner), 'Green Functions in Statistical Mechanics', *Repts. on Progr. in Phys.*, **27**, 23 (1964).

(R) Patterson, J. D., 'Modern Study of Solids', *Amer. J. Phys.*, **32**, 269 (1964).

(T), (C) Pines, D., *The Many-Body Problem*, Benjamin, New York (1961).

(T) Pines, D., *Elementary Excitations in Solids*, Benjamin, New York (1963).

Platzman, P. M. (with N. Tzoar), 'X-Ray Scattering From an Electron Gas', *Phys. Rev.*, **139**, A410 (1965).

Platzman, P. M., 'Experimental Aspects of X-ray Scattering from Electrons in Matter', *Comments on Solid State Phys.*, **4**, 101 (1972).

(R) Rado, G. T. (with H. Suhl) (ed.), *Magnetism*, Vol. V, Academic Press (1973).

(T) Raimes, S., *Many-electron Theory*, North-Holland, Amsterdam (1972).

Raimes, S., *The Wave Mechanics of Electrons in Metals*, North-Holland, Amsterdam (1961).

Rikayzen, G., *Theory of Superconductivity*, Wiley, New York-London (1965).

Roulet, B. (with J. Gavoret and P. Nozières), 'Singularities in the X-Ray Absorption and Emission of Metals. I. First-Order Parquet Calculation', *Phys. Rev.*, **178**, 1072 (1969).

(C) Schrieffer, J. R., 'Theory of Superconductivity', p. 541 of *The Many-Body Problem* (Les Houches), Wiley-Dunod, New York-Paris (1959).

(T)   Schrieffer, J. R., *Theory of Superconductivity*, Benjamin, New York (1964a).

Schrieffer, J. R., 'Electron-Phonon Interaction and Super-conductivity', p. 343 in *Phonons and Phonon Interactions*, Ed. Thor Bak, Benjamin, New York (1964b).

(T)   Schultz, T. D., *Quantum Field Theory and the Many-Body Problem*, Gordon and Breach, New York (1964).

Schweber, S. S., *An Introduction to Relativistic Quantum Field Theory*, Row, Peterson and Co., Illinois (1961).

(T)   Stanley, H. E., *Phase Transitions and Critical Phenomena*, Oxford University Press (1971).

(R)   Ter Haar, D., 'Some Recent Developments in the Many-Body Problem', *Cont. Phys.*, **1**, 112 (1959-60).

(R)   Ter Haar, D., 'On the Use of Green's Functions in Statistical Mechanics', p. 119 in *Fluctuation, Relaxation and Resonance in Magnetic Systems*, Oliver and Boyd, Edinburgh (1962).

Theumann, A., 'Linearized Parquet Equations and Truncation Procedures', *Phys. Lett.*, **33A**, 204 (1970).

(T)   Thouless, D. J., *The Quantum Mechanics of Many-Body Systems*, 2nd Edition. Academic Press, New York (1972).

(R)   Thouless, D. J., 'Green Functions in Low Energy Nuclear Physics', *Repts. on Progr. Phys.*, **27**, 53 (1964).

Tinkham, M., 'Superconductivity', p. 149 in *Low Temperature Physics* (Les Houches), Gordon and Breach, New York (1962).

Tsuneto, T. (with Abrahams, E.), 'Skeleton Graph Expansion for Critical Exponents', *Phys. Rev. Lett.*, **30**, 217 (1973).

(T)   Van Hove, L. (with N. Hugenholtz and L. Howland), *Quantum Theory of Many Particle Systems*, Benjamin, New York (1961).

Vashishta, P. (with K. S. Singwi), 'Paramagnetic susceptibility of an Interacting Electron Gas at Metallic Densities', *Solid State Commun.*, **13**, 901 (1973).

Wilson, K. G., 'Renormalization Group and Strong Interactions', *Phys. Rev. D*, **3**, 1818 (1970).

Wilson, K. G., 'Renormalization Group and Critical Phenomena I, II', *Phys. Rev. B*, **4**, 3174, 3184 (1971a, b).

Wilson, K. G. (with M. E. Fisher), 'Critical Exponents in 3.99 Dimensions', *Phys. Rev. Lett.*, **28**, 240 (1971c).

Wilson, K. G., 'Feynman Graph Expansion for Critical Exponents', *Phys. Rev. Lett.*, **28**, 548 (1972).

(C)   Wilson, K. G., 'Solution of the Spin-$\frac{1}{2}$ Kondo Hamiltonian', in *Collective Properties of Physical Systems*, proceedings of the 24th Nobel Symposium, B. Lundqvist and S. Lundqvist (Eds.), Academic Press (1973).

Wilson, K. G. (with J. Kogut), 'The Renormalization Group in the ϵ-expansion', *Phys. Rpts.*, **12C**, 75 (1974).

Woolley, R. G. (with R. D. Mattuck), 'Effect of local order on electronic states of binary alloys', *J. Phys. F.*, **3**, 75 (1973).

(C) Zawadowski, A., 'Review of the Application of the Renormalization Group Method to Logarithmic Problems', in *Collective Properties of Physical System*, proceedings of the 24th Nobel Symposium, B. Lundqvist and S. Lundqvist (Eds.), Academic Press (1973).

Ziman, J. M., *Electrons and Phonons*, Oxford, Clarendon Press (1962).

Index

A CATALOG OF SELECTED
DOVER BOOKS
IN SCIENCE AND MATHEMATICS

A CATALOG OF SELECTED
DOVER BOOKS
IN SCIENCE AND MATHEMATICS

QUALITATIVE THEORY OF DIFFERENTIAL EQUATIONS, V.V. Nemytskii and V.V. Stepanov. Classic graduate-level text by two prominent Soviet mathematicians covers classical differential equations as well as topological dynamics and ergodic theory. Bibliographies. 523pp. 5⅜ × 8½. 65954-2 Pa. $10.95

MATRICES AND LINEAR ALGEBRA, Hans Schneider and George Phillip Barker. Basic textbook covers theory of matrices and its applications to systems of linear equations and related topics such as determinants, eigenvalues and differential equations. Numerous exercises. 432pp. 5⅜ × 8½. 66014-1 Pa. $9.95

QUANTUM THEORY, David Bohm. This advanced undergraduate-level text presents the quantum theory in terms of qualitative and imaginative concepts, followed by specific applications worked out in mathematical detail. Preface. Index. 655pp. 5⅜ × 8½. 65969-0 Pa. $13.95

ATOMIC PHYSICS (8th edition), Max Born. Nobel laureate's lucid treatment of kinetic theory of gases, elementary particles, nuclear atom, wave-corpuscles, atomic structure and spectral lines, much more. Over 40 appendices, bibliography. 495pp. 5⅜ × 8½. 65984-4 Pa. $11.95

ELECTRONIC STRUCTURE AND THE PROPERTIES OF SOLIDS: The Physics of the Chemical Bond, Walter A. Harrison. Innovative text offers basic understanding of the electronic structure of covalent and ionic solids, simple metals, transition metals and their compounds. Problems. 1980 edition. 582pp. 6⅛ × 9¼. 66021-4 Pa. $14.95

BOUNDARY VALUE PROBLEMS OF HEAT CONDUCTION, M. Necati Özisik. Systematic, comprehensive treatment of modern mathematical methods of solving problems in heat conduction and diffusion. Numerous examples and problems. Selected references. Appendices. 505pp. 5⅜ × 8½. 65990-9 Pa. $11.95

A SHORT HISTORY OF CHEMISTRY (3rd edition), J.R. Partington. Classic exposition explores origins of chemistry, alchemy, early medical chemistry, nature of atmosphere, theory of valency, laws and structure of atomic theory, much more. 428pp. 5⅜ × 8½. (Available in U.S. only) 65977-1 Pa. $10.95

A HISTORY OF ASTRONOMY, A. Pannekoek. Well-balanced, carefully reasoned study covers such topics as Ptolemaic theory, work of Copernicus, Kepler, Newton, Eddington's work on stars, much more. Illustrated. References. 521pp. 5⅜ × 8½. 65994-1 Pa. $11.95

PRINCIPLES OF METEOROLOGICAL ANALYSIS, Walter J. Saucier. Highly respected, abundantly illustrated classic reviews atmospheric variables, hydrostatics, static stability, various analyses (scalar, cross-section, isobaric, isentropic, more). For intermediate meteorology students. 454pp. 6⅛ × 9¼. 65979-8 Pa. $12.95

RELATIVITY, THERMODYNAMICS AND COSMOLOGY, Richard C. Tolman. Landmark study extends thermodynamics to special, general relativity; also applications of relativistic mechanics, thermodynamics to cosmological models. 501pp. 5⅜ × 8½. 65383-8 Pa. $12.95

APPLIED ANALYSIS, Cornelius Lanczos. Classic work on analysis and design of finite processes for approximating solution of analytical problems. Algebraic equations, matrices, harmonic analysis, quadrature methods, much more. 559pp. 5⅜ × 8½. 65656-X Pa. $12.95

SPECIAL RELATIVITY FOR PHYSICISTS, G. Stephenson and C.W. Kilmister. Concise elegant account for nonspecialists. Lorentz transformation, optical and dynamical applications, more. Bibliography. 108pp. 5⅜ × 8½. 65519-9 Pa. $4.95

INTRODUCTION TO ANALYSIS, Maxwell Rosenlicht. Unusually clear, accessible coverage of set theory, real number system, metric spaces, continuous functions, Riemann integration, multiple integrals, more. Wide range of problems. Undergraduate level. Bibliography. 254pp. 5⅜ × 8½. 65038-3 Pa. $7.95

INTRODUCTION TO QUANTUM MECHANICS With Applications to Chemistry, Linus Pauling & E. Bright Wilson, Jr. Classic undergraduate text by Nobel Prize winner applies quantum mechanics to chemical and physical problems. Numerous tables and figures enhance the text. Chapter bibliographies. Appendices. Index. 468pp. 5⅜ × 8½. 64871-0 Pa. $11.95

ASYMPTOTIC EXPANSIONS OF INTEGRALS, Norman Bleistein & Richard A. Handelsman. Best introduction to important field with applications in a variety of scientific disciplines. New preface. Problems. Diagrams. Tables. Bibliography. Index. 448pp. 5⅜ × 8½. 65082-0 Pa. $11.95

MATHEMATICS APPLIED TO CONTINUUM MECHANICS, Lee A. Segel. Analyzes models of fluid flow and solid deformation. For upper-level math, science and engineering students. 608pp. 5⅜ × 8½. 65369-2 Pa. $13.95

ELEMENTS OF REAL ANALYSIS, David A. Sprecher. Classic text covers fundamental concepts, real number system, point sets, functions of a real variable, Fourier series, much more. Over 500 exercises. 352pp. 5⅜ × 8½. 65385-4 Pa. $9.95

PHYSICAL PRINCIPLES OF THE QUANTUM THEORY, Werner Heisenberg. Nobel Laureate discusses quantum theory, uncertainty, wave mechanics, work of Dirac, Schroedinger, Compton, Wilson, Einstein, etc. 184pp. 5⅜ × 8½. 60113-7 Pa. $4.95

INTRODUCTORY REAL ANALYSIS, A.N. Kolmogorov, S.V. Fomin. Translated by Richard A. Silverman. Self-contained, evenly paced introduction to real and functional analysis. Some 350 problems. 403pp. 5⅜ × 8½. 61226-0 Pa. $9.95

PROBLEMS AND SOLUTIONS IN QUANTUM CHEMISTRY AND PHYSICS, Charles S. Johnson, Jr. and Lee G. Pedersen. Unusually varied problems, detailed solutions in coverage of quantum mechanics, wave mechanics, angular momentum, molecular spectroscopy, scattering theory, more. 280 problems plus 139 supplementary exercises. 430pp. 6½ × 9¼. 65236-X Pa. $11.95

CATALOG OF DOVER BOOKS

ASYMPTOTIC METHODS IN ANALYSIS, N.G. de Bruijn. An inexpensive, comprehensive guide to asymptotic methods—the pioneering work that teaches by explaining worked examples in detail. Index. 224pp. 5⅜ × 8½.   64221-6 Pa. $6.95

OPTICAL RESONANCE AND TWO-LEVEL ATOMS, L. Allen and J.H. Eberly. Clear, comprehensive introduction to basic principles behind all quantum optical resonance phenomena. 53 illustrations. Preface. Index. 256pp. 5⅜ × 8½.
65533-4 Pa. $7.95

COMPLEX VARIABLES, Francis J. Flanigan. Unusual approach, delaying complex algebra till harmonic functions have been analyzed from real variable viewpoint. Includes problems with answers. 364pp. 5⅜ × 8½.   61388-7 Pa. $7.95

ATOMIC SPECTRA AND ATOMIC STRUCTURE, Gerhard Herzberg. One of best introductions; especially for specialist in other fields. Treatment is physical rather than mathematical. 80 illustrations. 257pp. 5⅜ × 8½.   60115-3 Pa. $5.95

APPLIED COMPLEX VARIABLES, John W. Dettman. Step-by-step coverage of fundamentals of analytic function theory—plus lucid exposition of five important applications: Potential Theory; Ordinary Differential Equations; Fourier Transforms; Laplace Transforms; Asymptotic Expansions. 66 figures. Exercises at chapter ends. 512pp. 5⅜ × 8½.   64670-X Pa. $10.95

ULTRASONIC ABSORPTION: An Introduction to the Theory of Sound Absorption and Dispersion in Gases, Liquids and Solids, A.B. Bhatia. Standard reference in the field provides a clear, systematically organized introductory review of fundamental concepts for advanced graduate students, research workers. Numerous diagrams. Bibliography. 440pp. 5⅜ × 8½.   64917-2 Pa. $11.95

UNBOUNDED LINEAR OPERATORS: Theory and Applications, Seymour Goldberg. Classic presents systematic treatment of the theory of unbounded linear operators in normed linear spaces with applications to differential equations. Bibliography. 199pp. 5⅜ × 8½.   64830-3 Pa. $7.95

LIGHT SCATTERING BY SMALL PARTICLES, H.C. van de Hulst. Comprehensive treatment including full range of useful approximation methods for researchers in chemistry, meteorology and astronomy. 44 illustrations. 470pp. 5⅜ × 8½.   64228-3 Pa. $10.95

CONFORMAL MAPPING ON RIEMANN SURFACES, Harvey Cohn. Lucid, insightful book presents ideal coverage of subject. 334 exercises make book perfect for self-study. 55 figures. 352pp. 5⅜ × 8¼.   64025-6 Pa. $8.95

OPTICKS, Sir Isaac Newton. Newton's own experiments with spectroscopy, colors, lenses, reflection, refraction, etc., in language the layman can follow. Foreword by Albert Einstein. 532pp. 5⅜ × 8½.   60205-2 Pa. $9.95

GENERALIZED INTEGRAL TRANSFORMATIONS, A.H. Zemanian. Graduate-level study of recent generalizations of the Laplace, Mellin, Hankel, K. Weierstrass, convolution and other simple transformations. Bibliography. 320pp. 5⅜ × 8½.   65375-7 Pa. $7.95

THE ELECTROMAGNETIC FIELD, Albert Shadowitz. Comprehensive undergraduate text covers basics of electric and magnetic fields, builds up to electromagnetic theory. Also related topics, including relativity. Over 900 problems. 768pp. 5⅜ × 8¼. 65660-8 Pa. $17.95

FOURIER SERIES, Georgi P. Tolstov. Translated by Richard A. Silverman. A valuable addition to the literature on the subject, moving clearly from subject to subject and theorem to theorem. 107 problems, answers. 336pp. 5⅜ × 8½. 63317-9 Pa. $7.95

THEORY OF ELECTROMAGNETIC WAVE PROPAGATION, Charles Herach Papas. Graduate-level study discusses the Maxwell field equations, radiation from wire antennas, the Doppler effect and more. xiii + 244pp. 5⅜ × 8½. 65678-0 Pa. $6.95

DISTRIBUTION THEORY AND TRANSFORM ANALYSIS: An Introduction to Generalized Functions, with Applications, A.H. Zemanian. Provides basics of distribution theory, describes generalized Fourier and Laplace transformations. Numerous problems. 384pp. 5⅜ × 8½. 65479-6 Pa. $9.95

THE PHYSICS OF WAVES, William C. Elmore and Mark A. Heald. Unique overview of classical wave theory. Acoustics, optics, electromagnetic radiation, more. Ideal as classroom text or for self-study. Problems. 477pp. 5⅜ × 8½. 64926-1 Pa. $11.95

CALCULUS OF VARIATIONS WITH APPLICATIONS, George M. Ewing. Applications-oriented introduction to variational theory develops insight and promotes understanding of specialized books, research papers. Suitable for advanced undergraduate/graduate students as primary, supplementary text. 352pp. 5⅜ × 8½. 64856-7 Pa. $8.95

A TREATISE ON ELECTRICITY AND MAGNETISM, James Clerk Maxwell. Important foundation work of modern physics. Brings to final form Maxwell's theory of electromagnetism and rigorously derives his general equations of field theory. 1,084pp. 5⅜ × 8½. 60636-8, 60637-6 Pa., Two-vol. set $19.90

AN INTRODUCTION TO THE CALCULUS OF VARIATIONS, Charles Fox. Graduate-level text covers variations of an integral, isoperimetrical problems, least action, special relativity, approximations, more. References. 279pp. 5⅜ × 8½. 65499-0 Pa. $7.95

HYDRODYNAMIC AND HYDROMAGNETIC STABILITY, S. Chandrasekhar. Lucid examination of the Rayleigh-Benard problem; clear coverage of the theory of instabilities causing convection. 704pp. 5⅜ × 8¼. 64071-X Pa. $14.95

CALCULUS OF VARIATIONS, Robert Weinstock. Basic introduction covering isoperimetric problems, theory of elasticity, quantum mechanics, electrostatics, etc. Exercises throughout. 326pp. 5⅜ × 8½. 63069-2 Pa. $7.95

DYNAMICS OF FLUIDS IN POROUS MEDIA, Jacob Bear. For advanced students of ground water hydrology, soil mechanics and physics, drainage and irrigation engineering and more. 335 illustrations. Exercises, with answers. 784pp. 6⅛ × 9¼. 65675-6 Pa. $19.95

NUMERICAL METHODS FOR SCIENTISTS AND ENGINEERS, Richard Hamming. Classic text stresses frequency approach in coverage of algorithms, polynomial approximation, Fourier approximation, exponential approxima- tion, other topics. Revised and enlarged 2nd edition. 721pp. 5⅜ × 8½.
65241-6 Pa. $14.95

THEORETICAL SOLID STATE PHYSICS, Vol. I: Perfect Lattices in Equilib- rium; Vol. II: Non-Equilibrium and Disorder, William Jones and Norman H. March. Monumental reference work covers fundamental theory of equilibrium properties of perfect crystalline solids, non-equilibrium properties, defects and disordered systems. Appendices. Problems. Preface. Diagrams. Index. Bibliog- raphy. Total of 1,301pp. 5⅜ × 8½. Two volumes. Vol. I 65015-4 Pa. $12.95
Vol. II 65016-2 Pa. $12.95

OPTIMIZATION THEORY WITH APPLICATIONS, Donald A. Pierre. Broad- spectrum approach to important topic. Classical theory of minima and maxima, calculus of variations, simplex technique and linear programming, more. Many problems, examples. 640pp. 5⅜ × 8½. 65205-X Pa. $13.95

THE MODERN THEORY OF SOLIDS, Frederick Seitz. First inexpensive edition of classic work on theory of ionic crystals, free-electron theory of metals and semiconductors, molecular binding, much more. 736pp. 5⅜ × 8½.
65482-6 Pa. $15.95

ESSAYS ON THE THEORY OF NUMBERS, Richard Dedekind. Two classic essays by great German mathematician: on the theory of irrational numbers; and on transfinite numbers and properties of natural numbers. 115pp. 5⅜ × 8½.
21010-3 Pa. $4.95

THE FUNCTIONS OF MATHEMATICAL PHYSICS, Harry Hochstadt. Com- prehensive treatment of orthogonal polynomials, hypergeometric functions, Hill's equation, much more. Bibliography. Index. 322pp. 5⅜ × 8½. 65214-9 Pa. $9.95

NUMBER THEORY AND ITS HISTORY, Oystein Ore. Unusually clear, accessible introduction covers counting, properties of numbers, prime numbers, much more. Bibliography. 380pp. 5⅜ × 8½. 65620-9 Pa. $8.95

THE VARIATIONAL PRINCIPLES OF MECHANICS, Cornelius Lanczos. Graduate level coverage of calculus of variations, equations of motion, relativistic mechanics, more. First inexpensive paperbound edition of classic treatise. Index. Bibliography. 418pp. 5⅜ × 8½. 65067-7 Pa. $10.95

MATHEMATICAL TABLES AND FORMULAS, Robert D. Carmichael and Edwin R. Smith. Logarithms, sines, tangents, trig functions, powers, roots, reciprocals, exponential and hyperbolic functions, formulas and theorems. 269pp. 5⅜ × 8½. 60111-0 Pa. $5.95

THEORETICAL PHYSICS, Georg Joos, with Ira M. Freeman. Classic overview covers essential math, mechanics, electromagnetic theory, thermodynamics, quan- tum mechanics, nuclear physics, other topics. First paperback edition. xxiii + 885pp. 5⅜ × 8½. 65227-0 Pa. $18.95

CATALOG OF DOVER BOOKS

HANDBOOK OF MATHEMATICAL FUNCTIONS WITH FORMULAS, GRAPHS, AND MATHEMATICAL TABLES, edited by Milton Abramowitz and Irene A. Stegun. Vast compendium: 29 sets of tables, some to as high as 20 places. 1,046pp. 8 × 10½. 61272-4 Pa. $22.95

MATHEMATICAL METHODS IN PHYSICS AND ENGINEERING, John W. Dettman. Algebraically based approach to vectors, mapping, diffraction, other topics in applied math. Also generalized functions, analytic function theory, more. Exercises. 448pp. 5⅜ × 8¼. 65649-7 Pa. $8.95

A SURVEY OF NUMERICAL MATHEMATICS, David M. Young and Robert Todd Gregory. Broad self-contained coverage of computer-oriented numerical algorithms for solving various types of mathematical problems in linear algebra, ordinary and partial, differential equations, much more. Exercises. Total of 1,248pp. 5⅜ × 8½. Two volumes. Vol. I 65691-8 Pa. $14.95
Vol. II 65692-6 Pa. $14.95

TENSOR ANALYSIS FOR PHYSICISTS, J.A. Schouten. Concise exposition of the mathematical basis of tensor analysis, integrated with well-chosen physical examples of the theory. Exercises. Index. Bibliography. 289pp. 5⅜ × 8½. 65582-2 Pa. $7.95

INTRODUCTION TO NUMERICAL ANALYSIS (2nd Edition), F.B. Hildebrand. Classic, fundamental treatment covers computation, approximation, interpolation, numerical differentiation and integration, other topics. 150 new problems. 669pp. 5⅜ × 8½. 65363-3 Pa. $14.95

INVESTIGATIONS ON THE THEORY OF THE BROWNIAN MOVEMENT, Albert Einstein. Five papers (1905–8) investigating dynamics of Brownian motion and evolving elementary theory. Notes by R. Fürth. 122pp. 5⅜ × 8½. 60304-0 Pa. $4.95

NUMERICAL METHODS FOR SCIENTISTS AND ENGINEERS, Richard Hamming. Classic text stresses frequency approach in coverage of algorithms, polynomial approximation, Fourier approximation, exponential approximation, other topics. Revised and enlarged 2nd edition. 721pp. 5⅜ × 8½. 65241-6 Pa. $14.95

AN INTRODUCTION TO STATISTICAL THERMODYNAMICS, Terrell L. Hill. Excellent basic text offers wide-ranging coverage of quantum statistical mechanics, systems of interacting molecules, quantum statistics, more. 523pp. 5⅜ × 8½. 65242-4 Pa. $11.95

ELEMENTARY DIFFERENTIAL EQUATIONS, William Ted Martin and Eric Reissner. Exceptionally clear, comprehensive introduction at undergraduate level. Nature and origin of differential equations, differential equations of first, second and higher orders. Picard's Theorem, much more. Problems with solutions. 331pp. 5⅜ × 8½. 65024-3 Pa. $8.95

STATISTICAL PHYSICS, Gregory H. Wannier. Classic text combines thermodynamics, statistical mechanics and kinetic theory in one unified presentation of thermal physics. Problems with solutions. Bibliography. 532pp. 5⅜ × 8½. 65401-X Pa. $11.95

CATALOG OF DOVER BOOKS

ORDINARY DIFFERENTIAL EQUATIONS, Morris Tenenbaum and Harry Pollard. Exhaustive survey of ordinary differential equations for undergraduates in mathematics, engineering, science. Thorough analysis of theorems. Diagrams. Bibliography. Index. 818pp. 5⅜ × 8½. 64940-7 Pa. $16.95

STATISTICAL MECHANICS: Principles and Applications, Terrell L. Hill. Standard text covers fundamentals of statistical mechanics, applications to fluctuation theory, imperfect gases, distribution functions, more. 448pp. 5⅜ × 8½. 65390-0 Pa. $9.95

ORDINARY DIFFERENTIAL EQUATIONS AND STABILITY THEORY: An Introduction, David A. Sánchez. Brief, modern treatment. Linear equation, stability theory for autonomous and nonautonomous systems, etc. 164pp. 5⅜ × 8¼. 63828-6 Pa. $5.95

THIRTY YEARS THAT SHOOK PHYSICS: The Story of Quantum Theory, George Gamow. Lucid, accessible introduction to influential theory of energy and matter. Careful explanations of Dirac's anti-particles, Bohr's model of the atom, much more. 12 plates. Numerous drawings. 240pp. 5⅜ × 8½. 24895-X Pa. $5.95

THEORY OF MATRICES, Sam Perlis. Outstanding text covering rank, non-singularity and inverses in connection with the development of canonical matrices under the relation of equivalence, and without the intervention of determinants. Includes exercises. 237pp. 5⅜ × 8½. 66810-X Pa. $7.95

GREAT EXPERIMENTS IN PHYSICS: Firsthand Accounts from Galileo to Einstein, edited by Morris H. Shamos. 25 crucial discoveries: Newton's laws of motion, Chadwick's study of the neutron, Hertz on electromagnetic waves, more. Original accounts clearly annotated. 370pp. 5⅜ × 8½. 25346-5 Pa. $9.95

INTRODUCTION TO PARTIAL DIFFERENTIAL EQUATIONS WITH AP-PLICATIONS, E.C. Zachmanoglou and Dale W. Thoe. Essentials of partial differential equations applied to common problems in engineering and the physical sciences. Problems and answers. 416pp. 5⅜ × 8½. 65251-3 Pa. $10.95

BURNHAM'S CELESTIAL HANDBOOK, Robert Burnham, Jr. Thorough guide to the stars beyond our solar system. Exhaustive treatment. Alphabetical by constellation: Andromeda to Cetus in Vol. 1; Chamaeleon to Orion in Vol. 2; and Pavo to Vulpecula in Vol. 3. Hundreds of illustrations. Index in Vol. 3. 2,000pp. 6⅛ × 9¼. 23567-X, 23568-8, 23673-0 Pa., Three-vol. set $41.85

ASYMPTOTIC EXPANSIONS FOR ORDINARY DIFFERENTIAL EQUA-TIONS, Wolfgang Wasow. Outstanding text covers asymptotic power series, Jordan's canonical form, turning point problems, singular perturbations, much more. Problems. 384pp. 5⅜ × 8½. 65456-7 Pa. $9.95

AMATEUR ASTRONOMER'S HANDBOOK, J.B. Sidgwick. Timeless, compre-hensive coverage of telescopes, mirrors, lenses, mountings, telescope drives, micrometers, spectroscopes, more. 189 illustrations. 576pp. 5⅜ × 8¼. (USO) 24034-7 Pa. $9.95

SPECIAL FUNCTIONS, N.N. Lebedev. Translated by Richard Silverman. Famous Russian work treating more important special functions, with applications to specific problems of physics and engineering. 38 figures. 308pp. 5⅜ × 8½.
60624-4 Pa. $7.95

OBSERVATIONAL ASTRONOMY FOR AMATEURS, J.B. Sidgwick. Mine of useful data for observation of sun, moon, planets, asteroids, aurorae, meteors, comets, variables, binaries, etc. 39 illustrations. 384pp. 5⅜ × 8¼. (Available in U.S. only) 24033-9 Pa. $8.95

INTEGRAL EQUATIONS, F.G. Tricomi. Authoritative, well-written treatment of extremely useful mathematical tool with wide applications. Volterra Equations, Fredholm Equations, much more. Advanced undergraduate to graduate level. Exercises. Bibliography. 238pp. 5⅜ × 8½. 64828-1 Pa. $6.95

CELESTIAL OBJECTS FOR COMMON TELESCOPES, T.W. Webb. Inestimable aid for locating and identifying nearly 4,000 celestial objects. 77 illustrations. 645pp. 5⅜ × 8½. 20917-2, 20918-0 Pa., Two-vol. set $12.00

MODERN NONLINEAR EQUATIONS, Thomas L. Saaty. Emphasizes practical solution of problems; covers seven types of equations. ". . . a welcome contribution to the existing literature. . . ."—*Math Reviews.* 490pp. 5⅜ × 8½. 64232-1 Pa. $9.95

FUNDAMENTALS OF ASTRODYNAMICS, Roger Bate et al. Modern approach developed by U.S. Air Force Academy. Designed as a first course. Problems, exercises. Numerous illustrations. 455pp. 5⅜ × 8½. 60061-0 Pa. $8.95

INTRODUCTION TO LINEAR ALGEBRA AND DIFFERENTIAL EQUATIONS, John W. Dettman. Excellent text covers complex numbers, determinants, orthonormal bases, Laplace transforms, much more. Exercises with solutions. Undergraduate level. 416pp. 5⅜ × 8½. 65191-6 Pa. $9.95

INCOMPRESSIBLE AERODYNAMICS, edited by Bryan Thwaites. Covers theoretical and experimental treatment of the uniform flow of air and viscous fluids past two-dimensional aerofoils and three-dimensional wings; many other topics. 654pp. 5⅜ × 8½. 65465-6 Pa. $16.95

INTRODUCTION TO DIFFERENCE EQUATIONS, Samuel Goldberg. Exceptionally clear exposition of important discipline with applications to sociology, psychology, economics. Many illustrative examples; over 250 problems. 260pp. 5⅜ × 8½. 65084-7 Pa. $7.95

LAMINAR BOUNDARY LAYERS, edited by L. Rosenhead. Engineering classic covers steady boundary layers in two- and three-dimensional flow, unsteady boundary layers, stability, observational techniques, much more. 708pp. 5⅜ × 8½.
65646-2 Pa. $15.95

LECTURES ON CLASSICAL DIFFERENTIAL GEOMETRY, Second Edition, Dirk J. Struik. Excellent brief introduction covers curves, theory of surfaces, fundamental equations, geometry on a surface, conformal mapping, other topics. Problems. 240pp. 5⅜ × 8½. 65609-8 Pa. $6.95

ROTARY-WING AERODYNAMICS, W.Z. Stepniewski. Clear, concise text covers aerodynamic phenomena of the rotor and offers guidelines for helicopter performance evaluation. Originally prepared for NASA. 537 figures. 640pp. 6⅛ × 9¼.
64647-5 Pa. $14.95

DIFFERENTIAL GEOMETRY, Heinrich W. Guggenheimer. Local differential geometry as an application of advanced calculus and linear algebra. Curvature, transformation groups, surfaces, more. Exercises. 62 figures. 378pp. 5⅜ × 8½.
63433-7 Pa. $7.95

INTRODUCTION TO SPACE DYNAMICS, William Tyrrell Thomson. Comprehensive, classic introduction to space-flight engineering for advanced undergraduate and graduate students. Includes vector algebra, kinematics, transformation of coordinates. Bibliography. Index. 352pp. 5⅜ × 8½. 65113-4 Pa. $8.95

A SURVEY OF MINIMAL SURFACES, Robert Osserman. Up-to-date, in-depth discussion of the field for advanced students. Corrected and enlarged edition covers new developments. Includes numerous problems. 192pp. 5⅜ × 8½.
64998-9 Pa. $8.95

ANALYTICAL MECHANICS OF GEARS, Earle Buckingham. Indispensable reference for modern gear manufacture covers conjugate gear-tooth action, gear-tooth profiles of various gears, many other topics. 263 figures. 102 tables. 546pp. 5⅜ × 8½. 65712-4 Pa. $11.95

SET THEORY AND LOGIC, Robert R. Stoll. Lucid introduction to unified theory of mathematical concepts. Set theory and logic seen as tools for conceptual understanding of real number system. 496pp. 5⅜ × 8¼. 63829-4 Pa. $10.95

A HISTORY OF MECHANICS, René Dugas. Monumental study of mechanical principles from antiquity to quantum mechanics. Contributions of ancient Greeks, Galileo, Leonardo, Kepler, Lagrange, many others. 671pp. 5⅜ × 8½.
65632-2 Pa. $14.95

FAMOUS PROBLEMS OF GEOMETRY AND HOW TO SOLVE THEM, Benjamin Bold. Squaring the circle, trisecting the angle, duplicating the cube: learn their history, why they are impossible to solve, then solve them yourself. 128pp. 5⅜ × 8½. 24297-8 Pa. $3.95

MECHANICAL VIBRATIONS, J.P. Den Hartog. Classic textbook offers lucid explanations and illustrative models, applying theories of vibrations to a variety of practical industrial engineering problems. Numerous figures. 233 problems, solutions. Appendix. Index. Preface. 436pp. 5⅜ × 8½. 64785-4 Pa. $9.95

CURVATURE AND HOMOLOGY, Samuel I. Goldberg. Thorough treatment of specialized branch of differential geometry. Covers Riemannian manifolds, topology of differentiable manifolds, compact Lie groups, other topics. Exercises. 315pp. 5⅜ × 8½. 64314-X Pa. $8.95

HISTORY OF STRENGTH OF MATERIALS, Stephen P. Timoshenko. Excellent historical survey of the strength of materials with many references to the theories of elasticity and structure. 245 figures. 452pp. 5⅜ × 8½. 61187-6 Pa. $10.95

GEOMETRY OF COMPLEX NUMBERS, Hans Schwerdtfeger. Illuminating, widely praised book on analytic geometry of circles, the Moebius transformation, and two-dimensional non-Euclidean geometries. 200pp. 5⅜ × 8¼.
63830-8 Pa. $6.95

MECHANICS, J.P. Den Hartog. A classic introductory text or refresher. Hundreds of applications and design problems illuminate fundamentals of trusses, loaded beams and cables, etc. 334 answered problems. 462pp. 5⅜ × 8½. 60754-2 Pa. $8.95

TOPOLOGY, John G. Hocking and Gail S. Young. Superb one-year course in classical topology. Topological spaces and functions, point-set topology, much more. Examples and problems. Bibliography. Index. 384pp. 5⅜ × 8¼.
65676-4 Pa. $8.95

STRENGTH OF MATERIALS, J.P. Den Hartog. Full, clear treatment of basic material (tension, torsion, bending, etc.) plus advanced material on engineering methods, applications. 350 answered problems. 323pp. 5⅜ × 8½. 60755-0 Pa. $7.50

ELEMENTARY CONCEPTS OF TOPOLOGY, Paul Alexandroff. Elegant, intuitive approach to topology from set-theoretic topology to Betti groups; how concepts of topology are useful in math and physics. 25 figures. 57pp. 5⅜ × 8½.
60747-X Pa. $2.95

ADVANCED STRENGTH OF MATERIALS, J.P. Den Hartog. Superbly written advanced text covers torsion, rotating disks, membrane stresses in shells, much more. Many problems and answers. 388pp. 5⅜ × 8½. 65407-9 Pa. $9.95

COMPUTABILITY AND UNSOLVABILITY, Martin Davis. Classic graduate-level introduction to theory of computability, usually referred to as theory of recurrent functions. New preface and appendix. 288pp. 5⅜ × 8½. 61471-9 Pa. $6.95

GENERAL CHEMISTRY, Linus Pauling. Revised 3rd edition of classic first-year text by Nobel laureate. Atomic and molecular structure, quantum mechanics, statistical mechanics, thermodynamics correlated with descriptive chemistry. Problems. 992pp. 5⅜ × 8½. 65622-5 Pa. $19.95

AN INTRODUCTION TO MATRICES, SETS AND GROUPS FOR SCIENCE STUDENTS, G. Stephenson. Concise, readable text introduces sets, groups, and most importantly, matrices to undergraduate students of physics, chemistry, and engineering. Problems. 164pp. 5⅜ × 8½. 65077-4 Pa. $6.95

THE HISTORICAL BACKGROUND OF CHEMISTRY, Henry M. Leicester. Evolution of ideas, not individual biography. Concentrates on formulation of a coherent set of chemical laws. 260pp. 5⅜ × 8½. 61053-5 Pa. $6.95

THE PHILOSOPHY OF MATHEMATICS: An Introductory Essay, Stephan Körner. Surveys the views of Plato, Aristotle, Leibniz & Kant concerning propositions and theories of applied and pure mathematics. Introduction. Two appendices. Index. 198pp. 5⅜ × 8½. 25048-2 Pa. $6.95

THE DEVELOPMENT OF MODERN CHEMISTRY, Aaron J. Ihde. Authoritative history of chemistry from ancient Greek theory to 20th-century innovation. Covers major chemists and their discoveries. 209 illustrations. 14 tables. Bibliographies. Indices. Appendices. 851pp. 5⅜ × 8½. 64235-6 Pa. $17.95

THE FOUR-COLOR PROBLEM: Assaults and Conquest, Thomas L. Saaty and Paul G. Kainen. Engrossing, comprehensive account of the century-old combinatorial topological problem, its history and solution. Bibliographies. Index. 110 figures. 228pp. 5⅜ × 8½. 65092-8 Pa. $6.95

CATALYSIS IN CHEMISTRY AND ENZYMOLOGY, William P. Jencks. Exceptionally clear coverage of mechanisms for catalysis, forces in aqueous solution, carbonyl- and acyl-group reactions, practical kinetics, more. 864pp. 5⅜ × 8½. 65460-5 Pa. $19.95

PROBABILITY: An Introduction, Samuel Goldberg. Excellent basic text covers set theory, probability theory for finite sample spaces, binomial theorem, much more. 360 problems. Bibliographies. 322pp. 5⅜ × 8½. 65252-1 Pa. $8.95

LIGHTNING, Martin A. Uman. Revised, updated edition of classic work on the physics of lightning. Phenomena, terminology, measurement, photography, spectroscopy, thunder, more. Reviews recent research. Bibliography. Indices. 320pp. 5⅜ × 8¼. 64575-4 Pa. $8.95

PROBABILITY THEORY: A Concise Course, Y.A. Rozanov. Highly readable, self-contained introduction covers combination of events, dependent events, Bernoulli trials, etc. Translation by Richard Silverman. 148pp. 5⅜ × 8¼. 63544-9 Pa. $5.95

THE CEASELESS WIND: An Introduction to the Theory of Atmospheric Motion, John A. Dutton. Acclaimed text integrates disciplines of mathematics and physics for full understanding of dynamics of atmospheric motion. Over 400 problems. Index. 97 illustrations. 640pp. 6 × 9. 65096-0 Pa. $17.95

STATISTICS MANUAL, Edwin L. Crow, et al. Comprehensive, practical collection of classical and modern methods prepared by U.S. Naval Ordnance Test Station. Stress on use. Basics of statistics assumed. 288pp. 5⅜ × 8½. 60599-X Pa. $6.95

DICTIONARY/OUTLINE OF BASIC STATISTICS, John E. Freund and Frank J. Williams. A clear concise dictionary of over 1,000 statistical terms and an outline of statistical formulas covering probability, nonparametric tests, much more. 208pp. 5⅜ × 8½. 66796-0 Pa. $6.95

STATISTICAL METHOD FROM THE VIEWPOINT OF QUALITY CONTROL, Walter A. Shewhart. Important text explains regulation of variables, uses of statistical control to achieve quality control in industry, agriculture, other areas. 192pp. 5⅜ × 8½. 65232-7 Pa. $6.95

THE INTERPRETATION OF GEOLOGICAL PHASE DIAGRAMS, Ernest G. Ehlers. Clear, concise text emphasizes diagrams of systems under fluid or containing pressure; also coverage of complex binary systems, hydrothermal melting, more. 288pp. 6½ × 9¼. 65389-7 Pa. $10.95

STATISTICAL ADJUSTMENT OF DATA, W. Edwards Deming. Introduction to basic concepts of statistics, curve fitting, least squares solution, conditions without parameter, conditions containing parameters. 26 exercises worked out. 271pp. 5⅜ × 8½. 64685-8 Pa. $7.95

CATALOG OF DOVER BOOKS

DE RE METALLICA, Georgius Agricola. The famous Hoover translation of greatest treatise on technological chemistry, engineering, geology, mining of early modern times (1556). All 289 original woodcuts. 638pp. 6¾ × 11.
60006-8 Pa. $17.95

SOME THEORY OF SAMPLING, William Edwards Deming. Analysis of the problems, theory and design of sampling techniques for social scientists, industrial managers and others who find statistics increasingly important in their work. 61 tables. 90 figures. xvii + 602pp. 5⅜ × 8½.
64684-X Pa. $15.95

THE VARIOUS AND INGENIOUS MACHINES OF AGOSTINO RAMELLI: A Classic Sixteenth-Century Illustrated Treatise on Technology, Agostino Ramelli. One of the most widely known and copied works on machinery in the 16th century. 194 detailed plates of water pumps, grain mills, cranes, more. 608pp. 9 × 12. (EBE)
25497-6 Clothbd. $34.95

LINEAR PROGRAMMING AND ECONOMIC ANALYSIS, Robert Dorfman, Paul A. Samuelson and Robert M. Solow. First comprehensive treatment of linear programming in standard economic analysis. Game theory, modern welfare economics, Leontief input-output, more. 525pp. 5⅜ × 8½.
65491-5 Pa. $13.95

ELEMENTARY DECISION THEORY, Herman Chernoff and Lincoln E. Moses. Clear introduction to statistics and statistical theory covers data processing, probability and random variables, testing hypotheses, much more. Exercises. 364pp. 5⅜ × 8½.
65218-1 Pa. $9.95

THE COMPLEAT STRATEGYST: Being a Primer on the Theory of Games of Strategy, J.D. Williams. Highly entertaining classic describes, with many illustrated examples, how to select best strategies in conflict situations. Prefaces. Appendices. 268pp. 5⅜ × 8½.
25101-2 Pa. $6.95

MATHEMATICAL METHODS OF OPERATIONS RESEARCH, Thomas L. Saaty. Classic graduate-level text covers historical background, classical methods of forming models, optimization, game theory, probability, queueing theory, much more. Exercises. Bibliography. 448pp. 5⅜ × 8¼.
65703-5 Pa. $12.95

CONSTRUCTIONS AND COMBINATORIAL PROBLEMS IN DESIGN OF EXPERIMENTS, Damaraju Raghavarao. In-depth reference work examines orthogonal Latin squares, incomplete block designs, tactical configuration, partial geometry, much more. Abundant explanations, examples. 416pp. 5⅜ × 8¼.
65685-3 Pa. $10.95

THE ABSOLUTE DIFFERENTIAL CALCULUS (CALCULUS OF TENSORS), Tullio Levi-Civita. Great 20th-century mathematician's classic work on material necessary for mathematical grasp of theory of relativity. 452pp. 5⅜ × 8½.
63401-9 Pa. $9.95

VECTOR AND TENSOR ANALYSIS WITH APPLICATIONS, A.I. Borisenko and I.E. Tarapov. Concise introduction. Worked-out problems, solutions, exercises. 257pp. 5⅜ × 8¼.
63833-2 Pa. $6.95

TENSOR CALCULUS, J.L. Synge and A. Schild. Widely used introductory text covers spaces and tensors, basic operations in Riemannian space, non-Riemannian spaces, etc. 324pp. 5⅜ × 8¼.   63612-7 Pa. $7.95

A CONCISE HISTORY OF MATHEMATICS, Dirk J. Struik. The best brief history of mathematics. Stresses origins and covers every major figure from ancient Near East to 19th century. 41 illustrations. 195pp. 5⅜ × 8½.   60255-9 Pa. $7.95

A SHORT ACCOUNT OF THE HISTORY OF MATHEMATICS, W.W. Rouse Ball. One of clearest, most authoritative surveys from the Egyptians and Phoenicians through 19th-century figures such as Grassman, Galois, Riemann. Fourth edition. 522pp. 5⅜ × 8½.   20630-0 Pa. $10.95

HISTORY OF MATHEMATICS, David E. Smith. Nontechnical survey from ancient Greece and Orient to late 19th century; evolution of arithmetic, geometry, trigonometry, calculating devices, algebra, the calculus. 362 illustrations. 1,355pp. 5⅜ × 8½.   20429-4, 20430-8 Pa., Two-vol. set $23.90

THE GEOMETRY OF RENÉ DESCARTES, René Descartes. The great work founded analytical geometry. Original French text, Descartes' own diagrams, together with definitive Smith-Latham translation. 244pp. 5⅜ × 8½.
                                                            60068-8 Pa. $6.95

THE ORIGINS OF THE INFINITESIMAL CALCULUS, Margaret E. Baron. Only fully detailed and documented account of crucial discipline: origins; development by Galileo, Kepler, Cavalieri; contributions of Newton, Leibniz, more. 304pp. 5⅜ × 8½. (Available in U.S. and Canada only)   65371-4 Pa. $9.95

THE HISTORY OF THE CALCULUS AND ITS CONCEPTUAL DEVELOPMENT, Carl B. Boyer. Origins in antiquity, medieval contributions, work of Newton, Leibniz, rigorous formulation. Treatment is verbal. 346pp. 5⅜ × 8½.
                                                            60509-4 Pa. $7.95

THE THIRTEEN BOOKS OF EUCLID'S ELEMENTS, translated with introduction and commentary by Sir Thomas L. Heath. Definitive edition. Textual and linguistic notes, mathematical analysis. 2,500 years of critical commentary. Not abridged. 1,414pp. 5⅜ × 8½.   60088-2, 60089-0, 60090-4 Pa., Three-vol. set $29.85

GAMES AND DECISIONS: Introduction and Critical Survey, R. Duncan Luce and Howard Raiffa. Superb nontechnical introduction to game theory, primarily applied to social sciences. Utility theory, zero-sum games, n-person games, decision-making, much more. Bibliography. 509pp. 5⅜ × 8½.   65943-7 Pa. $11.95

THE HISTORICAL ROOTS OF ELEMENTARY MATHEMATICS, Lucas N.H. Bunt, Phillip S. Jones, and Jack D. Bedient. Fundamental underpinnings of modern arithmetic, algebra, geometry and number systems derived from ancient civilizations. 320pp. 5⅜ × 8½.   25563-8 Pa. $8.95

CALCULUS REFRESHER FOR TECHNICAL PEOPLE, A. Albert Klaf. Covers important aspects of integral and differential calculus via 756 questions. 566 problems, most answered. 431pp. 5⅜ × 8½.   20370-0 Pa. $8.95

CHALLENGING MATHEMATICAL PROBLEMS WITH ELEMENTARY SOLUTIONS, A.M. Yaglom and I.M. Yaglom. Over 170 challenging problems on probability theory, combinatorial analysis, points and lines, topology, convex polygons, many other topics. Solutions. Total of 445pp. 5⅜ × 8½. Two-vol. set.
Vol. I 65536-9 Pa. $6.95
Vol. II 65537-7 Pa. $6.95

FIFTY CHALLENGING PROBLEMS IN PROBABILITY WITH SOLUTIONS, Frederick Mosteller. Remarkable puzzlers, graded in difficulty, illustrate elementary and advanced aspects of probability. Detailed solutions. 88pp. 5⅜ × 8½.
65355-2 Pa. $3.95

EXPERIMENTS IN TOPOLOGY, Stephen Barr. Classic, lively explanation of one of the byways of mathematics. Klein bottles, Moebius strips, projective planes, map coloring, problem of the Koenigsberg bridges, much more, described with clarity and wit. 43 figures. 210pp. 5⅜ × 8½. 25933-1 Pa. $5.95

RELATIVITY IN ILLUSTRATIONS, Jacob T. Schwartz. Clear nontechnical treatment makes relativity more accessible than ever before. Over 60 drawings illustrate concepts more clearly than text alone. Only high school geometry needed. Bibliography. 128pp. 6⅛ × 9¼. 25965-X Pa. $5.95

AN INTRODUCTION TO ORDINARY DIFFERENTIAL EQUATIONS, Earl A. Coddington. A thorough and systematic first course in elementary differential equations for undergraduates in mathematics and science, with many exercises and problems (with answers). Index. 304pp. 5⅜ × 8½. 65942-9 Pa. $7.95

FOURIER SERIES AND ORTHOGONAL FUNCTIONS, Harry F. Davis. An incisive text combining theory and practical example to introduce Fourier series, orthogonal functions and applications of the Fourier method to boundary-value problems. 570 exercises. Answers and notes. 416pp. 5⅜ × 8½. 65973-9 Pa. $9.95

THE THEORY OF BRANCHING PROCESSES, Theodore E. Harris. First systematic, comprehensive treatment of branching (i.e. multiplicative) processes and their applications. Galton-Watson model, Markov branching processes, electron-photon cascade, many other topics. Rigorous proofs. Bibliography. 240pp. 5⅜ × 8½. 65952-6 Pa. $6.95

AN INTRODUCTION TO ALGEBRAIC STRUCTURES, Joseph Landin. Superb self-contained text covers "abstract algebra": sets and numbers, theory of groups, theory of rings, much more. Numerous well-chosen examples, exercises. 247pp. 5⅜ × 8½. 65940-2 Pa. $6.95

Prices subject to change without notice.
Available at your book dealer or write for free Mathematics and Science Catalog to Dept. GI, Dover Publications, Inc., 31 East 2nd St., Mineola, N.Y. 11501. Dover publishes more than 175 books each year on science, elementary and advanced mathematics, biology, music, art, literature, history, social sciences and other areas.